Sheep Production in the Tropics and Subtropics

Related Books

Title	Author	ISBN
• Advances in Microbiology	*P.C. Trivedi*	9788172333324
• Applied Microbiology (Agriculture, Environmental, Food and Industrial Microbiology)	*S.M. Reddy, S. Girisham , G. Narendra Babu & B. Vijaypal Reddy*	9789386347084
• An Introduction to Population Genetics Theory	*J.F. Crow*	9788172336240
• Animal Biodiversity Patterns and Process	*T.N. Ananthakrishnan*	9788172334303
• Animal Husbandry Research, Education and Development	*P.N. Bhat*	9789386652218
• Chara Utpadan Evam Sanrakshan (Hindi)	*B.S. Meena*	9788172339517
• Comparative Anatomy of the Vertebrates	*T.H. Eaton*	9789383692286
• Comprehensive Laboratory Manual of Life Science	*Jyoti Saxena, Mamta Baunthiyal Indu Ravi*	9789387893931
• Comparative Anatomy of the Vertebrates T. H. Eaton	T. H. Eaton	9789383692286
• Essentials of Virology	*S.R. Reddy*	9788172334734
• Feeding and Management of Livestock During Drought and Scarcity	*N.V. Patil*	9788172336479
• Forage Legumes	*J.V. Singh*	9788172336387
• Grassland Techniques and Management	*L.N. Singh, Antaryaml Singh Janardan Singh*	9788172332280
• Glossary of Dairy Technology	*Deepak Mudgil*	9788172339289
• Managing Manure	*Mark Kopecky*	9789383692910
• Microbial Diversity: Exploration & Bioprospecting	*S.R. Reddy*	9788172337827
• Objective Animal Physiology **for JRF, SRF, ARS, NET, *Civil Services & Other Competitive Examinations***	*Arpita Mohapatra*	9789386652461
• Objective Type Questions and Answers in Veterinary Immunology	*T.R. Kannaki*	9788172335212
• Pathological Investigation with Analysis	*M.M. Dashora*	9788172338541
• Principles of Animal Taxonomy	*G.G. Simpson*	9788172337636
• Principles of Systematic Zoology	*E. Mayr*	9789383692279
• Questions and Answers in Environmental Science Practical	*R.K. Kamble*	9789387893849
• Scientist: A Unique Species and the Backbone of a Nations Economic and Intellectual Prosperity	*B.K. Tyagi*	9788172335830
• Traditional Folk Veterinary Medicines	*S.S. Katewa*	9788172336653

Sheep Production in the Tropics and Subtropics

S. K. Kaushish

SCIENTIFIC®
PUBLISHERS

Published by
SCIENTIFIC PUBLISHERS (INDIA)
5 A, New Pali Road, P.O. Box 91
Jodhpur 342 001 (INDIA)
E-mail: info@scientificpub.com
Website: https://www.scientificpubonline.com

Print: 2022

Trademark Notice: Product or corporate names may be trademarks or registered trademarks, and are used only for identification and explanation without intent to infringe.

ISBN: 978-81-72330-72-9
eISBN: 978-93-88449-22-9

Print in India

PREFACE

The aim of this book is to collect all available informations about sheep production to improve traditional methods. Most of the previous sheep books have concentrated on wool, but the emphasis in the present book is more in lamb with trend of the increasing importance of meat and milk from sheep.

It is hoped that this book will be able to fill a gap in the literature for the students of sheep production, working for their graduate and post-graduate degrees in the field of veterinary and agricultural sciences. It is hoped that the book should also meet the need of the great number of those with veterinary and agricultural degrees in every walk of life-farmers and advisors alike. It is also hoped that it will provide reference materials and additional self study guide for research workers and veterinary practitioners.

I am indebted to many colleagues who have made the writing of this book possible, in particular to staff colleagues who have shared the burden of some of the development work reported. Parts of the draft at various stages were read by kind colleagues Dr. J. P. Mittal, Dr. B. D. Sharma, Dr. N. D. Yadav, Sh. Shiv Prasad and Sh. P. M. Singh.

I am finally thankful to my wife and my children who provided assistance in all sorts of ways.

Bikaner **S. K. KAUSHISH**

FOREWORD

Agriculture and Animal Husbandry are two fundamental basis of our economy. Advances have been made in agricultural sciences, which resulted in increase in income. This improvement increased problems in animal husbandry. Sheep husbandry was given maximum emphasis during fourth-five year plan. Central Sheep and Wool Research Institute was established in the state of Rajasthan in 1962. Some exotic breeds of sheep were imported in sixties for cross breeding native sheep for increased mutton and wool production. All India Coordinated Research Projects for fine wool and mutton were started. The objective of AICRP on sheep for mutton was to produce sheep which would weigh 30 kg at 6 months of age. Karakul was imported for increasing pelt production.

The subject of sheep production and management was introduced in many agricultural universities at graduate and post-graduate levels. Many books have been written by foreign authors and we have been copying them. The climate of this region being different from that of those countries, these are far away from realities under our conditions. Very few books have been written on sheep management under tropical and subtropical conditions. These are not according to the syllabi being followed in agricultural universities. The author has written this book keeping in view the requirements of undergraduate and post-graduate students of sheep production which is a rare combination.

This book will be useful not only for students but also for the teachers farm managers and planning officers alike. Dr. Kaushish had been associated with sheep research and teaching for the last 20 yrs. It is hoped that this book will be able to help students, teachers and researchers alike.

R. M. ACHARYA

Deputy Director General (Animal Sciences)
Indian Council of Agricultural Research
New Delhi.

CONTENTS

1

General View

History

Sheep were domesticated long before the dawn of recorded history. Wool fibers have been found in the remains of primitive villages in Switzerland that date back to an estimated 20,000 years. Sheep and goat were first domesticated by man between 7500 and 6500 B.C. for meat and milk (Harlan, 1976). Mention is made in the Bible of sheep flocks and lambs.

Very little information is available about the original selection and domestication of sheep, they are thought to have descended from wild type like the Moufflon sheep. Selection practices not only removed most of the wild instincts, leaving the species completely dependent on man for management and protection, but the tail also lengthened. As weaving began to develop as an important advancement, more definite type and breeds of sheep began to emerge. A breed was developed in Spain for fine wool. The English also developed many breeds that would adapt to their varying climate. Spain and England developed as two super powers in wool production between 1000 and 1500 A.D. Most of the breeds we come across in U S are from these two stocks.

Columbus introduced domestic sheep in West Indies during, his second voyage in 1493. With the invasion of Spain by Napoleon, Merino spread to other parts of the world. Some along with many English breeds found their way to the United States, where their number increased.

Classification

All the sheep have several characteristics in common. As a member of animal kingdom, they belong to the Phylum *Chordata* (back bone), class *Mammalia* (suckle their young ones), order *Artiodactyla* (hooved, even footed), family *Bovidae* (ruminants), genus *Ovis* (domesticated and wild sheep) and species *Ovis aries.* Within this species many different breeds exist. All the pure breds and crossbreds have their place.

Class	Mammalia
Subclass III	Eutheria (females possessing placenta)
Order IV	Ungulata (hoofed mammals)

Suborder I. Artiodactyles
(even-toed animals)
A. Suina
Family C : Suidae
(true pigs)
Genus: *Sus*
Sus Domesticus (domestic pig)
C. TYLOPODA Family: Camilidae
Genus: *CAMELUS* Linn
Camelus dromedarius Linn
(Arabian camel)
2. *Camelus bactrianus* Linn
(two-humped Bactrian camel)
D. Pecora (true ruminants)
Family d : Bovidee (hollow horned)
Genus: *Bos*
1. Taurinc group :
Bos taurus (humpless cattle)
Bos indicus (humped or zebu cattle)
2. Bubaline group:
Bos (Bubalus) bubalis (Indian water buffalo)
Genus: Ovis
Ovis aries (domestic sheep)
Genus: *Capra*
Capra hircus (domestic goat)

Suborder 2. Perissodactyles
(uneven-toed animals)

Family C: *Equidae*
(hoirses, donkeys and zebras)
Genus: *Equs*
1. *Equs caballus* (horse)
2. *Equs asinus* (donkey)
(Note. Mule is a cross between male ass and female horse. Jennet is a cross between female ass and male horse. Both are sterile)

Sheep are now being used for wool, hairs, pelt, apparel, blankets, carpets etc. Endocrine glands of slaughtered sheep are used for making some medicines, fat for making candles and intestines for sausages. Sheep and goat husbandry is the main occupation in arid and semi arid areas. Availability of water is poor in these areas. Watering sheep once in two days does not affect production. They can climb steep mountains easily.

Veriation is found in size, wool production, reproduction and adaptation. The sheep population of different countries are given in Table 1.1. These figures are according to FAO (1978). Sheep population per hactare in Newzealand, South Africa, Australia and England is highest in the world.

Table 1.1 : World sheep population in millions (FAO, 1978).

Country	Population	Country	Population
Afghanistan	2.20	Mongolia	1.35
Algeria	0.95	Mauritania	0.44
Argentina	3.40	Mexico	0.79
Australia	13.50	Medagasker	0.07
Bangladesh	0.12	Morrocco	1.43
Bolivia	0.78	Namibia	0.50
Botswana	0.84	Nepal	0.23
Brazil	1.73	NewZealand	0.59
Bulgaria	0.97	Niger	0.25
Camroun	0.21	Nigeria	0.81
Chile	0.57	Pakistan	1.97
China	7.60	Peru	1.45
Chad	0.24	Poland	0.35
Columbia	0.21	Portugal	0.37
Egypt	0.19	Rhodesia	0.08
England	2.80	Romania	1.43
Equador	0.22	South Africa	3.12
Ethiopia	2.31	Spain	1.56
France	1.09	Sudan	1.82
Ghana	0.15	Syria	0.68
Greece	0.81	Tanzania	0.30
Hungary	0.23	Tunesia	0.36

Country	Population	Country	Population
India	4.03	Turkey	4.15
Indonesia	0.33	Uganda	0.11
Iran	3.50	Uruguay	1.78
Iraq	1.14	Upper Volta	0.13
Ireland	0.35	USA	14.00
Italy	0.94	USSR	1.25
Ivory coast	0.10	Yemen North	0.33
Kenya	0.39	Yemen South	0.09
Liberia	0.02	Venezuala	0.01
Libya	0.30	Yugoslavia	0.75
Mali	0.44		

World Sheep Population

There are 1030 million sheep in the world, out of which, 14 and 13 percent are found in Australia and USSR, respectively. Twenty six percent sheep (268 million) of the world are found in Asia. They are found in larger numbers in China, Turkey, India and Iran. Seventy percent of the total sheep population of Asia is found in these countries. Their number increased by 13 percent between 1955 and 1975 (Table 1.1).

Sheep population increased by 3 percent in Europe during this period. Largest number of sheep in Europe are found in Bulgaria, France, Romania and Spain.

Of the total 113 million sheep found in South America, 23 and 32 percent alone are found in Brazil and Argentina, respectively. Their number decreased by 4 percent. During 1955 and 1975 the sheep number decreased by 44 percent in North and Central America whereas it increased by 38 percent in USSR during the same period.

There were 155 million sheep in Africa according to 1975 census. The sheep population registered an increase of 35 percent between 1955 and 1975. Fifty percent of the total sheep population of this continent is found in South Africa, Ethiopia, Morrocco and Sudan alone.

Australia is the biggest wool producing and exporting country of the world. About half of the world merino wool is produced here. Australia produces about one third of the world wool (Table 1.2).

Table 1.2 : Mutton, milk and wool production in various countries of different continents from sheep.

Continents/ countries	% of total milk production	Washed wool (kg) per sheep	Meat production from sheep & goats per thousand (Tones)
Asia			
Afghanistan	41.50	0.75	5.93
China	21.50	0.71	5.11
India	–	0.52	0.67
Iran	30.20	0.36	7.49
Iraq	45.50	0.48	0.85
Mongolia	19.90	0.94	64.84
Nepal	–	0.99	1.44
Pakistan	–	0.71	1.18
Syria	41.9	1.15	6.96
Turkey	21.5	0.71	5.11
Yemen	20.2	–	8.89
Africa			
Algeria	21.7	0.75	2.86
Ethiopia	0.8	0.01	3.67
Lesotho	–	0.67	3.69
Mali	16.3	0.04	6.18
Morrocco	29.2	0.41	5.14
Nigeria	–	–	1.84
South Africa	–	1.89	8.05
Sudan	7.0	0.22	6.24
Tunesia	22.2	0.71	2.16
Australia			
Australia	–	2.79	58.86
Newzealand	–	3.81	190.44
South America			
Argentina	–	2.00	7.60
Bolivia	38.1	0.53	3.88
Brazil	–	0.82	0.59
Chile	–	1.67	2.15
Uruguay	–	2.25	11.84
Europe			
Albania	25.40	1.13	12.88
Bulgaria	18.32	1.52	9.31

Continents/ countries	% of total milk production	Washed wool (kg) per sheep	Meat production from sheep & goats per thousand (Tones)
England	–	1.19	3.90
France	2.90	0.84	2.41
G.D.R.	–	2.10	0.63
Greece	33.40	0.58	10.30
Hungary	2.90	1.89	0.48
Italy	4.40	0.68	1.00
Ireland	–	2.33	16.38
Poland	–	1.69	0.75
Portugal	12.50	0.60	1.84
Romania	9.54	1.28	4.00
Spain	3.93	0.72	3.97
USSR	0.10	1.80	3.95
Yugoslavia	5.20	0.82	2.41

According to 1982 census, there are 48.2 million sheep in India (Table 1.3) and ranks sixth in the world. According to 1985-86, Ministry of Agriculture report, India produces 38.2 million kg wool, which is 8.3 percent of the total wool produced in Asia (Chaudhry, 1988).

Indian Sheep Population

Sheep husbandry remained in the hands of poor, landless and economically backward farmers. They graze sheep on common unproductive land and migrate over long distances with their stock, in search of fodder. Therefore, it is difficult to adopt breeding on scientific lines. The state wise sheep population trend is given in Table 1.3. Native ewes and rams on average weigh 20 to 30 and 30 to 40 kg, respectively. They produce medium to low quality wool varying from 0.25 to 1.75 kg. Some of the exotic breeds weigh 90 to 120 kg and produce much more wool.

We get about 3.43 thousand tonne wool, 10.1 thousand tonne mutton and 101 million pieces of hides in addition to manure and milk. The demand for mutton is increasing day by day. It is nutritious. The manure produced from sheep is rich is potash and nitrogen.

Table 1.3 : State wise sheep population (in million) in different years.

States	Years						
	1951	1956	1961	1966	1972	1977	1982
Andhra Pradesh	10.19	7.85	8.30	8.00	7.06	8.34	8.36
Assam	0.03	0.17	0.05	0.07	0.59	0.59	0.50
Bihar	0.91	1.05	1.15	1.25	1.12	1.18	1.19
Gujarat	3.61	3.70	1.49	1.65	1.59	1.72	1.90
Maharashtra	Combined	–	2.09	2.20	0.54	0.45	0.50
Himachal Pradesh	0.63	0.70	0.66	0.81	1.05	1.04	0.70
Jammu & Kashmir	0.98	1.47	1.16	1.51	1.21	1.07	1.00
Karnataka	4.35	4.06	4.77	4.75	4.53	4.66	4.50
Kerala	0.43	0.10	0.02	0.01	0.03	0.01	0.01
Madhya Pradesh	0.69	0.90	1.51	1.01	0.96	1.00	0.80
Orissa	0.68	1.08	0.96	1.18	1.43	1.36	1.20
Punjab	0.85	1.23	0.93	0.52	0.43	0.38	0.37
Haryana	Combined	–	–	–	0.54	0.45	0.51
Rajasthan	5.37	7.37	7.34	8.81	9.93	8.55	13.68
Tamil Nadu	7.93	7.04	7.15	6.62	5.28	5.39	5.80
Uttar Pradesh	1.64	2.18	2.46	2.62	2.05	1.95	2.38
West Bengal	0.62	0.62	0.54	0.64	0.79	0.79	0.60

Source : Livestock census of various states

During third five year plan more emphasis was given to sheep production in India. Many extention centers were set up. Central Sheep & Wool Research Institute was established in 1962 in Rajasthan for sheep and wool development and for imparting higher training. Wool classification work was started first of all in Rajasthan.

In collaboration with Australian government Central Sheep Breeding Farm was established in Hisar and Corriedale sheep were imported. Sheep farms were also set up in Andhra Pradesh, Bihar, Jammu & Kashmir, Karnatak, Madya Pradesh and Uttar Pradesh. Units of All India Coordinated Research Project on fine wool were set up in Avikanagar (Raj.), Sandinalha (Tamil Nadu) and Dan-tiwada (Gujarat) and for mutton production in Avikanagar (Raj.), Palamner (Andhra Pradesh) Makhdoom (Uttar Pradesh), Rahuri (Maharashtra) and Kuttupattam (Tamil Nadu).

Breeds of Sheep

The import of different breeds of sheep started in nineteenth century. At that time South Down and Cape Merino were imported in small numbers. Romney Marsh and Merino sheep were im-ported in Uttar Pradesh, Punjab and Karnataka in the beginning of this century. Cross breeding of Deccani sheep with Merino rams was started in 1938. Some Cheviot, Border liecester and Merino sheep were imported in 1940. Rambouillet and Dallen Merino sheep were imported in Jammu & Kashmir from USA in 1951. Cross breeding with these gave good encouraging results therefore Merino, Rambouillet, Corriedale, Polworth, Somalia, South Down Soviet Merino etc. were imported.

The details of some exotic, native and crossbred sheep are given below :-

Exotic Breeds

1. *Dorset* : This breed is found in England. It is good for mutton production. It is medium sized breed. The males have curved horns but females are hornless. Their ears, legs and face are white and skin and lips are pink in colour. A sheep yields 2.5 to 3.5 kg wool annually. Fiber length varies from 7 to 10 cm. The wool is white and rough. The animals are shorn once in a year. The growth rate in case of lambs is good. Good pastures are necessary for their growth.

2. *Suffolck* : This breed is found in Great Britain. This is also a cross which is good for mutton production. Their face, ears and legs are black in colour. Male and female are hornless. Adult males and females weigh 70 to 100 and 100 to 135 kg, respectively. Their legs are short and straight. The growth rate is high in this breed. Yearly wool production varies from 2.0 and 3.0 kg. Staple length is 5 to 6 cm. The wool is white but may be black in some places.

3. *Corriedale* : This breed is found in Australia and is good for mutton and wool production. This is a cross between Merino and Lincolon. The legs are short and the body heavy, wool is lusterous and of medium quality. The fiber length varies from 8 to 12 cm. Yearly wool production per sheep is 2.5 to 3.5 kg. They are white in colour but may have black spots. The head is of medium size. Both sexes are hornless. The adult males and females weigh 80 to 110 and 60 to 75 kg, respectively (Hultz, 1943; Link, 1942).

4. *Rambouillet* : This breed was developed from Merino in France. It is very sturdy breed and can adapt itself in different types of climates. If fed well it can produce good quality of wool. Its body is heavy, head is carried high. The rams have horns but the ewes are hornless. They have wool on face and legs, their skin is pink in colour and ears are small. Adult males and females weigh 65 to 95 and 100 to 120 kg, respectively. Annual wool yield per animal is 4.5 to 5.5 kg. Staple length is 5.5 to 7.5 cm. Lambs at birth weigh 3.2 to 4.5 kg. (Lagrix, 1949).

5. *Karakul* : The origin of this breed is in South Central Asia. It got its name from a village of the same name in USSR. The rainfall is scanty and fodder production is less in this area. The sheep are reared for pelt production. This breed is not good for wool or mutton production. Adult males and females weigh 60 to 70 and 80 to 100 kg, respectively. The neck is long, the body thin, hind quarters are high, the ribs are not wide spread, they have fat deposit at the base of their tails. The males have horns but the females are hornless. The wool is inferior which may be black or white or brown in colour which is used for carpet making. Yearly wool yield per animal is 3.7 kg (Florich, 1931).

6. *Awassi* : It is known as Baladi or Syrian. This breed is native of Syria and is also found in Central and West Iraq. The wool is rough and the breed is good for milk and mutton production. The sheep can yield 40 litre milk in one lactation. Male and females on

an average produce 2.50 and 1.75 kg wool annually, respectively. The sheep of this breed found in Central East Asia have fat on their humps. Their face is black may be brown also. Only males have horns. (Hirsch 1933; Schuler, 1937).

7. *Ossimi* : It is found in Egypt. It has a fat ceposite on its hump. The wool may be white or brown in colour. Adult sheep may weigh from 45 to 55 kg. The annual wool yield per sheep varies from 1.4 to l.8 kg (Sedky, 1948).

8. *Cheviot* : This breed is found in Scotland. Nothing definite can be said about its origin (Frazer, 1948). It is good for mutton production. Both sexes have horns. This breed has some blood of Border Cheviot (Phillips, 1947).

9. *Barbary* : This breed is found in Libya, Tunisia and Eastern Algeria. The wool is rough and the breed is good for milk and mutton production. The wool is white in colour but may be black and brown also. Only rams have horns The lambs at birth weigh 2.5 to 3.0 kg. Adult rams and ewes weigh 40 to 45 and 35 to 40 kg, respectively. The wool is white. The sheep produce two lambs in one year or 3 lambs in two years. The ram produces 4.0 to 5.0 kg wool annually (Andreoni 1938; Pissaloux 1943).

10. *Border Liecester* : Thisbreed was produced by crossing Cheviot with Liescester. It is found in Southern Scotland and Northern England. The wool is of medium quality. Both the sexes are hornless (Young, 1950).

11. *Clunforest* : This breed is found in England and is good for mutton production. Horns are absent. Lambs at birth weigh 3.2 kg. Adult rams and ewes weigh 74 to 86 and 55 to 62 kg, respectively. The annual wool yield is 3.4 to 3.7 kg. The average staple length is 7-8 cm (Builow and Dzhadaridze, 1975).

12. *Kurdi* : It is found in Northern Eastern part of Iran. The wool is rough and the breed is good for mutton production. Head and legs are black. Some sheep may have black and brown colour. Only rams have horns.

13. *Waziri* : This breed is found in North West Frontier region of Pakistan. It is good for mutton production. The ears are long (Lal, 1947).

Native and Crossbreds

Different types of climates are found in our country. The highest rainfall is found in one part and no rainfall for several years in another part of this country. The atmospheric temperature is several degrees below freezing point in some parts and it may reach as high as 50° C in other parts. The country may be divided into four regions on the basis of climate and sheep rearing practices. (Acharya,1982).

(A) Northern Temprate Region

This includes Jammu and Kashmir, Himachal Pradesh and Hilly areas of Uttar Pradesh. Dense forests are found in this area. Good quality grasses grow under trees from March to November. 27 and 87 percent flocks go on migration in Himachal Pradesh and Uttar Pradesh, respectively in April. One or two family members take the flock for grazing. Animal are not given supplemental feeding in addition to grazing. About 60 percent of the sheep are sheared thrice and the rest only twice. The prominent breeds of this region are :-

1. *Bhakerwal* : It drives its name from a tribe of the same name. It is found in the districts of Udampur and Rajauri Tehsil of Jammu & Kashmir. They are heavy but good in climbing mountians. They weigh 28 to 35 kg. The ewes are hornless but rams have horns. The wool is white but may be coloured sometimes. They are shorn thrice in a year. Annual wool production per sheep varies from 1.5 to 2.0 kg. The fiber length varies from 10 to 15 cm and the diameter is 37μ. The wool is of 40's count (Lal, 1947).

2. *Biangi* : It is found in Kangra, Lahul and Spiti. It resembles Gaddi. It is short in size and reared for wool and mutton. Its wool is used for making garments. They can tolerate cold climate very well.

3. *Gaddi* : This is also known as Bhaderwa. It is found in the districts of Hamirpur and Rampur of Himachal Pradesh, Bhaderwa Tehsil of Jammu & Kashmir and districts of Dehradun, Nainital, Teh-rigarhwal of Uttar Pradesh. The sheep, on an average weigh 21 to 30 kg. The height of these sheep is short and can climb mountains with ease. Rams have horns but the ewes are hornless. Their ears are long, the tail is thin and of medium length. Their face is brown. They give birth to single lambs, but 15 to 25

percent sheep give birth to twins. They can be milked upto 3 months. Average staple length is 6 cm and fiber diameter 25 to 30μ. They are shorn thrice in a year. They are found in Kangra, Lahul and Spiti.

4. *Kashmir Valley* : Its height is short, wool is coloured and of the inferior quality.

5. *Gurej* : It is found in Gurej tehsil of Kashmir. They are white in colour but some may be brown or black or brown with blackspots. The ears are short and tail is thick. It is shorn twice in year. This breed is good for milk production. The wool is hairy and rough. Annual wool yield per animal is 1.3 to 1.8 kg. The fiber may reach length upto 15.0 cm.

6. *Rampur Bushair* : This breed is found in the districts of Mahasu, Simla, Nahan, Solan, Kinnore and Bilashpur of Himachal Pradesh and districts of Dehradun and Rishikesh and Nainital of Uttar Pradesh. The average temperature of this region varies from 0 to 37° C. The annual rainfall of this region is 149 mm. The body weight of adult sheep varies from 22 to 28 kg. Their face is romanose and ears are long, the rams have horns and the ewes are hornless. Their wool is white in colour but may have brown or blackspots. The wool is absent from legs, face and belley. Annual wool production per sheep is 1.3 to 1.9 kg. The fiber length varies from 15 to 23 cm and diameter from 30 to 40μ. The medullation is 25 to 30 percent.

7. *Karnah* : It is found in Karnah tehsil of Kashmir. They weigh from 28 to 38 kg. The horns of the rams are curved but ewes are hornless. Their wool is of good quality but of shorter fiber length. They are shorn twice in a year. Annual wool yield per sheep is 1.0 to 1.5 kg. Their staple length is 4 to 6 cm and fiber diameter is 28 to 32μ. It is of 58 to 60's count.

8. *Punchi* : It is found in Punch and Rajauri districts of Jammu & Kashmir. It resembles Gaddi but its height is more. They weigh 25 to 30 kg and the horns are absent in both the sexes. The ears are of medium length and the tail is thin and short. The wool quality varies from medium to high and is white. They are shorn thrice in a year. The annual wool production per sheep varies from 1.4 to 1.6 kg. The fiber length varies from 14 to 15 cm and diameter is 26μ.

9. *Changthangi* : It is found in Changthag region of Ladakh they are sturdy and have wool all over their body. They weigh from 28 to 36 kg. The wool is good for apparel making. They are shorn twice in a year but at places, they are shorn annually. The annual wool yield per sheep varies from 1.0 to 1.5 kg.

10. *Kashmir Merino* : This breed was evolved by crossing Gaddi, Bhakerwal and Punchi sheep with Russian Merino and Rambouillet rams. They have 50 to 75 percent exotic blood. Annual wool production per sheep varies from 2.0 to 2.5 kg. The staple length varies from 12 to 16 cm and fiber diameter is 20μ.

(B) Arid Western Region

This region includes Punjab, Haryana, plains of Uttar Pradesh, Rajasthan, Madhya Pradesh and Gujarat. Sheep from Rajasthan migrate to Uttar Pradesh, Madhya Pradesh and Haryana. The sheep from Rajasthan usually go on migration in November and return by the beginning of rainy season. The major breeds found in this region are:-

1. *Chokla* : This is also known as Chhaper and Shekhawati. It is found in Churu, Jhunjhunu and Sikar districts and parts of Bikaner, Nagaur and Jaipur districts. The average body weight of rams and ewes is 35.0 and 23.1 kg, respectively. The tail is thick and sheep are medium in height. The ears are of medium length. Both the sexes are hornless. The age of maturity for male and female lambs is 15 and 18 months, respectively. The wool is of medium quality and is good for carpet and apparel making. The adults on an average yield 2.75 kg. wool per annum. The average staple length is 4.7 cm, diameter 28μ and medullation 24 percent. The wool is of 54 to 60's count. The wool shorn after rainy season is yellow. Milk production is less. They give birth to one lamb at a time (Bhasin and Desai 1965; Arora *et al.,* 1975 c; Mathur 1977).

2. *Pugal* : It is found in Pugal region of Bikaner and in Northern parts of Jaisalmer district. Its face is black and have yellow band running from eyes towards nose. The tail is short. The age at maturity in case of male and female is 15 and 18 month, respectively. The average body weight of adult males and females is 31.8 and 27.0 kg respectively. The wool is white in colour and of medium quality. The annual wool production varies from 1.4 to 2.3 kg. Average staple length is 5.7cm, fiber diameter 35μ and medullation 80 to 86 percent (Acharya, 1982).

3. *Marwari* : Marwar is the native place of this breed. It is found in the districts of Jodhpur, Pali, Nagaur and Sirohi. It is also found in the adjoining districts of Udaipur, Ajmer, Bikaner and Jaisalmer. Average body weight of rams and ewes are 30.7 and 26.2 kg, respectively. The face is black, ears are small, tail is thin and of medium size. Male and female lambs mature at the age of 12 and 18 months, respectively. They are not good for milk production. Average wool yield per animal per annum is 1.78 kg. The staple length in 6.6 cm average fiber diameter is 36.93μ and medullation percentage is 65.2. The wool is of 46 to 50's count.

4. *Jaisalmeri* : In addition to Jaisalmer district, it is also found along the western border of Jodhpur and Barmer districts. It is tallest of all the breeds found in Rajasthan. The average body weight of ewe varies from 33 to 36 kg. The face is black but sometimes may be brown also. Its ears are long and horns are absent in both the sexes. Male and female lambs mature at the average age of 16 and 18 month, respectively. Annual wool yield per sheep is 1.5 kg. The average staple length is 6.5 cm, fiber diameter 39.1μ and medullation 56.1 percent. The wool is of 52 to 54's count.

5. *Magra* : It is found in Churu, Nagaur, and Bikaner districts. The average body weight of adult rams and ewes is 26.84 and 24.6 kg, respectively. Its face is white and has short brown circle around the eyes. The colour of skin in pink. The tail is of medium length and horns are absent in both the sexes. The wool is lusterous, white and of medium quality. Annual wool yield per animal is 2.18 kg. The average staple length and fiber diameter are 5.8 cm and 32.45μ, respectively. The medullation percentage is 48.3. It is shorn thrice in a year.

6. *Nali* : This breed is found in the districts of Ganganagar and Churu and in the northern parts of Jhunjunu district of Rajasthan and Hisar and Rohtak districts of Haryana. The average body weight of rams and ewes is 34.6 and 24.3 kg, respectively. They are of medium size, the face is brown and covered with wool. The ears are long and the legs are short. The rams and ewes of this breed mature at the age of 12 and 18 month, respectively. The wool is present on head, abdomen and legs. The wool is pale yellow in colour. The annual wool yield per animal is 2.29 kg. The average staple length is 7.4 cm, fiber diameter, 34.9μ medullation 45 to 50 percent. The wool is of 42 to 50's count (Arora *et al.*, 1975c).

7. *Sonadi* : It is found in Udaipur district of Rajasthan and northern parts of Gujarat. The average body weight of ram and ewe is 38.6 and 21.2 kg., respectively. Its face is white and neck is light in colour. Their tail is thin and long. It is quite tall. The ears are long. Male and female lambs mature at the age of 12 and 16 month, respectively. They are also milked and good for mutton production. The wool is absent from legs and abdomen. Annual wool yield per sheep is 0.90 kg. The average staple length, fiber diameter and medullation percentage are 4.98 cm,52.7μ and 8.16, respectively. It is of 30 to 36'count (Arora *et al.,* 1977).

8. *Malpura :* This breed is found in Jaipur, Tonk and Sawaimadhopur districts and parts of adjoining districts of Ajmer, Bhilwara and Bundi. The breed got its name from a tehsil of same name in Tonk district. The average adult body weight of the rams and ewes are 41.6 and 24.5 kg, respectively. The frame of breed is strong, the face is light brown, tail is thin and long and ears are short. Both the sexes are hornless. The male and female mature at the age of 10 and 16 mouth, respectively. The wool is white and of inferior quality. A sheep may yield 21.0 kg milk in a lactation of 3 ½ months. The average staple lenght is 6.5 cm, fiber diameter 51.4μ and medulla-tion 84.1 percentage. The wool is of 30 to 35's count (Arora *et al.,* 1975 a).

9. *Patanwndi* : This is also known as Kuchi, Deshi and Kadhiawadi and Charotry. It is found in Kutch, Jamnagar, Bhavnagar and Junagarh districts of Gujarat. The average body weight of adult rams and ewes is 33.34 and 25.53 kg respectively. The legs are long and are of medium size. Its face is brown, ears long, neck short and tail thin and short. It gives milk up to 3 months of lactation. The wool is white in colour and of medium quality. Average wool production is 1.0 kg per sheep per annum. Average staple length, fiber diameter and medullation percentage are 8.5 cm, 40.0μ and 29.9 percentage, respectively.

10. *Lohi :* This breed is found in Ferozepur and Amritsar districts of Punjab. It has a good height. Its head is big and brown in colour. The tail is thick and short, udder well developed, teats long. It usually gives birth to one lamb but twins are also not uncommon. Though its wool is thick and long, it is one of the best breeds of Asia for milk and mutton. The average body weights of rams and ewes are 45 and 70 kg., respectively (Nanda, 1942).

11. *Jalauni* : It is found in Jalaun and Jhansi districts of Uttar Pradesh. Average body weight of sheep varies from 28 to 32 kg. Their height is average, nose flat, tail thin and of medium length and ears are long. They are usually white but there may be black or brown spots on face and around eyes. The wool is inferior and of short staple length. It is absent from belly and legs. They are milked and are good for mutton. Annual wool production per sheep is 2.0 kg. Average fiber diameter is 41μ.

12. *Bundelkhandi* : It is also known as Jabalpuri. It is found in Bundelkhand. The wool is coloured and inferior. They are hornless.

13. *Avikalin* : This breed has been evolved by crossing Malpura with Rambouillet rams at Central Sheep and Wool Research Institute Avikanagar (Rajasthan). This breed is good for carpet making. Rams may or may not be horned. Wool is lustrous and soft. Annual wool production per sheep varies from 2.0 to 2.5 kg. The wool is absent from face but present on legs and abdomen. Average staple length and fiber diameter are 7.0 cm and 24.6μ respectively (Anon 1976).

14. *Avivastra* : It has been produced by crossing Chokla with Rambouillet rams. The average body weights of rams and ewes vary from 35 to 45 and 40 to 50 kg, respectively. Average unwashed wool production is 2.4 kg per sheep per annum. Average staple length, fiber diameter and medullation percentage are 3.0 cm, 20μ and 21, respectively (Mani Mohan *et al.,* 1979).

15. *Hisar dale* : It was evolved at Govt. Livestock Farm by crossing Bikaneri with Merino rams. Average body weights of ram and ewes are 55 to 60 and 33 to 37 kg, respectively. Its lips are thin, nose pink and ears triangular and hanging. They may or may not be horned. The wool is of superior quality. Annual wool production per sheep is 2.3 to 2.8 kg. Average staple length, fiber diameter and medullation percentage are 6 to 10 cm, 24μ and 0.0, respectively. It is of 58's count (Nanda, 1947)

(C) Eastern Region

This includes Bihar, Bengal, Orissa, Assam, Arunachal Pradesh, Mizoram, Manipur, Tripura, Meghalya, Sikkim, and Nagaland. Most of the flocks of plains go on migration. Flocks of hilly areas migrate to heights. Flocks start migrating in Sikkim with the onset of summer season and return with the onset of rainy season. The important breeds of this area are :-

1. *Balangir* : It is found in Balangir and Sambalpur districts of Orissa. Their weight varies from 18 to 20 kg. Their ears are small. Horns are present in rams only. Their wool is white or brown but may be black sometimes and is rough.

2. *Bonpala* : It belongs to Sikkim. They are very high, weigh 33 to 60 kg and their face is white or black in colour. The wool is white, rough and absent on belly and legs. A sheep produces 1.5 kg wool annually, whose fiber diameter varies from 28 to 35μ and medullation 77 to 84 percent. They go on migration (Katiyar *et al.*, 1981).

3. *Chhotanagpuri* : This breed is found in Chhotanagpur, Ranchi, Sinhbhoom, Palamau, Hajaribag, Dhanbad and Santhal Pargana districts and parts of Bankura district of West Bengal. They are small in size (15 to 20 Kg). Ears are small and legs are thin, but the tail is short but thick. They are hornless and their wool is rough, inferior, grey or brown in colour. The wool production is 0.2 to 0.5 kg per animal per annum. The fiber diameter is 50μ and medullation is 80 percent.

RAMBOUILLET RAM

SUNFFOLK AND DORSET RAMS

GADDI

MAGRA RAM

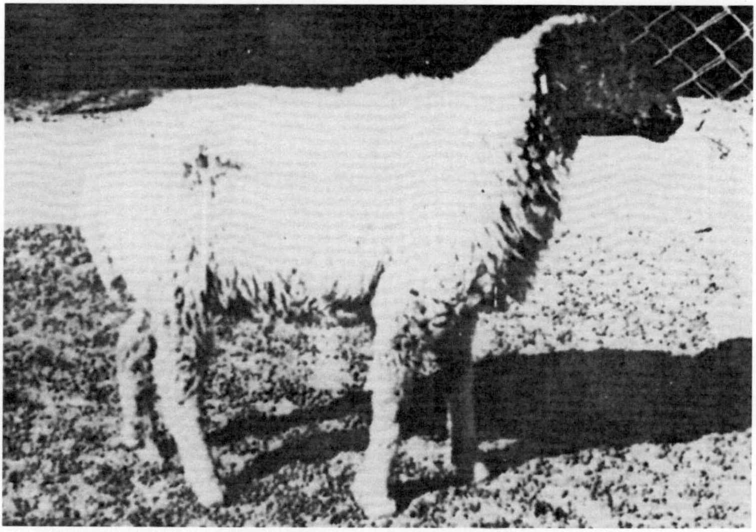

MARWARI EWE

MALPURA EWE

SONADI

AVIKALINE RAM

AVIVASTRA RAM

MANDYA

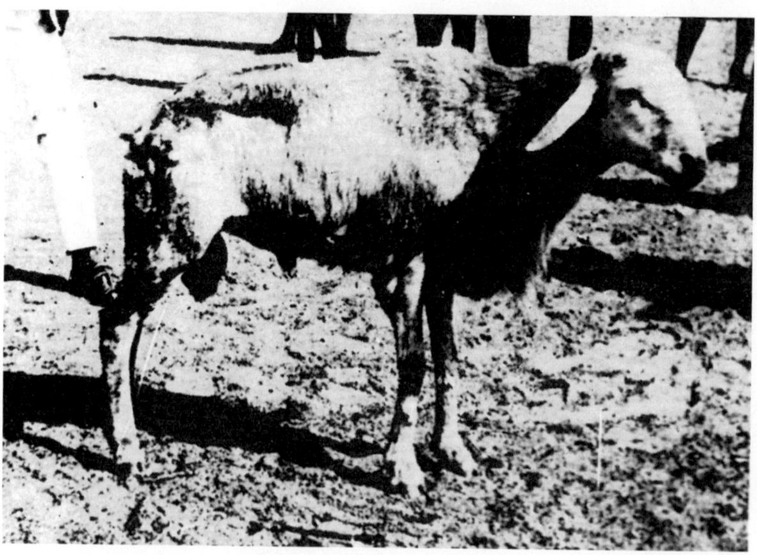

MECHHERI RAM

DECCANI FLOCK

KILAKARSAL EWE

MADRAS RED EWE

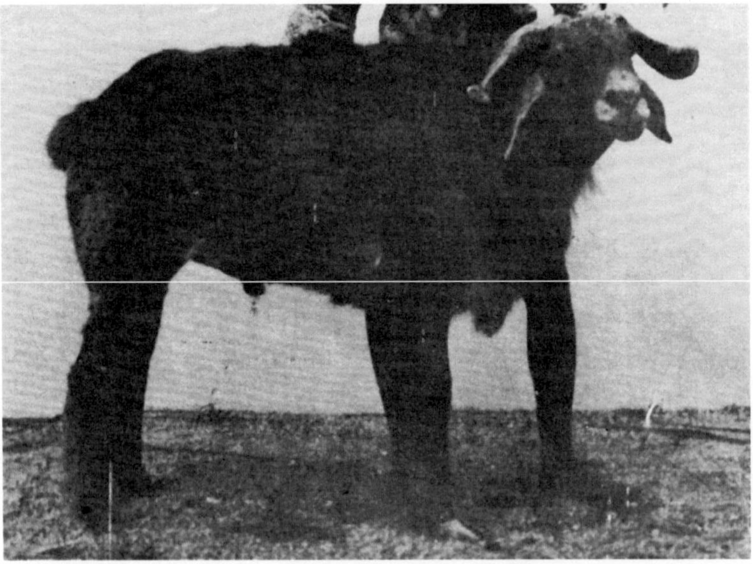

NELLORE

NILGIRI

VEMBURE EWE

CHHOTANAGPURI

4. *Ganjam* : It is native of Phulbani and Koraput districts of Orissa. It is of medium size and weigh 20 to 28 kg. There may be spots on body and face. It has hair on its body and good for mutton produc-tion. Only rams are horned.

5. *Shahbadi* : It is found in Patna, Gaya and Shahbad districts of Bihar. They weigh 20 to 27 kg, their ears are short but tail is long. They don't have horns. They are of medium size, with long legs and have spots on their bodies. Annual wool production per sheep is 250 g. The wool is hairy and medullation is 80 to 85 percent.

(D) Southern Plateau Region

It encompasses Andhra Pradesh, Karnataka, Kerala and Tamil Nadu. About 20 percent sheep of this area go on migration. They are not given concentrate in addition to grazing, but occasionally they are given ground nut leaves. Peaks in heat are observed between May and August and November and February.

1. *Bellari* : It is found in the districts of Dharwar, Raichur and Bellari of Karnataka and Kurnool district of Andhra Pradesh. Their colour is white, black or grey. Their average weight varies

from 25 to 30 kg. Their ears are of medium size and the tail is thick. The wool is mixed with hair. The wool is rough. Annual wool production is 0.4 to 0.9 kg and medullation percentage varies from 40 to 50 percent (Shahi, 1941).

2. *Coimbatore :* It is found in Coimbatore and Selam districts of Tamil Nadu and weigh 20 to 25 kg. Their face is white, black or black spotted. Rams may bear horns, but generally they are hornless. It is good for mutton production. A sheep may produce 0.7 to 0.9 kg inferior wool annually. The staple length varies from 5 to 8 cm and has 55 to 60 percent medullation.

3. *Deccani* : It is found in the districts of Kolhapur, Nasik, Pune, Sangli, and Satara of Maharashtra, Kurnool, Malgouda, Nizamabad and Vidar of Karnataka. Rams may or may not have horns sometimes ewes also have small horns. On an average they weigh 26 to 30 kg. Annual wool production per sheep is 0.7 kg. Wool is of inferior quality and is used for making blankets.

4. *Hasan* : It is native of Chikmangloore and Hasan districts of Karnataka. They are small sturdier, strong with bright eyes, medium sized ears and thin tail. They are usually white but may have brown or black spots on their bodies. Rams are horned. The sheep on an average weigh 20 to 25 kg. The wool is white, rough and inferior. The annual wool production per sheep varies from 0.4 to 0.7 kg.

5. *Kilakarsal* : Also known as Kilakresal, Karuvi and Ramanad. They are predominently found in Madurai, Ramanathpuram, Ramanad and Thanjavoore districts of Tamil Nadu. They are brown in colour but may have black spots on head and belly, Rams and ewes weigh 25 to 35 and 18 to 25 kg, respectively. Their ears are long but tail is short.

6. *Mandya* : It is also called Banure and is found in the districts of Banglore, Kolar and Mandya of Karnataka. Their average body weight varies from 21 to 25 kg. Face is brown, ears are long, leaf like. They are hornless and kept for mutton and milk production. They are hairy.

7. *Machheri* : It is also called Melavadi. It is found in Salem and Coimbatore districts of Tamil Nadu. They are brown in colour, ears are of medium size and the tail is short. Rams and ewes weigh 30 to 35 kg and 20 to 25 kg, respectively. They are hairy (Acharya, 1982).

8. *Madras Red:* It is found in Madras and Chingalpet districts of Tamil Nadu. They are brown in colour. Rams and ewes weigh 20 to 25 and 32 to 35 kg, respectively. Rams are hornless.

9. *Nellore* : Found in Guntur, Nellore and Nalgonda districts of Andhra Pradesh. They weigh 29 to 34 kg, ears are long, tail thin and small and their body is hairy (Chakravarty and Haq, 1946).

10. *Nilgiri* : It is found in Nilgiri hills of Tamil Nadu. It originated in ninteenth century because of crossing of native sheep with Tasmanian Merino, South Down and Cheviot rams. They weigh 21 to 26 kg. Rams have small horns. A sheep produces 1.0 to 1.6 kg wool annually, whose fiber diameter is 25 to 30μ. The medullation percentage is very low (10 to 12 percent).

11. *Tenguri* : Also know as Benguri and are found in Tingsagar, Gangradi and Sathnore talukas of Raipur districts of Karnataka. They weigh 20 to 28 kg. They are dark brown but may have white or black spots. Rams are horned.

12. *Trichuri Black* : It is found in Paramboore and Arivaloore tehsils of Trichuri district and weigh 15 to 20 kg. They are black in colour and only rams are horned. Their ears and tail are small. The hair are rough.

13. *Vemburi* : It is found inTirunavelley district of Tamil Nadu and weigh 25 to 32 kg. They are white with brown spots. Their ears are of medium size and tail is small. Rams have horns. They have small hair on their body.

2

Housing

Of all livestock, sheep are best fitted by nature to withstand cold weather. They have their own natural protection from cold. During cold weather sheep prefer to remain out of doors, except during severe storms, if they are protected from strong winds. Moreover, sheep are usually associated with the system where grazing is the major source of feed throughout the year. Consequently, for a portion of the year, little or no housing is necessary. Thus the shelter for sheep need not be expensive or particularly warm. During lambing season, warm quarters should be provided as the lambs are susceptible to chilling during the first few days of their lives. Hence, housing of sheep is considered as a less important feature than other livestock species. However, the housing system depends upon the breed, age, physiological status and climatic conditions for better production performances as well as to avoid predation during night.

The natural climate is not continuously optimum for livestock production throughout the year. Since man began to domesticate the animals, he has progressively restricted their freedom of selecting the best environment suitable for them during different adverse environment conditions, which affect their development, response and growth. Therefore, the provision of suitable shelter to the sheep during the adverse environmental factors will buffer the extremes of these climatic factors and reduce the peak stress on them.

The following major points should be considered before constructing shed for the animals :

1. *Topography of the land* : The building should be situated at a higher elevation than the surroundings to offer a good slope for rainfall and easy, economic drainage of the waste products, manure, urine etc.

2. *Soil type* : Clay soil which considerably swells during rainy season may not be suitable for construction. Soil composed of gravel and sand is best suited for foundation.

3. *Exposure to sun and protection from wind* : Shed should be constructed to obtain maximum sunlight. The long axis should always be in north-south direction. For protection from strong winds fodder trees may be planted at least 10 feet apart from the sheds.

4. *Accessability* : The sheds should be connected to good roads.

5. *Water supply* : Fresh, clean and soft water should be ensured.

6. *Durability and attractiveness* : Sheds should be constructed with durable materials and may be designed in a proper manner i.e. round shape, L-shape or T-shape to give a scenic view.

7. *Surroundings* : Area infested with wild animals, poachers and industries may not be selected.

8. *Labour* : Farm should be at an approachable distance from the locality to ensure in time availability of labours.

9. *Marketing* : The market should be at an easy approachable distance and the consumers should be in a position to accept the products at a comparatively reasonable price.

10. *Electricity* : The site may be at a reasonable distance for electrification.

11. *Fodder production* : There should be sufficient area surround-ing the shed for fodder cultivation and its preservation as hay and silage.

12. *Scope for future extension* : Sufficient area may be available round the shed for future expansion.

13. *Limited scope for spread of diseases* : The site should be at a place where grazing of local animals is generally avoided.

Types of Sheds

(a) *Open shed* : In this system there is no other special construction except for roof and fencing. This is good for healthy and adult stock throughout the year. During severe winter and summer, artificial protection with grass or bamboo, may be provided on the sides.

(b) *Semi open shed* : In this one side of the shed is open.

(c) *Enclosed shed* : This type of shed is fully enclosed with controlled environment, light and ventilation.

Farm Plan

It is necessary to provide in the yard for housing sheep different types of housing for sheep kept for wool and mutton production. The general plans may be as given in Fig 2.1 The number of units of each type to be provided in the yard would depend on the number of animals to be looked after. The area of each unit shall depend upon the minimum floor space specified for each animal and the number of animals to be kept in each unit. The minimum floor space for each type of animal shall be as given in table below.

Table : Minimum floor space requirements per animal.

Type of Animal	Floor space per animal (Sq. m)
Ewe	1.0
Ram	3.3
Lamb	0.4

Location

The housing for sheep shall be located on a dry, elevated and well drained place, not exposed to strong hot or cold winds, A place where there is a longer period of shade in the day is to be prefered.

Units

A yard meant for keeping sheep for wool and mutton production may have the following units :

(a) Sheds

(b) Shearing and store room, and

(c) Shephereds house

Sheds

The sheds of the following types are required for housing the animals.

(a) Flock shed

(b) Ram shed

(c) Lamb

(d) Lambing and

(e) Sick shed

Flock shed : The flock shed shall be used for housing ewes kept for breeding purpose. The shed shall be 15 × 4 × 3 m high and accommodates not more than 60 ewes.

Ram shed : The ram shed shall be used for housing rams kept for breeding purpose. The shed shall be 4 × 2.5 × 3 m high and accommodate not more than 8 animals. The shed shall be partitioned length wise to form four equal compartments as shown in Fig 2.1.

The partitions may be either of wooden planks or half cut ballies. The partitions shall be not more than one metre high from the floor.

Lamb shed : The shed be used for housing lambs until they attain maturity and are weaned and also those animals which are weaned but have not attained maturity. The shed shall be 7.5 × 4 × 3 m high to accommodate not more than 75 animals. The shed shall be partitioned bredth wise dividing it into two compartments. The compartments having the dimensions 5 × 4 m shall be used to keep the unweaned animal and the other compartment having dimensions 2.5 × 4.0 m shall be used for keeping the weaned but immature animals.

Lambing shed : The shed shall be used as maternity room for ewes. The shed shall be 1.5 × 1.2 × 3 m high. A manger for holding feed hay and a bucket for keeping water be provided in the shed.

Sick shed : There shall be a sick shed for segregating sick and disabled animals. The shed be 3 × 2 × 3 m high. There shall be a door one metre wide and two metre high and one of the lower half of the door may be made of wooden planks, and window 0.7 m broad and 1.2 m high with a wire net covering.

Shearing and store room

The shearing and store room may consist of two compartments with a dividing wall. One room may be exclusively meant for storing wool and shearing equipement and the other for keeping feed and medicines. The room may be 6 × 2.5 × 3 m high. There shall be a door one metre wide and two metre high in the front side of the room. The door leaf may be made of wooden battens. It may also have two windows one on each side of the long sides of the room. Each window may be 0.7 m wide and 1.2 m high. The window may be covered with wire netting.

Shephard's House

The shephard's house meant for the caretaker shall be located at a convenient place in the yard. The house may be 6 × 4 × 3 m high. There shall be a door one metre wide and two metre high. The door leaf may be of wooden planks. There may be four windows one of these facing the passage of the yard and the other three facing outside (See Fig 2.1). Each window may be 0.7 m broad and 1.2 m high and covered with wire netting.

Constructional Details of Units

Floor : The flooring may be either of *moorum* or of strong wooden battens and where the rainfall is quiet heavy, the latter type of flooring may be prefered. In the case of wooden batten flooring the width of each plank shall vary from 75 to 100 mm and the thickness between 25 mm and 40 mm. The sides of the planks shall be well rounded and the clearance between two planks shall range between 10 mm and 15 mm to facilitate the disposal of dung and urine. The wooden batten flooring shall be constructed at a hight of at least one metre above the ground level. In this case a suitable ramp or steps of wooden planks shall be provided. In the case of *moorum* flooring, a plinth wall between 150 mm and 300 mm high shall be provided, for the shearing and store room of and shephard house the flooring may be of moorum or brick in cement morter and the floor shall be levelled properly.

Walls: The walls may be of bricks in lime mortar up to two thirds of their height from the common burnt clay building bricks. The upper one third part of the walls may be of chain link netting of galvanized steel barbed wire. The chain link netting may be supported by angle irons of suitable section and brick masonry

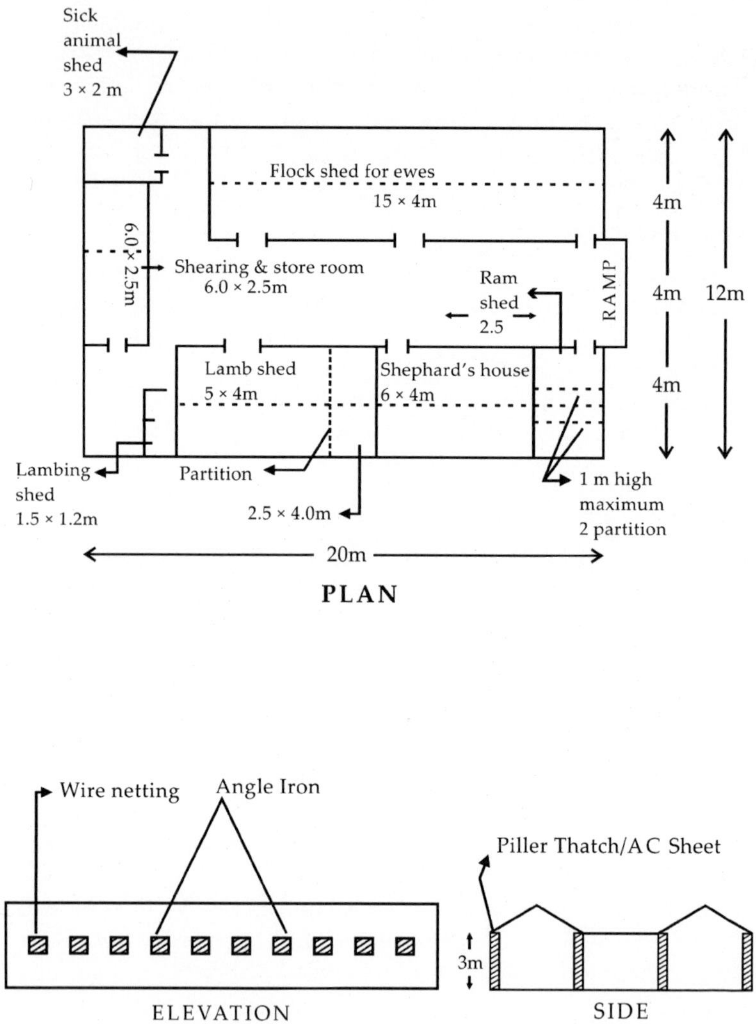

Fig. 2.1 Ground plan for a sheep shed

Improved Sheep Shed

placed alternatively at 1.5 m centre to centre. In the case of lamb shed the upper one third part may be of wire netting supported by angle irons. The walls on the three sides of the shearing and store room may be of brick in cement mortar. The front wall (One of the broad sides of the room) may be of brick in cement mortar up to one half the height from the ground level and the upper one half portion may be of wooden battens or wire netting.

Roof : The roof may be made up of corrugated steel sheet, or asbestos cement sheet and where the rainfall is not heavy it may be of thatch.

Gate : Each shed may be provided with one or more gates on the long or broad sides of the sheds. Depending upon the dimensions of the shed the dimension of each gate may be 0.8 m broad and one metre high, the gate leaf and frame may be made of wooden battens.

Manger: The manger may be either of cement concrete or of wood with two compartments for providing food and hay. A separate hay rack may also be provided by fixing at level or slightly before the heads of the animals. With the half of clamp the manger may be raised within the height ranging between 450 and 600 mm from the ground. The water trough may be of cement concrete or and palls or buckets and may be fixed or hung from a hook fixed to the walls. The manger may also be of portable type. The numbers of mangers and water troughs in each shed may vary according to the number of animals.

Dipping Tank: To protect the animals from infection, the dipping tank may be made either of galvanized steel sheets or constructed of stone or brick in cement mortar. If a galvanized steel tank is used, it shall be well bedded down and the soil tight against it to prevent the sides of the bath from bulging when it is filled. If the base of the soil is unstable the tank may be bedded in cement concrete. The dipping tank may be at one side of the yard.

Foot-Bath : A foot bath made of galvanized steel sheet or brick in cement mortar shall be provided at the entrance to the yard to protect the animals from foot rot disease. These baths may be embedded in the soil suitably.

Providing Shelter for the Farm Stock

Construct the shed large enough to provide 1.0 to 1.2 square metre of floor space per ewe. Make the door ways wide to prevent crowding and injury to pregnant ewes. Provide doors so that the building can be closed during lambing season and periods of severe storm. Slotted flooring is preferred which should be 1 metre above the ground level. The walls are made of brick in lime mortar upto 1.0-1.25 m height. The upper 0.5-0.7 m may be chain link netting of galavanised steel or barbed wire. The height of the shed at the centre should be 2.7 m and on the sides 1.5 m. Thatched roofing should be provided to prevent against severe heat and cold.

For lambing pens, construct panels about 1.2 m in length and fasten them in the building, providing about 1.5 square metre of floor space for the ewe and lamb. The panels can be hinged together in pairs.

Allow approximately 0.1 square metre of window space for each 2.0-3.0 square metre of floor space so that the interior will be reasonably lighted. An earthern floor is satisfactory if dry and kept well bedded. Gable type of building with sufficient space overhead for storage of bedding materials, hay and grain. This adds to the cost of the structure, but less labour is required in carrying to the stock.

Shelter for the Range Flock

For range flocks, shelters are primarily at lambing time. Central lambing quarters are provided. Moveable lambing equipment is used on some ranches. For most or all of the year, the typical range flock is provided by moving sheep to winter ranges, where the temperature is more moderate and where there is some natural shelter from intense winds. As a protection for sheep in open range, provide 1.5 metre high thorny wall and gate of 2.0 metre height to protect against wolves, predators etc.

Materials for Shed Construction

These are straw, asbestos, galvanised iron sheets, angle iron, tubular frame, cement, lime, sand and bricks.

Farm Equipments

These are spring balance, clincher, ear-tags, tag printing dieset, hoof cutter, shearing scissors, measuring tape, feed and water troughs, hand troley, axe, sickle, wooden scraper, buckets, torch, tattooing forceps, hay racks etc.

Managemental Practices

The success of farming system lies in observing certain routine practices which help in proper upkeep of the livestock and their records. The valuable information thus collected can be utilised for the improvement in the system and for better production from the animals. The following are certain managemental practices being routinely undertaken on a sheep farm.

Cleaning and Disinfection of the Shed : In slotted flooring, the dung and urine are accumulated below, which may be cleaned once in a week, the rest of the flooring should be cleaned daily. Lime or bleaching powder may be sprinkled in the shed once in a week. The walls, floors and roofs may be disinfected with flame gun once in a fortnight. The walls may be white washed with lime water to which 5% phenol is added. Sodium hydroxide at 4% may be used as spray for disinfection. The frequency and type of disinfection can be decided according to necessity.

Ideal Shed under Village Conditions

The methods of housing the sheep differ according to climatic conditons. In tropical regions open sheds or semi open sheds are suitable for providing shelter to sheep in order to protect them from severe rains, heat and predators. For open sheds and semi open sheds, no special construction is required. Only fencing and roofing is essential. The materials required for construction of such sheds should be chosen depending upon the availability of articles locally at cheaper rate. Iron angles, wire, asbestos sheets, thatch, galvanised iron sheets, poles, bamboo, bushes, wooden planks, slates etc. are the materials required for fencing and roofing and one can select the articles for construction of such sheds according to his economic situation. The space required varies from breed to breed and individual to individual as such dimension of the shed also varies.

Open sheds and semi open sheds are good for healthy adult livestock throughout the year. However, during severe winter and summer artificial protection made up of grass or bamboo may be provided on the sides. The difference between open shed and semi open shed is slight. The semi open shed is open only on one side.

In temperate regions viz. Jammu and Kashmir, Himachal Pradesh and Hilly districts of U.P. majority of notable sheep breeders are nomads. They carry their sheep flock to lower hills during winter season and again migrate their livestock to alpine pastures during summer months. They remain always on migration and do not construct any shed for their sheep. However, some sheep breeders do not take their sheep flock to winter abode and are kept at their native place even during the season when heavy snowfall is ensuing. Under such conditions permanent sheep sheds are required to be constructed for protecting the animals from biting chills and to minimise the deaths from pneumonia. For construction of permanent sheep sheds, it is desired that it should be cheap and hygienic and warm. The materials needed are stones or bricks, mud plastering, wire mesh netting, wooden beams and plants, bhojpatra, bushes, earth etc. and if financial position permits iron sheets or slates can be used for better roofing. During very severe cold, the doors, windows are kept closed and covered with curtains made up of gunny bags. The animals remain entirely on stall feeding till summer months set in.

The design of a modern shed at S.B.F. Tal (H.P.) or Central sheep Breeding Farm, Hisar is as under :-

In this type of shed, special attention has been given to the flooring system. The cement and stone flooring stain the wool and unhealthy to sheep. Similarly the 'Katcha' flooring is also not suitable. Therefore, to avoid all this, wooden batten flooring has been used in these ideal sheds.

In this type of flooring, the floor is constructed about 2 metres above the ground level with the wooden battens 4-6 cm wide fitted at a distance of 1.0 cm from which urine and droppings can fall down easily. These sheds have been constructed in the shape of big halls 50 m × 30 m of size, sub-divided into 8 to 10 compartments of required size for different age groups of animals and different purposes. The walls of 2 to 2.5 m height from this wooden flooring with sufficient number of windows and ventilators have been provided. The roof is of slanting type and with the beam height from floor is 3.0 to 3.5 m. Entrance is provided from both the ends. The compartments are interlinked in such a manner that minimum manual handling of the animals is needed in performing various operations in the shed. The space for shearing and treatment is also provided along with feed storage facilities. The urine and droppings slip from the batten flooring which keeps the flooring clean and animals don't soil themselves. The droppings etc. are collected from the ground floor at regular intervals and converted into a high quality manure. In this way minimum labour is involved in addition to the healthy atmosphere for the animals and production of good quality wool.

A shed of 50 m × 30 m is sufficient for keeping 1000 to 1200 sheep of different ages and sizes and to perform different operations.

3

Establishment, Migration and Culling

Before starting a sheep farm, the purpose of starting sheep rearing should be clear in the mind of the sheep farmer. Normally sheep farming is done for, mutton, wool or pelt production. Different breeds of sheep are reared for different purposes. It is very essential to have primary information about these breeds. Sheep farmer should be aware about the demand of the particular product in the market e.g. wool, mutton or pelt. It is better to start sheep farming with the breed predominant in that area.

After knowing these points, ewes and rams should be selected carefully for the flock. Ewes should be 2-4 toothed, active, good size and wider pouch girth. Non of the teats should be blind. The wool should be uniform on the body. This can be done by moving hand on the back. The sheep which have lambed for the first time should be selected. They should be free from diseases.

The maxim that "the ram is half of the flock," because the quality of the lamb born depends upon the ram. Ram should have good quality wool on its body. It should be healthy, sturdy and high quality breedability. Its legs should be strong, back wide and age between 2-3 years, if is selected for wool, it should have uniform cover of wool all over its body. The wool fiber should be thin and fine. If it is selected for mutton, in that case, its body should be uniformly meaty, legs small and straight. Following points are to be kept in mind while selecting a breeding ram :

(a) *Look for twinning history* : In order to achieve a high lambing rate, one should search a ram with known history of twinning.

(b) *Select heavier rams* : The selection of a ram should be based on body weight basis because body weight is correlated with fleece weight and density. Normally the rams in private flocks are selected at the weaning stage as the weight at weaning is correlated with mature body weight.

(c) *Select healthy rams* : Selected breeding ram should be vigrous, free from diseases, internal and external parasites.

(d) *Buy early* : Do not wait until the last minute to find a ram. It is a desirable practice to get new rams accustomed to new environmental and feed conditions so they will be ready for the breeding season. A good rule is to purchase rams a month or two before turning them, with the ewes.

(e) *Obtain rams according to the quality of ewes* : Always keep in mind the poorer points of your ewe flock and buy rams that way to be able to correct these faults

(f) *Seek advice of others* : Experienced breeder or other similar individuals can assist materially in helping the inexperienced to obtain good rams.

(g) *Selection based on production records* : The surest way of selecting rams for breeding purpose is by a production test. Production testing includes (i) the type and finish at weaning time (ii) prolificacy (iii) weight gains upto weaning and (iv) the weight and quality of fleece. A ram whose offspring has shown up well in production test is known as proven sire.

Precautions for using the selected ram

Points which are to be kept in mind before using the selected ram for breeding purpose are

(a) *Feeding of rams* : Well balanced ration should be fed to breeding rams at least a month earlier to start breeding and should be continued till breeding operation is over.

(b) *Number of ewes per ram* : Under range conditions, three rams per 100 ewes is a desirable ratio. Yearling rams can handle 25 to 40 ewes. However, a mature strong ram of 2-5 years old may handle upto 75 ewes, although 35 to 50 is the most desirable number of ewes for breeding season under farm conditions.

(c) *Shearing of rams* : Shear the breeding rams all over if they are to be used early. Atleast shear legs, neck, brisket, belly and scrotum. Ram will be more active and it is easier for them to cover a ewe if they are shorn.

(d) *Care of the feet*: Rams and ewes should have their feet well trimmed so that they can walk naturally.

(e) *Cryptorchidism*: Ram with one testicle retained within the body cavity/abdomen should not be used for breeding purpose as there is tendency to occur in succeeding generations, although it will not appear in all cases.

Age

Normal age of sheep varies from 10-15 years. Its production capacity starts decreasing after 6 years of age. It is very essential to have knowledge about their age for purchasing or selling. The age of sheep can be judged from the number of rings on the horns in horned breeds, but it is not possible in most of our sheep. To judge the age of the sheep, condition of the teeth is helpful. The age of a sheep is indicated by the number and condition of teeth. Like other mammals, sheep have two sets of teeth, viz. the temporary set called the milk teeth and permanent set.

The lamb has small, narrow teeth that are known as milk teeth. At twelve to fourteen month of age, the two center incisors are replaced by two large, broad permanent teeth. Each succeeding year, an additional pair of permanent teeth appears, one on either side of the first pair, until at four years of age when there is a full mouth. A mature sheep has 32 teeth, of which 24 are molars and 8 are incisors. All incisors are in the lower jaw. The two central incisor teeth are called pincers; the adjoining ones intermediates; the third are, second intermediates, and the outer ones, corners.

Time at which these teeth come, varies with different breeds e.g. in Nellore sheep 1st, 2nd, 3rd and 4th pairs of incisors erupt on an average at 1 year 3 months, 1 year 8 months, 2 years 2 months and 2 years and 8 months, whereas the corresponding ages for Mandya breed are 1 year 5 months, 1 year 11 months, 2 years 9 months and 3 years and 4 months.

Table 3.1 : Relationship between age of sheep and dentition

Age	Number of permanent incisors	Sheep commonly called
6 month	None (8 milk teeth)	Lamb
1-1.5 years	2	2 teeth
1.5-2 years	4	4 teeth
2.5 - 3 years	6	6 teeth
3-4 years	8	full mouth

In this way all the incisors erupt by 4 years of age. This method is simply a method of estimation, because the eruption of incisors depends upon breed, condition of pasture etc. Fourth pair of molars erupt at 3-5 month of age and fifth and sixth pairs of molars develop by 1½ -2 years of age. The positioning of teeth at different ages are as follows:

$$\text{Formula for Temporary teeth : } \frac{0\ 0\ 3\ 0}{4\ 0\ 3\ 0}$$

$$\text{Formula for Permanent teeth : } \frac{0\ 0\ 3\ 3}{4\ 0\ 3\ 0}$$

It is difficult to tell the age upto which the teeth remain useful. They start falling at the age of 5 to 6 years. In some cases they may not fall even upto an age of 8 to 9 years. With the advancing age the gums start shrinking and the roots of the teeth are exposed. If the one or more than one incisors fall in a sheep, such sheep are called "broken mouth". The sheep whose all the incisors have fallen is called a "gummer". The sheep with broken mouth or gummer can not cut grass, with the passage of time they loose condition.

Implements needed for sheep farm

Following articles are needed at an organised farm :

1. Feeding troughs of 3.6 × 0.25 × 0.25 metre size can be used. These can be made up of wood, iron, cement, concrete and bricks.

2. Water troughs of 3.0 × 0.25 × 0.25 metre can be used for watering sheep.
3. Aluminium tags and clinchers.
4. Tatooing machine with letter and number set with tatooing ink.
5. Docking knife.
6. Rubber bands and elastrator for docking.
7. Burdizo castrator.
8. Dial type 5 kg Salter balance for weighing lambs
9. Drenching cup.
10. Dial type 20 and 100 kg, hanging type Salter balance.
11. Vaccination syringes and needles
12. Iron dipping tank.
13. Foot bath made up of aluminium
14. Scissors for shearing wool
15. Hoof pairing knife
16. Milk feeding bottles
17. Gas stove
18. A.I. kit.

Migration

The means of transporting sheep are not well developed. Normally, the sheep may be transported by road, rail, ship or aeroplane. Very little recorded information is available regarding the problems encountered during their transportation (Matharu, 1962, 1963a, b; Kale, 1966). Only trained hands should be selected for escorting sheep. Sick and lame animals should not be transported. Care should be taken to feed and water them before transporting. During transit, sheep should be given water twice a day and fed 1.0 to 1.5 kg of good quality hay or given *pala* or gram *bhoosa*. They should be fed at each resting point. In each truck or wagon there should be two attendants to lift the sheep whenever they fall. If sheep are to be transported by road, the journey should be started before dawn. If the journey is long and the day is hot, better undertake journey during night. Approximate space per sheep needed in a truck or Railwagon is given below :

Table 3.2 : Space needed while transporting sheep by rail or by road.

Body weight of sheep	Space per sheep in sq. fl.	
	In a truck	**In a rail wagon**
1. 25 kg wooled	2.2	2.3
2. 30 kg wooled	2.4	2.5
3. 25 kg shorn recently	2.3	2.4
4. 30 shorn recently	2.2	2.3
5. 40 kg wooled	2.6	2.7
6. Pregnant ewes	3.0	3.4

Small lambs should be separated from their mothers and provided safe place in a corner to save them from being trampled. It is better to keep them on upper deck.

Trucks and wagons should be thoroughly scrubbed, cleaned, disinfected and properly bedded. The CA type wagons with wooden floor should be preferred for transporting sheep. A broad gauge wagon of this type can be used for transporting 75 sheep and truck having a floor area of 4.8 × 1.2 m, can carry about 50 sheep. It is better to bring the sheep to loading. This will give them a chance to rest and empty their bowls to ease abdominal pressure. Migration of sheep also takes place in search of fodder. Most of the sheep farmers do not have land to graze their sheep. The time of migration as soon as the grass starts drying after rainy season. Infemine years 80-90 percent sheep move out on migration toother provinces. Some sheep go on migration for a short period and others remain on migration for a longer period. For the convenience it can be divided into two parts depending on the area from where the animals are migrating:

(1) *In plains* : Sheep migrate in plains due to failure of rains in winter season or due to being sheep in large numbers in that area. 1.5 million sheep migrate in Rajasthan every year. Out of these 0.6 million remain on migration throughout the year. Normally migration starts after Diwali. If there are good winter rains, the migration may start after the harvest-ing of fields in March/April.

    Flocks from Rajasthan migrate to Delhi, Gujarat, Maharashtra, Madhya Pradesh and Uttar Pradesh and flocks from Kutch region of Gujarat migrate to Jalore and Sirohi districts of Rajasthan. Eight to ten flocks of a village join to form a bigger flock. A man moves ahead of flock to know about the position of fodder and water in the next village. A flock covers 600-1000 km during one migration. A donkey and a camel cart moves with the flock; which is used for carrying newly born lambs and their luggage. Breeding in the migratory flocks is regulated by tieing prepuce of the ram with a woollen thread. Sheep return with the onset of rains. At that time they are 2-3 month pregnant

(2) *In Mountains* : Sheep move out on migration in mountains with the onset of summer season. While grazing, they climb upto 3000- 5000 m height. There are 400-500 sheep in each such flock. Six to eight week old lambs move with the sheep during migration. Sheep farmers cook their food themselves. They carry a piece of tarpaulin to make tent for resting at night. Sheep and lambs are given common salt once or twice a week. Sheep are not watered upto 2-3 days. Sheep farmers believe that by doing so they eat more. Rams are normally kept with the ewes. The flocks return to plains with the onset of rainy season.

Table 3.3 : Routes of migration of sheep in plains

Route	Period of migration	Months of shearing	Places of shearing	Wool quality
Nagada Region :				
a) Jaore, Sirohi Ratlam, Nagda	November to July	Mid December to mid February.	Nagda Agra	Medium to inferior quality
b) Jodhpur, Pali, Falna, Neemuch, Ratlam.	November to July	Mid December to mid February.	Nagda Agra	Medium to inferior quality
c) Ajmer, Bundi, Jhalawar, Ujain Indore	November to July	November	Kota Jhalawar	Medium to inferior quality
Gujarat Region :				
a) Jaisalmer, Barmer, Jodhpur Palanpur.	November to June	February to March	Diga Palanpur	Medium to inferior with burrs.

Eastern U. P.				
a) Jaisalmer, Barmer, Jodhpur, Ajmer, Dudu, Jaipur, Bharatpur, Agra	Mid November to December	February to April.	Jaipur Dausa Bharatpur.	Usually unwashed
b) Ajmer, Niwai, Chaksu, Dhaulpur, Agra.	Mid November to December	February to April.	Niwai Dhaulpur	Usually unwashed
c) Jalore, Pali, Jodhpur, Ajmer, Jaipur, Chaksu, Dhaulpur, Lalsot, Gangapur, Gwalior, Etawa, Mainpuri	November to July	January to May.	Jaipur Chaksu Lalsot Etawa Mainpuri	Medium to inferior
d) Ajmer, Jaipur, Dausa, Bharatpur, Agra, Hathrus.	November to july	January to May.	Jaipur Chaksu Lalsot Etawa Mainpuri	Medium to inferior
Kuchhi Region :				
a) Junagarh, Dhar, Indore, Ujain, Jhansi	May to June.	July to August	Kota Jhalawar	Medium to inferior.
b) Dhanduka, Baroda, Indore, Abu Road.	May to June.	July to August	Shivganj Sirohi.	Medium to inferior.
c) Ajmer, Radhapur, Sanchor, Barmer	May to June.	July to August	Balotra Barmer.	Medium to inferior.

Sheep Farm Calendar and Culling

The success of farming system lies in observing certain routine practices which help in proper upkeep of livestock and their records. The valuable information thus collected can be utilized for the improvement in the system and better production from the animals. We will be discussing certain operations being routinely undertaken in a farm as a package of practices for progressive sheep farming.

According to season, the operations followed at farm can be divided into:

1) Spring (February to March)

This part of the year is neither very hot nor very cold. The grasses start sprouting at the end of the winter season. In addition to these stubble grazing is also available in gram harvested fields. The sheep start coming in heat. Sporadic cases of enterotoxaemia are also noted. The sheep should be vaccinated against this disease during this period. The sheep are dipped in D.D.T., B.H.C., Gerathion or Smithion solution, 8-10 days after shearing, if there are no shearing wounds on the body. This is done to kill ticks, lices etc.

2) Summer (April to June)

The climate becomes hot and grasses start drying, but sheep survive by stubble grazing in wheat harvested fields. They get *Acacia* pods, *'Khejari loong'* and *'pala'* leaves during this period. They also come in heat during this period. These leaves and pods are excellent source of nutrition for pregnant sheep. They should be given supplemental ration during scarcity periods.

3. Rainy season (July to August)

With first showers, grass starts sprouting. The sheep start gaining weight, due to pregnancy and getting good grazing. The lambing starts by the end of the rainy season. The sheep should be drenched against internal parasites. They should be protected against foot rot.

4) Autumn (September to October)

Shearing operation starts during this period. Culling is also done at this time. Some of the left sheep also become pregnant. Lambing of sheep continues. Tail docking and castration is also done during this period. Sheep get sufficient fodder in Maize, *Jowar* and *Bajra* harvested fields.

5) Winter (November to January)

The grass starts drying and scarcity period starts. Sheep gone on migration on mountains start returning to plains. The lambs should be protected from cold.

Culling

In addition to improved management and health practices, culling of sheep is also very essential to keep the flock in good condition. It is better to keep a few sheep giving more and good quality wool than a large number giving small amount of inferior wool. Best time for culling is at shearing because at this time wool quality can be judged and body can be examined thoroughly (Buch and Jayaraman, 1953).

All those lambs who do not represent the breed, have black spots on the body, bent legs and suffering from contagious diseases should be culled.

Rams who produce inferior quality semen and wool and their age is more than five years should be culled.

Sheep which are black in colour and have black spots on the body, the wool is yellow inferior, and small in quantity and weigh less than the average of the flock can be culled. They should not have mastitis and giving small amount of milk. They should not have broken mouth and represent the breed. The ewes whose shoulders are not broad and have over or under shot jaws should not be retained in the flock.

4

Sheep and Lamb Management

Normally sheep come in heat throughout the year, but they come in heat in larger numbers after the harvesting of *Rabi* crops (Kaushish, 1980). Soviet Merino and Rambouillet come in heat in larger numbers in Autumn than in Spring season. If these are bred in the beginning of the breeding season, the conception rate is poorer than when bred in the middle or at the end of breeding season. The twining percentage is lower during first lambing (Kaushish and Sahni, 1977a). For the convenience, management of sheep can be divided into three parts:

Dry Ewes

The sheep continue to come in heat every 17-18 days unless they become pregnant. Many shepherds forget about the ewes after they are mated and start looking forward to her delivering a lamb after 5 months. The care of pregnant ewe is important as the lamb develops during this period and this development inturn determines, the rate of growth of the lamb after birth. A well looked after and properly maintained ewe during pregnancy will have a good flow of milk, when the lamb is born and will be able to develop and maintain good health later in life. On the other hand if the pregnant ewe is not cared for, the lamb born may be weak or develop poorly and die or there may be loss due to abortion and the ewe herself may be so weakened that she may be inclined to disown her lamb and may require much more attention at lambing time.

Pregnant Ewes

The control over the nutrition of the pregnant ewes is essential for economic reasons. The nutrition of the pregnant ewe must be concerned not only with the definition of nutrient requirements but also with an assessment of the effects on production of levels of feeding which are lower than their full requirements. It is convenient to divide the period of gestation as follows :-

(i) The first 3½ months (early pregnancy)

(ii) The last 1½ months (late pregnancy)

The first 3½ Months

Ewes in good body condition at mating shed more ova and have lower rates of embryo mortality than ewes in poor condition. The effects of current nutrition on embryonic mortality are less clear. Embryo survival can be adversely affected by severe under-nutrition even of very brief duration. It is a good practice to avoid undernutrition during breeding season and also in the subsequent 2-3 weeks. During the second and third months of gestation the various tissues and organs of the foetus are undergoing rapid development, although the absolute weight of the foetus is under-going rapid development, although the absolute weight of the foetus is still relatively low. By the end of third month the foetus weighs about 300 g or approximately 14% of its birth weight. Placenta attains about 95% its final weight of about 250 g per foetus by the 90th day of gestation.

The literature does not present a clear picture of what constitutes an optimum level of nutrition this time. In attempting to resolve these apparently anomalous results it is necessary to take into account a number of factors including the age and condition of the ewe and the level of late pregnancy nutrition. It is clear that severe undernutrition during mid pregnancy can reduce the weight of the foetus and of the placental material at 90 days. Compensatory growth of cotyledonary tissue and particularly of that part associated with the foetus, is known to occur during late pregnancy, but this will of course require a relatively high level of nutrition at that later stage. One must take into account the age and body condition of the ewe at mating in making any recommendation regarding nutrition during mid pregnancy. An acceptable body weight loss during 2-3 months of pregnancy is about 5% of mating weight.

The last 1½ months

By this time the placenta is almost fully developed and major foetal tissues and organs are formed and all that remains is an incarease in size and weight. In addition to the nutrients needed for foetal growth and development plus those required to support the other products of conception, nutrients are needed for the satisfactory development of mammary tissue and of course to meet the maintenance requirements of the sheep herself. Care should be taken that pregnant ewes do not lose weight during this period. It is very essential to give supplemental feeding to sheep. If good quality fodder is fed during this period, the lambs born will be heavier at birth and sheep will be able to produce sufficient quantity of milk (Sehgal and Kaushish, 1989).

Merino Ewe with Lamb

Lactating Ewes

During the earlier weeks of life, the growth of the lamb is largely determined by the amount of milk it obtains from its mother. The methods of estimating yield have been reviewed by Boyazoglu (1963). Although the effects of nutrition on lactation in the ewes are well established in principle, the diversity of conditions under which the experimentation has been carried out makes it difficult to describe relationship between feed inputs, milk production and weight changes which are of general validity.

Energy requirements for ewes during lactation have been published in UK by MAFF and Boaz *et al.* (1973) and the USA by NRC (1975). There are wide variations in the estimates of both the energy requirements for maintenance in sheep and the efficiency of conversion of metabolizable energy to milk energy. The efficiency of utilization of body tissue for the production of milk energy has not been measured in ewes and a value etimated in the dairy cow (Moe *et al.*, 1971) has to be used.

The response of the lactating ewes to metabolizable energy intake has not been well defined as most experiments have not had treatments spread over adequate range of nutrition for the response cruve to be established. In some experiments (Owen and Ingleton, 1963; Gardner and Hogue, 1964), the differences in milk yield were approximately proportioal to the levels of intake but in others (Barnicoat *et al.*, 1949) the responses were less than the differences in intake. In the first 4-6 weeks of lactation, reduction in live weight usually occur and these losses may be large. Maintenance or gains in weight in early lactation are usually associated with low milk yields (Guyer and Dyer, 1954).

Robinson *et al.*, (1974) and Calderton cartes *et al.*, (1977) have demonstrated rapid responses to dietary protein concentration in ewes in negative energy balance in early lactation.

Management

Pregnant ewes should not be subjected to fright. They should not be crowded together, nor allowed to jump over hedges or ditches. They should not be driven through narrow passages, nor taken to long marches during the hotter part of the day. They should be provided shade adjacent to midday watering points during hot summer days.

Preparation for lambing

The pregnant ewes should be sheared 4 weeks before the onset of lambing. If the ewes are not shorn, atleast crutching should be done on hind quarters, around the tail, vulva and udder to avoid soilage at the time of parturition. Lambing of ewes should be done in two ways (1) Open air or range lambing - in which the ewes have their lambs on pastures in warm climate. The shepherd should pay more attention in protecting the lamb and ewe from predators and assist the ewe in difficult lambing

(2) Shed lambing or pen lambing - in this the shepherd provides clean and neat pen for lambing of ewes. The shepherd should remove droppings every day in the morning. The pen should be in a position to protect the lamb and ewe from predators. A light may also be provided in the lambing pen.

If the size of the flock is large, the pregnant ewes may be separated and grazed separately so that the individual ewe receives due attention. Two teeth ewes need to be watched more closely as they sometime run away after giving birth to the lamb. The shepherd should remain closer to the flock but move slowly without disturbing the ewes. The incidence of the difficult births are higher among young ewes. The shepherd should assist the ewe at the time of lambing.

If the breed involved is a poor milker, it may be ensured that sufficient number of milking goats are available in the flock when the lambing is to commence.

When the lamb is born it is a general practice in the villages to give 250g of sweet oil and handful of molasses to each ewe. It is a belief of most of the farmers that the feeding of the oil and the mollasses produces sufficient quantity of milk in addition to helping them in recouping their health. The oil and molasses, both provide the high energy needed by the lactating ewes to cater to the needs of their offspring.

In young ewes and malnutritioned ewes, the placenta may not be thrown fully due to weakness, at that time it is better to give pitutrine to the ewe to help in letting down of milk and throwing the placental remanents to avoid metritis and poor milking. Sometimes the young ewes disown their own lambs, to rectify the disowning of the lambs, use different techniques such as (1) tying the ewe several times a day along with lamb and (2) tying a dog out side the pen (3) putting the ewe milk on the body of her lamb, (4) placing the skin of the dead lamb over the orphan lamb.

Lactation

When the lamb is bom, first thing that the shepherd does is to ensure that within half an hour or forty five minutes it begins to suckle and if he finds that the lamb is unable to suckle, he should assist it in doing so. Each lamb should receive the attention of shephred. In a number of cases, it may happen that the milk supply of the ewe is insufficient. Ewes those are normal otherwise and free from fever or infection, may be brought in milk to feed their milk to such lambs. Succulent fodders such as silage or root crops are excellent for this purpose. In the meantime the lamb will have to be fed either by foster mother (she Goat), cow milk or milk taken from other ewes. The of lambs should be allowed to remain with the ewes to get milk they may produce, as frequent nursing encourages milk production. Hand feeding should be used as a supplement to the ewes milk. When ewes fail entirely to come to their milk the lambs will have to be hand fed entirely unless they can be placed on another ewe.

How to treat plugged teats

Ewes teats often become plugged either by thickened colostrum or from skin growing over the teat opening due to an injury.

Soon after the lamb is born a little milk should be drawn from each teat. This will remove any thickened ckened milk and assure the lamb of being able to get its food. If skin has grown over the teat opening, a piece of small pointed wire which has been disinfected by heating over a match flame may be used to open the teat canal.

Symptoms of parturition

Sheep tries to separate herself from the flock. If shed is in the shade it tries to search peaceful corner. She scratches ground with her fore hooves, gets up and sits down frequently, lies down on right side and stretches out its legs outward. She grinds her teeth frequently. Respiration rate increases (Kaushish and Arora, 1974a). Native sheep mostly lamb between 3-6 am and 3-6 pm (Kaushish *et al.*, 1973).

Normally pasture grasses are deficient in phosphorus and protein, in addition to these deficiency of vitamin A is also observed. To compensate for the deficiency of these, they should be given barley, gram, guar, jowar or cakes like groundnut, sarson, cotton seed cakes or leaves of Khejiri, Ardu, Ginga and Acacia pods. To compensate for protein and energy deficiency they can be given urea-molases-mineral bricks. If they are fed dry fodder with these, it becomes better than green fodder and concentrate mixture.

Table 4.1 : Components of a lick.

S.No.	Name of material	Quantity (kg)
1.	Water	2.0
2.	Molases	10.0
3.	Urea	1.5
4.	Mineral mixture	1.0
5.	Common salt	0.5

Dissolve urea in water then add molasses; mineral mixture and common salt. Add 10-151 water to this mixture and spray it on 100 kg chaffed fodder and feed to adult sheep. It can be fed to pregnant sheep also, lambs below 3 months of age should not be given this ration (Sehgal and Kaushish, 1989).

Table 4.2 : Method of feeding

Type of animal	April to June	July to August
1. Ram a) used for breeding	i. Urea molases mixed fodder 500 g	Concentrate mixture 300 g
	ii. Concentrate mixture 200 g	—
b) other rams	i. above treated fodder 750 g	—
2. Ewes a) Non pregnant and pregnant upto 3½ months	i. above treated mixture 1000 g	—
b) in last 1½ month of pregnancy	i. above treated mixture 750 g	Concentrate mixture 300 g
c) upto 1 month of lactation.	i. above treated mixture 700 g ii. Concentrate mixture 300 g	Concentrate mixture 450 g
d) in 2nd and 3rd month of lactation	i. above treated mixture 500 g ii. Concentrate mixture 200 g	Concentrate mixture 300 g
3. Lambs		
0-1 month	Green fodder only	green leguminons fodder
1-2 month	Concentrate Free choice	concentrate freechoice

Type of animal	April to June	July to August
2-3 month	Concentrate 50-250 g	concentrate freechoice
3-6 month	Concentrate 350 g	concentrate 350 g

Lambs

Normally native sheep give birth to one lamb at a time, but crossbred sheep give birth to one or two lambs at a time. To make profits from sheep husbandry it is essential that more lambs are produced and they survive (Jayaraman and Buch, 1953). For this it is essential that they get sufficient quantity of milk at appropriate time. The management of lambs can be divided into two parts for the sake of convenience :-

1) Upto 3 months

The face, eyes and rest of the body are covered with membranes at the time of birth, normally, ewe herself cleans the lamb, but if she does not do it, they should be helped. If the mother of a lamb dies or does not suckle its lamb, such lamb should be put to other ewe which has sufficient milk or her lamb has died. If the ewe does not adopt such orphan lamb, the placenta of such ewe should be rubbed on the body of the lamb. The skin of the dead lamb can be tied on such lamb. If she still does not adopt such lamb, then her fore legs should be tied and a dog may be tied near by. She may adopt lamb because of fear.

If she still does not adopt lamb, then it will have to be fed artificially. Lamb should be given warm milk mixed with cod liver oil. It should be protected from hot and cold winds. It should be given milk 6 times a day upto first fortnight, 4 times a day in second fortnight and twice a day upto 3 months. Water should be provided to lamb in the shed. The floor of the shed should be dry and covered with bedding so that it does not start licking soil. It should be provided soft leguminous succulent fodder. It should be given concentrate mixture and should be taken to pasture after is two weeks of age.

Table 4.3 : Composition of concentrate mixture for lambs and adult sheep

S. No.	Contents	For lambs	For sheep
1.	Maize/barley	50%	58%
2.	Groundnut cake	30%	10%

S. No.	Contents	For lambs	For sheep
3.	Wheat bran	16%	28%
4.	Trace elements	02%	07%
5.	common salt	02%	02%

Table 4.4 : Quantity of concentrate mixture for native and crossbred lambs.

S.No.	Age	Native lambs g/day	Crossbred lambs, g/day
1.	Upto 1 months	30-60	50-100
2.	1-2 months	50-100	100-200
3.	2-3 months	100-150	200-300
4.	3-6 months	150-250	300-400
5.	6-9 months	200-250	300-350
6.	9-12 months	200	300

Identification of lambs

Common methods of identification of lambs are

1. Ear marks

Identification for sheep can be provided by ear marks, made with either a sharp knife or a regular ear notcher. Such marks are permanent and easily recognized, but unattractive.

2. Metal ear tags

It is a common method in our country. Sheep are provided with a metal ear tag in each ear, bearing breed and number. Metal ear tags are easily attached but easily lost. Also, they frequently rub or scratch the skin, thus making openings for screw worm infestation.

3. Ear or Thigh tattoo

This methods of marking consists of piercing the skin with instruments equipped with needle points which form letters. This operation is followed by rubbing indelible ink into the freshly pierced area. On darkskinned animals, tattoos are difficult to read, therefore, it is avoided.

Docking or Tailing Lambs

The chief reasons for docking lambs are :

1. Docked lambs are cleaner.
2. Long tails interfere with breeding ewes.
3. Docked lambs and sheep presentdt a uniform and blocky appearance.
4. The dressed carcass has a more attractive appearance

Methods of docking or tailing lambs

1. In lambs under the age of three weeks the tail may be cut away by a firm stroke with a sharp knife. Some bleeding occurs but in young lambs it is seldom serious.
2. Tailing by means of docking iron (hot chiesel or hot pincer) should be carried out on older lambs, say at the age of eight to ten weeks. The iron is heated sufficient to sear the flesh, thus preventing excessive bleeding and reducing risk of the tail stump.
3. The modem method which is blood less, humane and avoids the risk of infection is placing a small, highly tensile rubber ring with the help of elastrator between the segments of the tail, causing atrophy and separation. It is prefered when lambs are seven to fourteen days old. The effect of docking on fat deposition and breeding performance has been studied by Qureshi (1968) and Qureshi and Shaw (1968).

Castration

Wether lambs produce more desirable carcass having a greater development in the regions of high priced cuts. Wether lambs are quieter and fatten better than ram lambs. Ram lambs also annoy the ewe lambs of the flock. These are the chief reasons for castrating male lambs.

Methods of castration of lambs

1. When sharp instrument like knife is used, the hands, instruments and site of operation should be thoroughly disinfected with good disinfectant. After disinfection, with the help of sharp knife remove lower third of scrotum, exposing the testicles which are then drawn out by a special forceps until the spermatic cords break. The method known as "drawing" can be safely used in case of lambs between age of three to six weeks.

2. For older lambs, it is advisable to use a clamp before severing the testes from the cord and to apply a hot iron to seal severed ends. This prevents serious bleeding. The method is usually employed in rams used in breeding flock where late castration is unavoidable because male lambs must be given adequate time for development before the best are selected for breeding purpose.
3. The bloodless method involves the use of a special instrument known as the "Berdizo" which can exert great pressure on the spermatic cord through the walls of scrotum. There is no cutting or loss of blood and the tissues of the cords are so damaged by pressure that the testes atrophy and disappear in the course of time. This method is more humane and avoids the risk of infection. But this procedure has two major disadvantages i.e. a small percentage of lambs may be imperfectly castrated if not operated carefully. Such lambs will show rather more masculine characters. Hence it is worthwhile to check up throughly a month after castration.
4. The elastrator method involves placing a small, highly tensile rubber ring around the neck of the scrotum as soon as the testes are present in the scrotum. The pressure of this ring causes atrophy of the spermatic cord and scrotum by cutting off blood supply.

Elastrater Docking

Hoof pairing

Particularly those sheep flocks which live on clay and loamy soils, there is a tendency for the feet of such sheep to become soft and to develop an abnormal growth of hoof. The front part of hoof usually grows too long, while the wall tends to curve inwards over the sole. Mountain flocks living on stony soils may need little attention to their feet because normal wear permits normal growth.

Failure to deal with excessive hoof growth creates a grave risk of foot rot, the majority of such cases being of an infectious nature and likely to cause widespread infection in a flock. Timely use of the pairing knife preserves a good shape and proper bearing surface to the hoof. This reduces the risk of foot rot to a minimum. Even if foot rot is not apparent, minute and detailed flock examination once a year is a part of good management.

More than 3 months

Lambs should be weaned when they are more than 3 months in age. If it is not done they continue to suckle upto 5-6 months of age. They can harm testis while doing so. Some sheep farmers apply dung to prevent them from suckling. This is unhygienic. This may cause mastitis also. It is better to graze them separately and keep them in a separate enclosure. At weaning they should be given sufficient amount of good quality fodder. In addition to grazing, they should be given concentrate mixture also. They should be given anthelmenthic drench immediately after weaning.

The weight of lambs increases by 11 kg if they are given 500 g of concentrate mixture in addition to 8 h grazing, it increases by 7 kg if they are fed only fo chule the feching or cowpea hay. The lambs can be given pelleted feed available in the market. It has been observed that lambs eat 6% more of such concentrate mixture as compared to ordinary concentrate mixture and produce 23 percent more meat.

5

Reproduction

Any amount of research on storage of semen without fertility tests on large scale will have little relevance to its application in sheep development. Reasonable conception rates have been obtained with freshly collected and diluted semen but there is a need to conduct further researches on storage, packing, transportation and utilization of ram semen and also on the effect of storage temperature, site of insemination and number of live and normal spermatozoa on fertility.

It is now fairly established that artificial insemination is a successful technique in sheep production. The importance of this technique for quickly improving productivity of native sheep in this country was emphasized by the investigations conducted under the able guidance of late Professor A. Roy. The factors which are considered favourable for adaptation of large scale artificial insemination in the country are non-seasonal nature of sexual activity in sheep, the lack of seasonality in semen quality of rams, plenty of cheap available labourers and sunshine for the location of cervix. Lack of organisational efforts are the strongest factors which come in the way of implementation of artificial insemination programme. The fertility of the ram semen drops seriously when it is stored in refrigerator for a period beyond 6-24 hr. Current experience with chilled and frozen semen is not satisfactory and, therefore, call upon new methods/alternatives which could be effectively used to minimise the drawbacks in the scientific application of artificial insemination (A.I.) programme.

However, the successful exploitation of A.I. depends upon gathering more knowledge in applied reproduction and management of sheep.

To evolve a sound basis of application of artificial insemination in sheep, it is necessary that researches on allied aspects of male and female reproduction are also carried out. It is only when basic information on reproductive behaviour of male and female is available in case of native and exotic sheep in this country that better conception rates could be expected from artificial insemination. The socialist system of agricultural organisation has greatly favoured large scale application of artificial insemination in USSR and other East European countries. The nonseasonal character of sexual activity in tropical sheep, the avialability of ample labour in this country together with the fact that all the ewes in a flock could be detected in heat within a span of 20-30 days, could make a reasonable case for adoption of artificial insemination in this species.

Semen Production

Studies on the age of puberty and maturity revealed that majority of indigenous and crossbred rams (50%) start showing sexual desire and start donating semen at an age of 8-9 months, when they attain 20-25 kg body weight. Good quality semen can be obtained at the age of 10-12 months. There was full establishment of spermatogensis and presence of matured leydig cells in testis and seminiferous tubules for secretory activity in seminal vesicles. Fine wool exotic breeds (Rambouillet and Russian Merino) and higher crosses (75%) attained puberty at the age of 18 months with 35.0 kg body weight. However, about 20 percent rams of these breeds ejaculated semen at 8-10 months with 22.0 kg body weight but good quality semen could be obtained only at the age of 25 months. Delayed puberty is observed in exotic and higher crosses (Tiwari and Sahni, 1982; Tiwari and Kalra, 1982). Age, body weight and testis dimensions were found positively correlated with semen production than the chronological age (Tiwari and Honmode, 1968; Mittal and Ghosh, 1981).

Because of variations in environmental temperature, rainfall and humidity, the seasonal changes in semen quality of exotic, indigenous and crosses have been reported by many workers (Singh and Roy, 1963; Ramamurthy, 1963; Mishra and Sengupta,

1965; Sahni and Roy; 1967a,b; Kaushish and Sahni, 1977b; Dhillon *et al.*, 1979; Sinha *et al.*, 1979; Mittal, 1980; Saxena and Tripathi, 1987).

Good quality semen suitable for breeding purposes could be obtained from indigenous and halfbred rams throughout the year (Sahni and Roy, 1972c,d; Honomode and Tiwari, 1974; Tiwari, 1978; Mittal, 1984, 1985). However, deterioration in semen quality was observed in exotic and higher crosses during summer months (Sahni and Roy, 1972a,b,c,d; Tiwari and Sahni, 1974b; 1975; Kaushish and Sahni, 1976, 1977b; Mittal, 1984) (Table 5.1). Provision of artificial cooling like under ground housing, electric fan and room coolers could not prevent seminal degenerations during these months in exotic breeds (Tiwari and Sahni, 1974a; Kaushish and Sahni, 1977a); (Table 5.2). Deterioration in semen quality in exotic rams due to season as extreme as summer, sterility are characterised by marked spermatogenic degenerations. The effect of high temperature coupled with increased day light has been responsible for production of inferior quality semen in exotic rams (Singh and Roy, 1963; Mishra and Sengupta, 1965; Sahni and Roy, 1972c; Tiwari and Sahni, 1975; Rai *et al.*, 1976; Sahni and Roy, 1967a).

Semen production in relation to frequency of semen collection revealed that semen collected upto four times daily either continuously at an interval of 10-15 minutes or twice in quick succession (3-5 months) during morning and evening hours for a period of 20 or 40 days did not bring any changes in semen quality of indigenous and half bred rams (Sharma *et al.*, 1969; Honmode and Tiwari 1974; Tiwari and Sahni, 1982). However, deterioration in semen quality with marked increase in number of refusals was observed in exotic rams when placed on frequency of once daily for a period of 20 days. When frequency was reduced to once only or twice in quick succession (3-5 minutes) on every alternate day, the semen of optimum quality suitable for breeding purpose was obtained upto 54 days during Autumn and Spring season from Rambouillet rams (Sahni and Tiwari, 1975a; Kaushish and Sahni, 1977a).

Considering the extent of reproductive wastage in exotic rams in terms of lack of libido (30%), erratic and poor quality semen production (30%) it was observed that about 40% rams could be effectively used for artificial insemination and natural

Table 5.1 : Monthly variation in the reproductive behaviour and semen quality of Russian Merino rams under semi-arid conditions.

Month	Volume (ml)	Motility (0-5)	Sperm Cone. $\times 10^9$/ml	Reaction time (Sec)	Max temp. (°C)	Min. temp. (°C)	Relative humidity (%)
January	1.00	4.20	2.233	68.70	26.2	5.3	60.0
	± 0.06	± 0.19	± 0.269	± 9.14			
February	0.93	4.00	3.734	32.30	24.7	6.9	54.0
	± 0.06	± 0.22	± 0.606	± 15.90			
March	1.01	4.10	2.993	72.30	34.2	14.0	50.0
	±0.67	± 0.21	± 0.184	± 11.33			
April	0.97	3.60	3.377	44.60	37.8	19.1	38.0
	±0.66	± 0.27	± 0.294	± 10.10			
May	0.93	3.40	2.475	54.50	42.9	24.4	19.0
	±0.07	± 0.27	± 0.294	± 10.10			
June	—	—	—	—	—	—	—
July	1.14	3.70	2.021	43.70	37.4	25.2	63.0
	±0.05	± 0.22	± 0.164	± 4.90			
August	1.16	3.30	2.117	57.70	35.9	23.9	84.0
	±0.06	± 0.16	± 0.174	± 6.90			
September	1.14	4.10	2.205	56.40	37.2	20.1	91.0
	±0.07	± 0.25	± 0.393	± 7.60			
October	1.14	4.10	1.377	47.10	33.9	15.1	33.0
	±0.08	±0.30	± 0.412	± 8.30			

Month	Volume (ml)	Motility (0-5)	Sperm Conc. $\times 10^9$/ml	Reaction time (Sec)	Max temp. (°C)	Min. temp. (°C)	Relative humidity (%)
November	1.00 ± 0.12	3.00 ± 0.27	1.693 ± 0.227	—	29.4	12.4	68.0
December	1.06 ± 0.06	3.40 ± 0.26	2.186 ± 0.231	108.3 ± 15.8	27.3	5.9	60.0

Table 5.2 : Libido and semen quality of Russian Merino rams maintained under various treatments during summer season.

Group	Volume	Motility^{xx}	Consis-tency**	Sperm** Concen-tration (million/ml)	pH	Reaction Time (Sec)	Thrust intensity	Mounts/ ejaculation
Control	0.85 ± 0.05 (43)	4.26 ± 0.19 (43)	3.23 ± 0.10 (43)	1358 ± 137.5 (37)	7.44 ± 0.06 (43)	25.2 ± 6.06 (43)	2.53 ± 0.10 (43)	2.10 ± 0.33 (43)
Coolled room	1.06 ± 0.05 (50)	3.51 ± 0.23 (50)	2.88 ± 0.11 (50)	1790 ± 181.9 (45)	7.60 ± 0.06 (50)	31.9 ± 7.23 (50)	2.78 ± 0.06 (50)	2.44 ± 0.34 (50)
Animal Protein	0.65 ± 0.05 (36)	1.31 ± 0.23 (36)	2.36 ± 0.14 (36)	948 ± 128.9 (31)	7.74 ± 0.05 (36)	24.9 ± 5.63 (36)	2.52 ± 0.10 (36)	2.19 ± 0.37 (36)
Trace mineral	0.97 ± 0.04 (44)	4.15 ± 0.21 (44)	3.14 ± 0.12 (44)	1300 ± 128.2 (39)	7.52 ± 0.05 (44)	28.6 ± 6.14 (44)	2.45 ± 0.11 (44)	2.48 ± 0.23 (44)

$P < 0.05$; $P < 0.01$.

service. About 20-30% of sexually active rams could be used for breeding with the use of electro-ejaculation (Tiwari and Sahni, 1974b).

Although with the passage of time and modified managerial practices the progenies of imported exotic parents (Rambouillet and Soviet Merino) appeared to have adapted in relation to semen production but were unable to sustain greater ejaculation stress as compared to native and halfbred rams (Tiwari, 1978; Mathur, 1985). Effect of scrotal wool cover has been studied by Rathore, (1969, 1970).

Sexual desire and semen production in rams was found to be affected by differential feeding regime. However, inspite of statistical differences in seminal attributes, the semen of optimum quality suitable for breeding operation was obtained in rams of different breeds and crosses, fed ration containing 14% protein in isoenergic ration.

Methods of semen collection in the Ram

There are three methods of collection. (1) Anoestrus ewe (2) Artificial vagina (3) Electrical stimulation.

Anoestrus ewe

In this method the rams are trained to serve anoestrus ewe and after the service the semen is aspirated from the vagina. It is unsafe to use the semen so collected for fear of contamination.

Artificial vagina

The pattern of A.V. is similar to that in the cattle, except that the A.V. is short. The outer hard rubber casing is 15.0 × 3.75 cm. The A.V. is prepared in the usual way by pouring hot water (45°C) through the inlet on the outer jacket. Necessary air is blown to provide cushioning effect and air valve tightened. Both the temperature and pressure are important. When the A.V. is ready, the inner temperature should be 40°C.

Kaushish and Sahni (1976) considered it advisable to warm the A.V. first with hot water (47°-50°C) initially and then pour out and refill the A.V. with hot water at 45° to 47°C, depending upon the season, in order to maintain the inner tube temperature between 40°-41°C at the time of collection.

A rubber cone is attached to one end of the A.V. at the end of which a graduated 5 ml test tube is fitted to serve as a receptacle for the semen. The open end of the A.V. should be properly lubricated preferably with liquid parafin. A glass collection cup may be directly fixed to the other end of the A.V. to obtain direct collection. Rams are very sensitive to temperature variations and therefore adjusting, proper temperature in the A.V. is important. Once the rams are exposed either to an A.V. which is too hot or cold, they may refuse to serve on subsequent occasions. A freshly prepared artificial vagina should be used for each collection from a ram to avoid contamination. It should be properly sterilised after the use and kept in dry sterile chamber ready for use.

Electrical stimulation:

The equipment for collection of semen from the rams by electrical stimulation as modified by Dziuk *et al.*, (1954) is said to be satisfactory. The power is obtained from 110 volt battery with 60 cycles which is reduced by transformer to 30 volt. This isolates the electrodes from the line of source. The probe for insertion into rectum is made up of rubber hose about 3/4 inch diameter with shoulder rings one inch apart. Alternate rings or electrodes are connected to provide the difference in electrical potential between adjacent electrodes. With this equipment the collection can be made in standing position. However, at times it is necessary to restrain the animal in an extended position on a table. The rectal probe is lubricated, moistened and inserted so that last ring is located just inside the rectum. The penis may be gently pulled out before stimulation. But on many occasions it generally protrudes with stimulation. The end of the penis should be held in a glass vial or a test tube. During the process the voltage is gradually increased and then reduced to zero during a five second period followed by an equal period of rest. Stimulus is gradually increased until the semen is ejaculated in the order of 2, 5, and 8 volts. The number of stimuli necessary to produce semen is variable.

Merits and Demerits of different methods of semen collection

Collection of semen by the A.V. is considered as the best method. Collection by electrical stimulation has an advantage that training of the rams is not necessary. When a large number of rams are to be tested for semen quality, collection by electrical stimulation is advantageous. The major disadvantage in the electrical stimulation is the likely injury to the animal which can be minimised by the use of the bipolar electrode. Other disadvantages are:

1. Slightly lowered spermatozoal density
2. The possibility of contamination of semen by urine
3. Difficulty in obtaining second collection within short time if the first one is not satisfactory
4. The refractoriness on the part of some rams. Experience shows that by collection in artificial vagina the volume of one ejaculate is smaller but denser as compared to an individual collection by electrical method. High quality of ejaculates is the main advantage in collection of semen by artificial vagina.

Frequency of collection

Rams are peculiar in their capacity for frequency of ejaculates with optimum quality. Ejaculates to the extent of even 20 to 30 a day followed by successive collections of equal number of ejaculates the next day have been obtained. Collections to the extent of 2 to 6 per day also have been obtained over a period of a week or 10 days (Chang, 1945). Salamon and Robinson (1962) observed that season and nutrition did not measurably affect frequency of service nor there was any apparent effect of season on volume, density or sperm count. However, marked effect was seen on these characters due to high protein ration. Satisfactory fertility can be maintained even with frequent ejaculations if the total number of spermatozoa is more than 80×10^6. Fall in lambing rate is observed when the sperm count is reduced.

Evaluation of Semen

Colour and consistency

The colour of ram semen is creamy. Variable grades are noticed according to frequency, season, age, rest and among individuals. The normal consistency is thick which is however, variable. It may be classified as dense, medium, thin and watery. It appears that there is very little direct evidence about the value of density and motility determinations in rams semen.

Volume

Considering the large size of the testicles in the ram, the volume of semen obtained at one ejaculate is relatively small, on an average from 0.5 to 2.0 ml. Wiggins *et al.,* (1953) observed that, volume of ejaculates gives a low but significant positive correlation with fertility.

Mass activity

When a fresh drop of ram semen is examined under the microscope at body temperature very brisk dark grey whirls are noticed. The swirling motion is very rapid and dense in appearance. The grading is done as for bull semen.

Initial motility

A drop of semen is diluted in Ringer's or Sodium citrate solution and examined under high power. Gradation of motility is done as for bull semen.

pH

The pH of ram semen varies between 5.9 and 7.3 with an average of 6.4 (Emmens, 1959).

Sperm concentration

The average sperm concentration in a single ejaculate of ram semen contains on an average 3 to 5 billion spermatozoa per ml. In a good sample about 90% of the sperms are alive. Ram semen with high sperm concentration is slightly acidic in reaction, while with low sperm concentration it is slightly alkaline. Semen with alkaline reaction is often associated with poor quality and low fertility (Anderson, 1945).

Morphology

A normal ejaculate may contain 5 to 15% of abnormal spermatozoa. A greater proportion of abnormal spermatozoa may lead to poor fertility. The most common forms of the abnormal type are tailless and deformed. Other morphological abnormalities found are: tapering heads, enlarged middle pieces, coiled or bent tails. It is usually observed that head abnormalities cause low fertility. The types of morphological abnormalities are of more importance than the total sperm count. High percentage of pyriforms or narrow heads are particularly associated with poor fertility.

Ram semen with sperm concentration of less than one billion per ml with a pH of 8.00 or higher, with slow sluggish motility and with more than 25% abnormal type and more than 1% abnormal sperm heads are of doubtful fertility and should not, therefore be used for insemination.

Percentage of live spermatozoa

The percentage of live sperms may be estimated by the vital stains (Lasley *et al.*, 1942) or by direct counting. A stain containing congo red and nigrosin gives satisfactory results for ram semen and it is preferred to Eosin and Nigrosin stain (Blackshaw, 1958). The percentage of live sperms is not significantly correlated with fertility in a ram but the percentage of live normal spermatozoa has a significant correlation (Wiggins *et al.*, 1953).

Biochemical Studies of Ram Semen

In the ram semen, 1/3 portion is made up of spermatozoa and 2/3 by the seminal plasma. There is marked difference in composition between the sperm and the seminal plasma. Nucleic acid is confined to spermatozoa. The spermatozoa contain more of iron, zinc, copper, haematin chiefly as cytochrome and plasmalogen than the seminal plasma (Table 5.3). Acid soluble phosphorus compounds are present in the sperm and also in the seminal plasma. Fructose and citric acid are present in large concentrations in ram semen and these are derived from the seminal vesicle secretions (Mann, 1964) (Table 5.3).

Table 5.3 : Ram semen components (Mann, 1964).

Attributes	mg/100 ml/g
Dry weight	14,820
Chloride	87
Sodium	103
Potassium	71
Calcium	9
Magnesium	3
Inorganic phosphorus	12
Total nitrogen	875
Non-protein nitrogen	57
Urea	44
Ammonia	2
Fructose	247

Attributes	mg/100 ml/g
Lactic acid	36
Citric acid	137
Ascorbic acid	5

Diluents for Ram Semen

The egg yolk diluent developed by Lardy and Philips (1939) is found satisfactory for ram semen. For this purpose equal volumes of fresh egg yolk and sterilised phosphate buffer (0.2 g dihydrogen potassium phosphate and 2.0 g disodium hydrogen phosphate per 100 ml of pyrogen free triple distilled water) are mixed thoroughly. The pH of the mixture is close to 6.75. If it is not, it should be adjusted by the addition of sterile M/15 citrate buffer or bicar-bonate can be used to replace the phosphate buffer without effect on the glycolysis (Moore, 1949). The replacement of the citrate buffer with glycine has improved the survival time of ram semen (Ahmed, 1955, Roy *et al.*, 1956 and Skolosvkaya *et al.*, 1956).

Milk appears to be the most useful dilutor for ram semen (Istvan, 1956; Foilmom *et al.*, 1956; Macpherson, 1957 and Hill *et al.*, 1958). Miller (1975) reported the use of diluents like egg yolk citrate, heated milk from ewes, cow skim milk and accessory secretions from vasectomised rams. Of the above diluents, cow milk or skim milk heated at 92°C for 10 minutes was found to give good results with dilution rate of 1:4.

Sahni (1967) recommended that at least 50 million spermatozoa per dose are necessary for conception in case the diluted semen is deposited in the cervix.

Semen Preservation

Successful storage with optimum fertility of ram semen is a world wide problem. Researcher have been carried out to evolve a suitable dilutor for ram semen in this country, but the results obtained are not very encouraging. The glycine egg yolk dilutor has been found to improve the spermatozoal survival of ram semen as compared to egg yolk citrate (E.Y.C.), later Roy *et al.*, (1962) observed successful storage of ram semen for a longer period in glycine yolk diluent in the 1:10 to 1:20 semen dilutions.

Sahni (1967) in his detailed studies on the storage problem of indigenous ram semen has reported a quite satisfactory survival in heated cow milk and has recommended that cow milk appears to be a dilutor of considerable merit in the preservation of ram semen and also observed it to be significantly superior to buffalo, goat and skim milk. The milk and yolk containing diluents enriched with additives did not show results better than milk diluent alone and Cornell University Extender. Keeping quality of ram semen in these diluents showed significant seasonal variations but generally superior values of motility could still be observed in spring, rainy and autumn season. Effect of higher dilutions (1:10 - 1:100) in these diluents upto 72 h of storage interval showed that there has been deterioration in keeping quality of ram semen with the increase of dilution rates beyond 1:10 and effect was more pronounced in egg yolk citrate as compared to glycine yolk and Cornell University Extender diluents on storage at 5-7°C upto 72 h. However, Tiwari, *et al.*, (1968) observed better survival upto 30 h in dilution rates varing from 1:3 to 1:10 depending upon the room temperature in different seasons it could be possible to utilise diluted semen within 3-4 h after collection and dilution for artificial insemination and glucose saline without egg yolk may be used for dilution of semen to be utilised quickly within 2 h of storage at 30°C.

Ram semen preservation studies (Sahni and Tiwari, 1972), were further carried out, on efficacy of the dilution rates, storage temperature, additives, semen container storage media, cooling rates etc. Results revealed higher spermatozoal survival in egg yolk citrate glucose diluent (EYCG) as compared to EYC, heated cow milk, milk + egg yolk at 8, 24, 48 and 72 h of storage interval. Irrespective of storage media, (refrigerator, semen shipper) and the diluents higher spermatozoal survival have been observed in glass screw capped vials. Out of all the diluents tested, EYCG gave higher satisfactory results at different storage interval which was mainly due to better buffering capacity of the diluent. Further studies were then confined to the EYCG diluents only. Significant differences were observed in percent spermatozoal survival and pH due to variation in number of live and normal spermatozoa ($25\text{-}400 \times 10^{6}$) in 0.1 ml of diluted semen. However, such differences were non significant upto 24 h of storage intervals and irrespective of number of live spermatozoa per inseminating dose (0.1 ml) about 70-80 per cent sperm motility have been observed upto 24 h of storage interval. This shows that concentrated ram semen could be stored successfully upto 24 h and could be utilized for breeding operations. Indigenous rams did not differ among themselves in relation to spermatozoal survival. However, differnce between breeds was found to be significant. Indigenous and halfbred rams showed relatively higher sperm survival than exotic rams at each storage interval. Differences between Rambouillet and Soviet Merino were nonsignificant. Addition of l, 3 and 5 percent gelatine and polyvenyle alcohol in EYCG diluent did not provide significant beneficial effect on spermatozoal survival and pH during storage.

In other studies carried out by Singh and Saxena (1965) and Saxena and Singh (1967) it has been observed that motility and percent live spermatozoa was highest in skim milk yolk followed by egg yolk glucose bicarbonate diluent than EYC and CUE upto 168 h of storage at 3°C. However, during storage there was no automorphological changes in spermatozoa as reported earlier in EYC diluent upto 72 h of storage, (Joshi and Singh, 1968a,b,c) in a comparative study on the efficacy of the diluents have reported higher sperm survival in yolk containing diluents enriched with fructose, tomato juice and glycine, skim milk diluent containing yolk, glucose or fructose. Beneficial effect of caproic acid on sperm motility during storage has been observed by Honmode (1972).

Sahni and Roy (1972c) studied the effect of deep freezing (-79°C) on post thawing revival of sheep and goat spermatozoa using various levels of glycerol, different equilibration time and rate of freezing in EYC and cow milk diluent. The post thawing survival varied from 10-14 percent in different trials and considerable variation between rams with respect to freezability was noted.

The efficiency of EYCG was further compared with diluents prepared from come known buffers by Tiwari *et al.,* (1978). Out of many, the percent motility and live spermatozoa was relatively better in diluents prepared from the Mcillvain and Tris buffer containing yolk and glucose similar to that of EYCG. Over and above in better survival, the most common characteristics in these diluents were the absence of sperm clustering (Sperm agglutination) which was about 20-30 percent in EYCG at 24 h of storage and also relatively better buffering capacity. This shows that these diluents might be able to give better fertility as compared to EYCG diluent.

Deep Freezing of Semen

Keeping in view, different aspects enumerated below, the strategy of developing the technology of ram semen shall have to be evolved and the scientists are working on these lines.

I. Sire individuality

The spermatozoa from all apparently healthy rams do not stand the cold shock, therefore rams have to be screened for this attribute and selected for donating semen for freezing.

II. Extenders

Neat semen can not be frozen and revived, it requires certain additives like nutrients, cryoprotectants, cell wall fortifying agents, buffers and antibacterial agents etc. to be provided. Semen has to be diluted to ensure the required number of spermatozoa per insemination dose to avoid wastage and to be able to cover more females from one ejeculate. Different workers have developed a number of extenders using different combination of various chemicals having qualities of a good extender but with varying results of conception rate. A good extender should consist of following:

1. *Buffering agents :* Ram semen has high number of vigorously motile spermatozoa which consume nutrients very quickly, collect the metabolites specially of the anaerobic respiratory pathway like lactate, pyruvate and lower the pH. Low pH disturbs the electric potential of the sperm cell wall leading to seepage of the enzymes and other material from within the spermatozoa and preventing some from entering into their body, causing toxicity, bursting of the cells and autolysis. To check this, it is important to provide suitable buffering material which may not allow the pH to go down drastically and keep it within the tolerable range. Alternately certain chemicals like cortisole and catalase etc which may avoid accumulation of lactate, pyruvate, H_2O_2 etc if provided may give good results. Milk has good buffering capacity but the results in the freezing experiments are not satisfactory. Egg yolk, blood serum albumin and different chemical buffers like tris, glycine, citrate, phosphate etc have given good results. Out of these egg yolk and tris have not become popular. Although electrolytes are helpful at the time of thawing they are unwanted at the time of freezing.

2. *Cryoprotactants :* Spermatozoa contain some water. On freezing the volume of frozen semen expands resulting in the bursting of spermatozoa cells and killing them. Cryoprotactants decrease the freezing point below zero degree celsius, allow the water to make smaller crystals and the chance to pass through that stage quickly. Glycerol, DMSO, ethylene glycol and propylene glycol have been used but varying percentage of glycerol has found favour with most of workers. Higher percentage of it is harmful because it is reported to deplete the ATP, the energy reserves of the spermatozoa. Cyroprotactants need some time to enter the sperm cell. It is called "equilibration time". Equilibration time of 1 h and 30 min have been tried but best results have been reported with 1, 4 and 5 h at 4-5°C temprature. Some workers have different opinion. It must, however be taken care that the cabinet temperature does not go below 2°C during equilibration.

3. *Cell wall fortification agents :* Egg yolk, egg yolk extract, lecithin, glutathione etc have been commonly used.

4. *Antioxidants:* Ram sperm cell wall lipids are oxidised quickly on coming in contact with oxygen in the medium or atmosphere, and lower the fertilizing ability. Alphatocopherol, BHT, etc have been used for the purpose. Boiling water just immediately before preparing the buffers also helps in removing free oxygen dissolved in it.

5. *Osmoregulators:* Living cells perform well in the media isotonic to the biological fluid they are used to. They should perform well in extender isotonic to semen plasma although certain workers have claimed better results with slightly hypertonic diluents.

6. *Nutrients:* Although spermatozoa utilize fructose and glucose in order of priority but maltose, raffinose, lactose, trebalose etc have given better results in freezing experiments.

7. *Antibacterial agents :* Bacteria are practically omnipresent. Spermatozoa like the living cells, have to be kept away from bacteria and be controlled by providing suitable bactericides antibiotics.

III. Dilution rate

It includes the number of live normal spermatozoa to be included per ml of the diluted semen per insemination dose when we are freezing in straws. Normally ram semen contains 3000-

5000 million sperms per ml of the neat semen and only 100-150 million sperms are enough to impregnate an ewe even though only one spermatozoa takes part in fertilization process. In certain extenders used for freezing, dilution rates of 1:2 and 1:3 have given better results than dilution rate of 1:1 which is general practice for short term preservation or for immediate use.

IV. Rate of heat exchange

It is the most important factor which needs monitoring. It is important not only to control the rate of exchange of heat at the time of cooling but also at the time of thawing revival. It is clarifed by some workers that it should be equal in both the stages and certain workers prefer high rate of cooling as well as thawing i.e. using high temperature water bath. There is a critical zone of temperature which is particularly harmful for the spermatozoa and the latter group in the absence of the knowledge of that zone advise the practice of fast rate of heat exchange. Now a days computer controlled programmable cooling and thawing cabinets are available and the day is not far off when the critical temperature zone will be identified and the measures to pass through it will be devised for successfully freezing the ram semen. Some workers have advised thawing the pelleted frozen semen in media like citrate solution others however, have found no difference between the results of thawing with or without the medium. Some prefer freezing pellets while others prefer freezing in straws to give wider surface area for quicker heat exchange. Pellets are prepared by dropping 0.2 ml of diluted equilibrated semen in the smooth pits carved on the surface of plastic tray floating on liquid nitrogen allowing it to freeze for 3-5 minutes, collecting the pellets into the goblets and lowering down into the liquid nitrogen for storage. Allowing the liquid nitrogen in direct contact or not has made no difference. Straws have to be filled in the clean cabinet atmosphere of 4-5°C temperature after equilibration, sealed, exposed to certain temperature below 0°C for a particular length of time and then only lowered into liquid nitrogen (-196°C) for storage. These days small size liquid nitrogen containers are also available for transporting smaller quantities of frozen semen. During thawing certain workers have claimed good results with as low temperature of water bath as 37°C, while some advise as high as 75°C. Similarly the time for exposure may be 10-15 seconds for straws while 1-2 minutes for pellets.

V. Additives

Addition to the semen of certain chemicals like cortisol, amylase, glucoronidase which play part in longivity of the spermatozoa in the female reproductive tract and are reported to be deficit under stressful conditions in the tract and their varying concentration in the ram semen with season, individuality etc, chelating agents like EDTA are also under study to see their effects *vis-a-vis* fertility rate.

VI. Miscellaneous factors

It is reported that semen from rams fed cotton seeds and phospholipids, freeze better. Similarly the level of certain minerals, vitamins and hormones also do play role on the freezing of the spermatozoa for successful revival. Certain workers are of the opinion that certain chemicals in the seminal plasma are harmful in the process of freezing and hence tried to freeze the washed spermatozoa in artificially prepared extenders but the results are equally disappointing. Feeding, housing and other managemental factors like vaccination, dipping, exercise, shearing, incidence of dieases like blue tongue, sheep pox etc affect the semen quality and the composition of seminal plasma and in turn the freez ability and fertility. Avoiding sudden change of temperature before equilibration, unnecessary exposure to oxygen and bright light, adhering to strict asepsis at every step using tripple glass distilled deionized water, use of pure and nontoxic chemicals are some factors which deserve attention in the venture of deep freezing of ram semen.

The studies on extention rates indicated that Rambouillet semen froze well to give optimum level of cryosurvival and motility, whereas 3/4 exotic semen showed very poor post thaw survival in both the concentrations and replicates (Mathur *et al.*, 1989).

Factors affecting semen quality

Season

Marked seasonal variations in density, percentage of abnormal sperms and glycolysis in semen of rams was observed by Comstock *et al.*, (1945). Poor quality of semen was recorded during summer and early autumn months i.e. July to September. Bell (1945) observed slight seasonal variations in Rambouillet rams in which semen quality was poor in spring (March-April). It improved in summer (May-July). Koger (1951) found that in autumn, ram semen has higher motility and longer viability on storage than in the semen collected in spring. Maqsood (1951) on his observations on young Suffolk rams reported that during the non-breeding season, there is a marked decrease in the libido, volume and spermatozoal density and increase in the number of abnormal sperms. Poor libido is believed to be due to interference with the production of androgens by the interstitial cells. Histological examination of testes of the young ram carried during spring and summer revealed arrested spermatogenesis in the majority of the seminiferous tubules. Waston (1952) observed that Merino rams show higher fertility in the autumn but there is uncertainty in the spring. Wiggins *et al.,* (1953) found a significant correlation between libido and fertility in the range rams. Volcani (1953) observed seasonal variations in spermatogenesis in Awassi sheep in Israel. In July and August when days are long and hot, histological examination of testes in rams indicated active spermatogenesis. Whereas from September to December, degenerative changes were noticeable. He considered that the temperature was not the causative factor and implicated availability of grazing facility.

Shukla and Bhattacharya (1952) observed that the reaction time, differed significantly between rams, but there was no seasonal trend and non relationship to quality of semen. Significant seasonal variations in the volume of semen, pH, sperm concentration, motility and percentage of abnormal spermatozoa were observed. Variations between months and within seasons were also significant. The quality of semen was poorest in autumn and best in spring.

Hafez *et al.,* (1955) observed that the quality of semen of Ausimi and Rahmani rams in Egypt was very good during spring and autumn than in summer and winter. Hulet *et al.,* (1956) found that ram fertility was 42.8% before September due to poor quality of semen and thereafter the fertility improved to 67.6%.

Sahni and Roy (1967b) observed in their studies on Romney Marsh rams brought in Uttar Pradesh (India) for improving the indigenous sheep, that in the month of January, the abnormal sperm count was 3.8% but during the months of February, as the climate begins to get warmer, the percentage of abnormal spermatozoa increased to 27.5%. The signs of seminal degeneration gradually developed, thereafter as the summer advanced the initial motility became poor, sperm concentration decreased, more than 90% of the sperms were found dead and the number of abnormal spermatozoa also increased. In the month of June, the wave motion was completely absent, during the months from July to September, there was complete cessation of spermatogenic activity. The semen was devoid of spermatozoa, ejaculates were watery and resembled those obtained from vasectomised rams. The restoration of normal spermatogenesis started from the end of October. There was remarkable improvement in November. The average initial motility was 3.2; the percentage of abnormal spermatozoa came down to 10.6% and the semen quality was excellent in December and January as in indigenous rams. The semen picture was normal and the abnormal sperm count was only 3 to 4%. These variations in semen quality are attributed to the variations in ambient temperature. It is evident that there is periodic rhythm for spermatogenesis. This pattern is influenced mainly by high temperature leading to deteriorating changes in semen quality which usually start from the mid summer. The semen quality gradually improves by mid- autumn. It is not clearly known to what extent the following factors such as shade available, winds, amount of wool on rams, availability of greens, feed, short term heat waves present in some summers and absent in others and frequency of service and/or collections, influence the semen production.

Decline in semen quality should invariably be accompanied by a gross reduction in the initial motility, sperm count and increase in the proportion of abnormal spermatozoa. It is on the basis of such changes in semen picture of rams in areas away from the equator that a gross decline in semen quality has been reported by Comstock *et al.*, (1945), Bogart and Mayer (1946) and Cupps *et al.*, (1960). Deshpande (1970) reported that in Bannur × Somali rams, the best semen samples were obtained during spring and winter.

Studies show that shifting indigenous rams even from far distant areas to other tracts which fall within the broad range of tropic and subtropics do not impair the reproductive efficiency of Mandya rams which were brought to North India from South, an approximate distance of about 1500 km, did not alter their reproductive efficiency. The variations in temperature or duration of daylight had no effect on the production of quality semen throughout the year (Sahni and Roy, 1969).

Day light

Ortavant (1958) studied interaction between temperature and photoperiodism on spermatozoal production. His P- 32 tracer studies in spermatogenesis show that the rate of division of primary spermatocytes and time of subsequent maturation are relatively unaffected by the photoperiodic environment but the number of spermatids which survive the complete maturation process are affected. The failure rate was observed to be high under conditions of increasing daily illumination. He observed that the effect of high temperature in the spring and early summer months superimposed on this basis photoperiodic effect is to destory a high proportion of the relatively few spermatozoa which have survived the maturation process. Deshpande (1970) observed that during spring and winter, the day length varied between 11.02 and 12.56 hours, whereas, the actual day light hours varied from 7.7 to 9.8 hours and the semen quality was best during this season.

Breed

Very little studies have been done on breed differences in semen production and on the interaction between breed. Season and temperature. Gunn *et al.,* (1942) observed that British breeds in Australia are susceptible to the high environmental temperatures similar to Merino. However, the period of minimum spermatogesis did not correspond with the high environmental temperature of mid and late summers but were observed in spring. Under Indian conditions autumn appears to be the season characterised with poor quality of ejaculates and summer provides most superior quality of semen, as has been reported in the indigenous breeds like Bikaneri and Mandya (Shukla and Bhattacharya, 1952; Sahni and Roy, 1969; Deshpande, 1970). This is in contradiction to the general trend in temperate breeds reared under temperate or tropical conditions.

Singh and Roy (1963) in their studies on Romey Marsh rams (U.P.) observed that similar to Bikaneri and Mandya breeds, the autumn season in characterised by poor quality semen but unlike these breeds which produce best semen in summer, it is in winter , that the semen of Romney Marsh rams has the best quality. As against this, Emmens and Robinson (1962) observed that Romney Marsh rams give most satisfactory semen in autumn in Australia. It is due to such differences in response of different breeds to different seasons that Webster (1952), Mies Filho and De Almeida Ramos (1956b), suggested that extreme care is required in choosing rams of right breed for a particular environment. It is not clear whether such limited data is adequate to prove breeds and environment interactions i.e. differences in sensitivity of breeds to environmental conditions, but gives a sufficient caution before any large scale import of exotic sires is made with an aim to improve the local breeds.

Altitude

High altitude and poor nutrition are deleterious to fertility in rams (Easley, 1951).

Disease

Any infection such as foot rot or abscess which might cause general febrile condition, will result in seminal degeneration (Gunn *et al.,* 1942). Morley (1948) reported that scrotal fly strike may not necessarily impair fertility. Engela (1948) reported incidence of epididymitis, varicocele, scrotal tumours and other diseases of male genitalia which cause semen degeneration. Andreevskii (1940) observed that the main cause of semen abnormality in 40, 976 adult rams was chronic orchidoepididymitis. Of these about 15% of rams showed morbid and degenerative changes in the testes, mostly in epididymis.

Shock

Webster (1952) observed that in New Zealand, a cold southernly wind during night rendered 18 rams infertile over night and the condition persisted for remaining period of the breeding season. He also observed from the past record that abnormally low lambing percentage was related to a sudden exposure to stormy weather during the mating season.

Dun (1956) recorded temporary infertility in rams, due to damp and marshy conditions which led to seminal degeneration associated with flabbiness of the testis. The condition was accompanied by warm humid weather and mosquito plague, shortly before mating.

Feeding of Thyroproteins

Feeding thyroproteins has not produced benefical results in rectifying the seasonal effect on rams semen.

Reproductive Behaviour in Ewes

Taneja (1966) has reported a fairly low level of female fertility in Marwari sheep during summer, whereas a higher fertility rate have been observed during autumn season. It has been suggested that the provision of shelter and off-grazing during noon in summer season may improve the situation. Honmode and Pachlag (1973) observed increase in chloride content during luteal phase (0.5-0.9%), it decreased (0.4-0.5%) after the heat was over.

Table 5.4 : Fertility rate in indigenous sheep during autumn and summer seasons.

Seasons	No. of ewes inseminated	No. of ewes lambed	Percent lambed	No. of inseminations per conception
Autumn	67	57	81.1	1.42
Summer	130	103	79.2	1.62

Sahni and Roy (1967a) and Sahni and Sahani (1973) in a study in the sexual activity in sheep have concluded that tropical sheep, Bikaneri (Magra) is non-seasonal in their sexual activity and about 80-90% come in heat within a span of 20-30 days (Table 5.4). The occurrence of estrous follows a similar trend both in dry and lactating ewes. Estrus length has been reported to be 16-18 days in 65% cases. Duration of estrous varied from 12-36 hours. 80 percent of lactating ewes came in heat within a post-partum interval of 50 days strongly suggest that neither suckling nor lactation had any influence on the onset of estrous in Bikaneri sheep.

Tiwari *et al.,* (1969d) have studied the incidence of estrous as well as lambing rate as a result of spring mating (April) in Rambouillet, Malpura. Chokla and their crossbred ewes. They concluded that estrous incidences were significantly lower in Rambouillet (4.5%) and crossbreds (30 to 45%) as compared to sexual behaviour of Malpura (85%) and Chokla (88%) sheep. Lambing rate has been observed to be of the same order. It was studied in exotic breeds by Kaushish and Sahni (1977a).

Srivastava and Roy (1969) have observed that fertility in adult Bikaneri sheep varied from 74 to 77 percent when rams were kept with breeding flock for 21 to 31 days. Three lamb crops could be harvested from the same flock within 2 years time by rationalization of management practicies.

In a study on the reproductive performance of different sheep during autumn season Tiwari *et al.*, (1969c) have reported that 80-85% ewes come in heat during first 18 days of start of heat detection, whereas 93-100% ewes exhibit their sexual desire in a span of 54 days. The duration of estrous varied between 24-36 hours in maximum number of cases in all the breeds studied (Rambouillet, Malpura, Chokla, Jaisalmeri, Rambouillet × Malpura and Rambouillet × Chokla). Estrous cycle length varied between 17-19 days with maximum concentration on 17th and 18th day. In case of Rambouillet and their crossbred there was a tendency for cycle length to occur between 16-17 days.

Tiwari and Sahni (1974b) and Tiwari *et al.*, (1969b) have observed that the mean post partum heat after 36.5 days in Malpura as compared to 41.6 days in Chokla sheep. The values for postpartum conception interval was 61.4 and 66.7 days in above breeds, respectively. It has been concluded that 50-60 percent of native sheep could give a second crop of lambs, if rebred between 45 to 90 days postpartum.

Tyagi and Lavania (1965) have reported a significant difference in the breeding season in different sheep breeds. Higher environmental temperature was more suitable to indigenous ewes and comparatively lower temperature to crossbred ewes, whereas cooler environment is necessary for the Polworth sheep. Relationship between monthly mean bright sunshine hours and percent ewes bred indicates that increase in reproductive activity followed by the decrease in bright sunshine hours. Rainy, autumn and early winter have been recommended for breeding sheep.

Honmode *et al.*, (1971c) in a study on the investigation into the effect of season, hormone and plane of nutrition on reproduction and characteristics of wool of indigenous sheep breed, have observed that thyroxine feeding showed a favourable effect on lambing percentage. Results on the reproductive behaviour of six breeds of indigenous sheep on low and high plane of nutrition revealed that nutrition had no effect on estrous cycle length (19-57 days). Ovulation time has been observed to vary from 28 to 36 h after the onset of heat.

Tiwari and Sahni (1974a) have observed that the provision of housing has no beneficial effect on the sexual behaviour of young cross bred ewes.

Sahni and Tiwari (1972) in a study on the reproductive behaviour in young ewes of various sheep breeds have observed that majority of native ewes showed their sexual desire at the age of one year as compared to 1.5 years and 2 years in cross bred and exotic ewes, respectively. Lambs born in autumn season attain maturity earlier than the spring born lambs. They made a comparative study of the reproductive behaviour of certain native, exotic and cross breeds of sheep under tropical conditions. High environmental temperature of summer had no adverse effect on fertility of native or cross bred sheep. The Corriedale when provided with some protection against intense heat of summer could be considered as the one exotic breed capable of satisfactory reproduction under tropics.

Biochemical changes in blood, vaginal mucus and oviduct have been observed by several workers. Rawat *et al.,* (1987) observed depression in thyroid status on day 3, but it was more on day 16 of the cycle than on any other day of the estrous cycle, these differences were not significant. Rawal *et al.,* (1986a) observed that the pH of vagina on 16th day and zero hour of estrous were close to neutrality, while it tended towards alkalinity at 8 h of estrous. The pH values followed increasing trend from zero to 8 h and then decreasing trend continued till 16th day of estrous. The values of ratio between vaginal mucus potassium and sodium and total sodium and potassium to chloride and pH value at 8th h of estrous showed that pre- estrual secretions are not quite conducive for sperm survival.

Fluctuations in the mean glucose, pyruvate, lactate, glycogen, total carbohydrates and activity levels of lactate dehydrogenase, glucose-6-phosphate dehydrogenase, phosphorylase a and b, acid and alkaline phosphatases and structural soluble and total proteins, DNA, RNA and total lipids in oviduct and vaginal tissue homogenate during various phases of roestus cycle were studied by Rao *et al.,* (1987).

Sahni and Roy (1972d) studied variation in postpartum conception interval in Bikaneri flock in which matings took place continuously over several years, gave an average conception period of 183.6 days indicating that, normally it is not possible to obtain more than one lamb from each ewe given by the one third of the flock within a period of 12-15 months is counter balanced by an equal proportion of ewes which take very long time to conceive again even when rams are run continuously.

Puberty

Puberty in the males is marked by production of viable sperms and a desire for mating influenced by hormone testosterone.. The age at puberty vary from 132 days for Tobasco (Pelibuey) × Dorset lambs on the high plane of nutrition in Mexico (Valencia *et al.*, 1977) to as old as 738 days for 3/4 Rambouillet 1/4 Malpura lambs in Rajasthan (Tiwari and Sahni 1981). The quantity and quality of ejaculates collected from puberscent rams are poor with a high incidence of dead and deformed spermatozoa (EL-Wishy, 1974). Ram lambs reach puberty about two months earlier than ewe lambs in each breed.

A similar desire for mating also takes place in the female. Puberty in the female is exhibited by the first estrous. The breed of the ewe affects the age and weight at puberty and sexual maturity. Ewe lambs in most of the mutton breeds reach puberty at the age of 10-12 months normally breeding during the first autumn. Larger, slower maturing breeds may mature at the age of 18-20 months.

Awassi lambs on a good diet first display estrous at an average age of 274 days (Younis *et al.*, 1978), but Rambouillet crossbred lambs in Rajasthan are about 615 days old before they display estrous (Kishore *et al.*, 1982). This difference is largely due to the different growth rates resulting from different nutrition. The average age at puberty in Ossimi and Barki ewe lambs reared on high plane of nutrition was 347 days in Egypt whereas, it was 366 days on a low plane of nutrition (El-Homosi and El-Hafiz, 1982). The variation in age at puberty occurs in sub-tropics where the ewes are reasonably anestrous so that if a lamb fails to cycle in her first season she will not cycle until the next, several months later.

Under Indian conditions, the ewes reach sexual maturity at 9 to 14 months of age, however, the full body growth is only attained between 18 to 24 months of age and the stage at which the mating is performed. Merino sheep mature more slowly than the Hampshire or Suffolk breeds, crossbred lambs generally mature earlier than pure bred lambs. Roy *et al.,* (1962) recorded first estrous in young Bikaneri ewes of less than one year of age. Majority of the Nellore ewes attain puberty at 630-810 days of age (Gupta *et al.,* 1987). The age at puberty has been reported to vary from 306-383 days in different exotic breeds (Martemucci *et al.,* 1983).

The Estrous Cycle

Rhythmic sexual behaviour patterns develop in females during puberty. This change in behaviour is called estrous and it occurs with each estrous cycle in non seasonal breeders unless pregnancy intervens. The combination of physiological events which begin at one estrous period and end at the next is termed as estrus cycle.

The ewes indigenous to temperate areas, during breeding season, estrous occurs every 14-19 days with a mean of 17 days (Terrill, 1974). Similar duration has been observed for ewes in tropics by Narayanaswamy and Balaine (1976); Gaillard (1979) and Yenekoye *et al.,* (1981). Cycle lengths of upto 70 days were recorded for Fulani ewes in Niger by Yenikoye *et al.,* (1982).

Visual signs of approaching estrous are, a swelling and redness of the vulva and restlessness or nervousness indicating a desire for company, but the most obvious sign is ridding and inturn being ridden. The breeding occurs only during estrous although the ram is capable of breeding at any time (Dhanda, 1970).

Three types of estrous cycles commonly observed are :

1) One type is seen in sheep, goat, cow and sow during breeding seasons. The infertile cycle in these species culminates in spontaneous ovulation of mature follicles; *corpora lutea* automatically formed, becomes functional and exists for a definite period of time.

2) Rat and mouse are example of another type of estrous cycle in which ovulation is spontaneous, but the *corpus leuteum* which is formed is not functional unless mating occurs.

3) In third type maturation and ovulation of follicles fail unless the male copulates with the female. Rabbit is an induced ovulator.

The main events of the estrous cycle of the ewe can be divided into those associated with the growth of follicle and those associated with the growth of the *corpus luteum.* The growth of the follicles takes place during proestrous and estrous. The period of the *corpus luteum* can be divided into two periods, metestros and diestrous. The estrous cycle is usually dated from the first day of estrous.

1. Estrous

Estrous is defined as the period of sexual receptivity during which ovulation occurs in most species and the corpus luteum begins to form. The psychic estrous is brought about by the action of estradiol on the central nervous system. Symptoms of heat are less pronounced in ewe than in cow or mare. The ewe in heat makes little effort to tease and mount. Becuase of weak symptoms, it is difficult to detect estrous in ewe. The duration of estrous in ewes varies from 12-72 h (Taparia 1972; Castillo Rojas *et al.,* 1977; Gaillard 1979; Yenikoye *et al.,* 1981) with an average of about 45 h. Breed of ewe does not affect the duration of estrous. Ewe lambs, usually have the shortest estrous, older ewes have the longest and the yearlings are intermediate. The duration of the estrous cycle is shorter at the beginning and at the end of the breeding season.

The first ovulation at the onset of puberty is frequently not associated with psychic estrous. It appears that a small amount of progesterone is necessary to trigger psychological estrus. In the cycling ewes this comes from the *corpus luteum* of previous cycle. Therefore, the first ovulation after the ovary is awakened from the quiescent state finds the ovary without an old *corpus luteum* to produce the small amount of progesterone necessary to bring about psychological estrous.

The ovulation takes places about 12 h before the end of estrous. Right ovary ovulates more frequently than the left, 60 and 40%, respectively.

During estrous swelling and reddening of vulva occurs. Ewes become restless. Pheromones begin to appear during proestrous. Ewes normally come in heat after the weaning of lamb. Some non-seasonal breeding ewes may show a post partum estrous a few days after delivery, but others wait for 4 to 6 weeks.

Two to 3 days before the onset of estrous grafian follicles destined for ovulation begin to enlarge rapidly and become turgid. The *theca interna* hypertrophies and the ovum, with attached cumulus oophorus, moves from the embedded position in the granulosa layer toward the enlarged fluid filled antrum of grafian follicle. A transparent area appears in the follicle membrane near the apex when ovulation is impending.

2. Metestrous

The ewe enters metestrous after estrous. This phase lasts for 2 days. The *corpus luteum* organises during this period and becomes functional. The progesterone level rises rapidly.

3. Diestrous

Diestrous is the period during which the influence of luteal progesterone on accessory sex structures predominates. This phase is referred to as the phase of the *corpus luteum.* Metestrous and diestrous are referred to as the luteal phase. Large amounts of progesterone enter general circulation resulting in the development of endometrial growth and also of mammary glands. The myometrium develops under the influence of progesterone and the uterine glands secrete a thick viscous material which will nourish the zygote. If a zygote reaches the uterus, the *corpus luteum* continues through out pregnancy, but if the ovum is not fertilized, the *corpus leuteum* remains functional only until day 12 or 13. The decline of progesterone permits a surge of FSH and growth of another follicle, resulting in initiation of new estrous cycle. In short the estrogen dominates for 3-4 days of cycle and progesterone dominates for about 13 days.

Development of corpus luteum : The *corpus luteum* is a temporary endocrine organ which functions for only afew days in cycling non pregnant ewes and throughout the duration of pregnancy once the ewe becomes pregnant. The *corpora lutea* of sheep reaches maximum size by day 3 and at this time it is reddish pink. *Corpora lutea* becomes paler as dioestrous approaches and after day 14 degeneration is rapid. Because the estrous cycle of the ewe is shorter, the *corpora lutea* of the ewe are devoid of the lipochrome pigment, hence the lighter colour. The circulating levels of progesterone in plasma during follicular and luleal phases are 0.25 and 3.7 ng/ ml in ewe.

4. Proestrous

It is the period after which the *corpus luteum* fails, when the progesterone level drops, FSH release stimulates follicle growth and rising estrogen levels lead to estrous. Proestrus and estrous are often referred to as the follicular phase. Estradiol increases the blood supply to the tubular genital tract from vulva to the uterine tube and causes oedema. The vulva swells, the vestibule becomes hyperemic and the glands of the cervix and vagina secrete a serous secretion which appears as a vaginal discharge.

Anterior pituitary and the gonads : The primary target organs of the pituitary gonadotropins are the ovary and testes. The growth and development of ovarian follicles in mammals are dependent upon FSH but LH is essential for follicle maturation. Both FSH and LH are essential for the synthesis of estrogen. Rising estrogen blood levels suppress release of FSH and facilitate release of LH.

Apparently FSH and LH are secreted continuously by the anterior pituitary throughout estrous cycle, but the proportions and levels of each, change during the different stages of the cycle.

Estrogen in large quantities inhibits FSH secretion. PGF_2 alpha from non pregnant progesterone dominated endometreum reaches ovary by a local route between day 14 and 15 and kills the *corpus luteum.* The progesterone level falls and FSH level increases. The peak of FSH can be seen just before estrous. FSH facilitates follicular growth and quick estrogen production. The rising estrogen induces behavioural estrous and trigers a release of an ovulation surge of LH on the days of estrous. Estrogen levels then decline but the new *corpus luteum* starts progesterone production which rises during the next few days and is sufficient to hold gonadotropin release to low level. On day 16 the *corpus luteum* is killed again and the cycle is completed.

Factors affecting corpus luteum : In polyestrous domestic animals, corporalutea develop after ovulation and function for 14-15 days unless pregnancy occurs to signal the *corpora lutea* to continue functioning. There appears to be marked species differences in the control of corpora luteal function. The pituitary is needed in most species to produce luteotropin at ovulation to form the *corpus luteum.* Nonpregnant uterus produces a substance probably PGF_2 alpha which has a luteolytic influence on the *corpus luteum.* The luteolytic effect of uterus is dominant to hypophysical control of the life of the *corpus luteum* in sheep. Presence of an embryo in the uterus will prevent luteolysis. Researches have shown that an embryo must be present in the uterus by the 12th day of the cycle in sheep if luteolysis is to be avoided. Luteal regression can be prevented in the ewe and cow by uterine infections. Presence of intra uterine devices (IUD) shorten, the estrous cycles in sheep and cow.

Artificial Insemination

During the recent years, it has become necessary to breed sheep by artificial insemination for making intensive use of superior sires. USSR is the foremost country in the world where thousands of sheep are annually bred by artifical insemination. Kenya, South Africa, Romania, Australia and Argentina are other countries where limited use of this method of breeding is being made in selected flocks. In India, it has been tried with success on sheep breeding farm at Poona by Sane *et al.,* (1952), at I.V.R.I, by Guha *et al.,* (1951); Shukla and Bhattacharya (1952) and Roy and his colleagues (1956) at Mathura on experimental basis.

Breeding by Artificial Insemination is now being practised in various sheep breeding farms in India. The main consideration in adopting this method of breeding in sheep is to use sires having the desirable traits specially for quality of wool and meat. The inherent danger of transmitting undersirable traits through sires which have not proved, should not be overlooked. The possibility of transmitting undersirable recessive characters due to widespread use of unproven sires through artificial insemination is a matter for consideration.

In flocks where it is not possible to synchronise the breeding season naturally it results in a prolonged lambing season with the consequent lowering of uniformity of the lamb crop. This method of breeding is very expensive but considering the large scale development in economic characters, adoption is most essential. The method may prove economical in case of large number of flocks where grading is desirable are concentrated in a locality and where there is dearth of quality rams for breeding purposes.

The age of sheep influences the fertility to a great extent in terms of ovulation rate and fecundity. Variation in fertility has been observed in different sheep breeds. Fertility percent in Muzaffar-nagari varied from 64.0 percent (1-2 year) to 85.3 percent (4-5 year) in different age groups (Sinha and Khan, 1989).

Selection of Rams

It is possible to select rams initially on their phenotypic characters coupled with judgement on qualitative requirements for wool and meat. However, most of the economically important characters of sheep are highly heritable and as such selection can be made on phenotypic characters. Selection has to be done at an early age as low as six months or so. They are usually ready for service at about nine months to a year. At this stage it is possible to study the sex behaviour and quality of semen. In case this is upto the mark, progeny testing should be done to evaluate their potentialities. Progeny tested rams can further be used on mass scale. Limited use should be made in case rams are not progeny tested for fear of transmission of undersirable traits especially lethal and sublethal characters. It is desirable to use rams with very high fertility rate so that the flocks can be settled in a reasonable period of about 60 days under natural mating.

Management of Rams

Ram sperms take 40 days to mature. The sperm cells which are required for A.I. programme are developed during two months prior to their collection.

Most of the insemination programmes are carried out from mid February till mid May, as this is the period when ewes are normally cycling. However, most of the semiarid areas experience very hot summers, resulting in summer infertility.

Rams, should have access to pasture or irrigated plots and before they are let out on pasture, should be inspected daily for fly strike, screw worm and foot abscess. Rams at two tooth stage are separated from adults to prevent fighting and homosexuality.

When the actual insemination programme starts, the rams are still housed and fed as before and walked to the shearing shed and back daily, to ensure exercise. When in use at the shearing shed, they are penned separately awaiting their turn as semen donors to avoid fighting with other males and to save time in handling.

The training, actually does not begin till the programme starts and it seldom takes more than two days to get all rams to serve an ewe and subsequently into an A.V. The number of collections to be obtained in a day varies from ram to ram. From most of the rams semen can be collected three times a day producing average ejaculates of 1.0 ml, sperm concentration of 35×10^8 per ml and live counts 85-90% (Miller, 1975). He further reported $15,000 \times 10^6$ sperm per day from two teeth rams in three collections (Table 5.5).

Table 5.5 : Results of rams semen examined on day 6th and 15th day of semen collected.

	Vol. (ml)			Sperm No. × 10^8			Live (%)		
	No.1	No.2	No.3	No.1	No.2	No.3	No.1	No.2	No.3
				DAY–6					
Ram A	0.7	1.4	0.8	39	45	45	90	92	88
Ram B	1.0	1.0	1.1	41	31	31	90	86	88
Ram C	0.8	0.6	0.5	34	45	45	82	95	95
Ram D	0.7	1.0	0.4	54	43	36	95	95	95
				DAY–15					
Ram A	0.6	1.0	0.8	40	40	35	90	90	90
Ram B	1.0	1.0	1.0	40	35	30	90	85	90
Ram C	0.0	0.6	0.4	32	40	45	80	90	95
Ram D	0.7	0.5	0.7	45	40	35	95	95	90

Semen could successively be collected up to 5 times daily for 17 days from mature rams without fall in semen quality (Miller, 1975).

Management of Ewes

Keast and Morley (1949) and Dun *et al.,* (1960) recommended overnight teasing and drafting raddled ewes off in the morning. This is quite manageable for small lots, but when large numbers have to be inseminated it takes lot of time as ewes cannot be presented for A.I. at proper time during estrous. Under the circumstances the rams have to be worked too frequently to provide necessary semen. When large number of ewes are to be handled, harnessed vasectomised rams are released with the ewes at about 5 p.m. and the mob turned into a paddock. They are mustered into the yards by 8.00 a.m. the following morning and marked ewes drafted off. With few modifications to most sheep yards, this can be done at the rate of 4,000 per hour.

Marked ewes are either inseminated to get the labour force acquainted with procedure and a few rams checked and trained or these are put aside and brought back into the mob at the end of the cycle. Generally the drafted ewes are inseminated on the acceptable type of estrous mucus. Obviously those ewes which are drafted late in estrous are rejected (Table 5.6).

After the morning draft the remaining flock is redrafted at 4.00 p.m. These ewes which are called as "A ewes" have commenced estrous between 9.00 a.m. and 4 p.m. The main flock is returned to the paddock with teasers and brought in for drafting again the following morning. The marked ewes which are called as "B ewes" have commenced estrous between 4.00 and 8.00 a.m.

The "A. ewes" are inseminated between 7.00 to 8.00 a.m. the following day. The "B. ewes" are inseminated between 10 a.m. and 3 p.m. the day they are drafted off.

All the "A. ewes" are inseminated between 15 and 23 hours after the onset of estrous.

All the "B. ewes" are inseminated between 2 and 23 hours after the onset of estrous.

Vasectomised Rams

The number of vasectomised rams used is 1% of the original number of ewes i.e. 5,000 ewes 50 teasers and even when there are only 1,000 ewes left in the mob 50 teasers should still be used. All vasectomised rams are tested for brucelosis. Two teams of teasers are used and alternated every 3rd day. The time to change teasers is when a large percentage of the ewes that are presented for insemination have creamy vaginal mucus. This indicates that estrous has terminated and this means that teasers did not detect estrous at the onset, either because there were too many ewes or they had lost interest.

Management calendar

Rams which are well fed and well grown may become functional as early as 8 months of age and produce viable sperms. It is difficult to assess their merit as sires for wool production under the age of 12-14 months. The two tooth performance is the best guide of suitability for use as sire. Selection for mutton characteristics can be done at an early age as growth rate is the important criteria. Waston *et al.,* (1956) observed that testicular development can be correlated more with live weight of the ram than with age.

Storage and transport of semen

Yoshika *et al.,* 1951) reported that ram semen can be successfully stored for seven days with high fertility rate. Temperature at 4°C is reported to be optimum for storage of ram semen (Moore *et al.,*1940). Storage of ram semen at 20°C in the presence of CO_2 is also considered as good (Blackshaw, 1958). Sudden temperature changes are considered harmful for the viability of the sperm. Semen should be cooled gradually for storage. Blackshaw (1958) reported that the presence of egg yolk in the diluent is largely responsible for preventing the cold shock.

The metabolism of ram semen is mainly glycolytic (Mann. 1964). Salisbury (1946) reported that the metabolism of ram semen is mainly glycolytic, oxygen may damage spermatozoa. Complete anaerobic condition is therefore necessary for storage. The viability of ram semen is prolonged with effective exclusion of air during storage. It is preferable to seal the ampules with minimum airspace as far as possible to avoid agitation during transport.

Insemination technique

Under the insemination technique there are a number of pre-requisites such as equipment for insemination, restraint of ewes, number of ewes for insemination and adjustment of inseminations during better part of estrous period end whether heterospermic inseminations are desirable or not.

Equipments

A speculum, a head light torch and a syringe with insemination pipette is all the equipment required. There are three types of specula in use (1) the duck bill type, (2) metal barrel, and (3) glass speculum. It is the usual experience that the duck bill type is easy for insertion and gives greater ease and freedom of movement. It can be easily dilated in the vaginal passage and removed in a closed condition after use. lt is also easy for sterilization. A glass speculum is also used.

A 2ml glass syringe attached to inseminating pipette with a rubber connection is most convenient for insemination. A fresh insemination pipette should be used for each insemination.

Restraint of ewes for insemination

With the help of two attendants, ewes can be inseminated at the rate of 100 per hour, if the operation is streamlined, using fresh neat semen. Various methods of holding ewes have been used and some methods used in Australia between 1950 and 1960 are :

1. Ewes held upside down in a cradle.
2. Ewes placed on a battery of bails on a raised plateform.

3. Ewes held on a rail, as for lamb marking.
4. The operator working in a pit and a ewe manoeuvred upto a hock bar at the edge of it.

Other methods used in Asia and in South America, include the use of various cradle and jacking devices to lift the hind quarters. These are not easy to operate. The method used in Australia, at present, is for the attendant to saddle the ewe facing the tail and with a hand under each flank to throw her hindquarters over a rail 24" above floor level. Attendant keeps the ewe's hind legs extended by pressing into her stifles. If the ewe's front feet are on the floor, she will not struggle. After insemination the ewe is released through a gate in the pen.

Table 5.6 : Percentage of ewes in heat according to duration of estrous.

Breed	12-24 hrs.	24-48 hrs.	48-72 hrs.	72 hrs & above
Jaisalmeri	45.00	40.00	13.00	1.80
Malpura	28.00	52.50	16.00	2.50
Chokla	20.60	49.00	31.70	4.70
Rambouillet × Chokla	40.80	40.30	12.50	1.80
Rambouillet × Malpura	36.60	45.80	13.70	3.60
Rambouillet	43.20	15.40	9.56	1.89

A false floor about 2 feet high is put into the catching pen to enable the inseminator to have the ewes presented at eye level.

When the ewe is placed over the rail, an assistant inserts a glass speculum whilst the inseminator loads his syringe. The inseminator then holds the speculum, locales and positions the cervical opening and inseminates the ewe. The semen is deposited in the cervix.

A simple wooden crate with sloping platform is considered for restraint. The rear portion is raised to a height of 3 feet so that the inseminator can conveniently inseminate in standing position (Sane *et al.*, 1952).

Time of insemination

During estrous mucus flows freely from the cervix into the vagina. Generally there is a small amount of clear mucus at about the time estrous commences, but as estrous advances the amount of mucus increases and its nature changes. It first becomes cloudy, then creamy and finally cheesy in consistency (Table 5.7). Although it is usually considered that changes occur progressively as estrous advances, these cannot be used to time different stages precisely. However, Morrant and Dun (1960) have demonstrated an association between fertility and the different types of mucus at the time of insemination.

Table 5.7 : The fertility of ewes different types of vaginal mucus.

Type of mucus	Grade	No. of Ewes inseminated	Fertility %
Clear and sparse	1	113	50
Clear and copious	2	369	62
Cloudy and copious	3	219	62
Cloudy	4	102	52
Creamy	5	92	33
Cheesy	6	34	15

Uterine contractions and motility of the spermatozoa are largely responsible for quick transport of spermatozoa from through the tubular genitalia to the site of fertilization in about 2 to 4 minutes. The ova are reported to survive from 12 to 23 h in the fallopian tubes but it is not very clear if they are fertilizable during this period. Generally, 8 to 10 h appears to be a reasonable time for the survival of tubal ova. Therefore, inseminations done any time within the estrous period may result in fertilization with the greater possibility of success to insemination done during mid estrous and least possibility in the later part of estrous i.e. close to ovulation time.

Sinclair (1957) confirmed this concept based on his studies on 1500 Merino ewes bred naturally or by artificial insemination in early, mid or late estrous. Conception by natural mating at each stage resulted in 55-65% fertility. Lower conception rate resulted due to inseminations in late estrous when the cervical mucus was becoming creamy. The observations of Corbonero Bravo (1955) show that the optimum time of insemination was mid estrous i.e. 12 to 18 h after onset of heat. Osikoski (1965) obtained highest lambing percentage 12 h after the onset of estrous even by natural service in 175 mountain ewes, 8 hours after in 122 Merino ewes. There was no difference in the conception rate of ewes (53.2%)

between morning and evening inseminations based on 45 day non returns (Mies Filho and De-Almeida Ramos, 1955 a).

Seasonal influence on successful percentage of inseminations

Season of the year has a marked influence on the successful percentage of insemination. Dun *et al.* (1960) observed significant differences in favour of inseminations done during autumn season. Salamon and Robinson (1962) observed 20-25% more of lambing crop in ewes inseminated during autumn as compared to those inseminated in spring. The difference is not due to differences in semen quality. Ewes returning to service during autumn are an evidence of reliable estimates of conception rates. On the other hand, percentage of returns in the spring can not be relied upon for a true estimate. Difference in the percentages of lambing drop are apparently visible between autumn and spring. Salamon and Robinson (1962) observed that of the estimations done on the sum of returns to service, 92% of successful single insemination as compared to 71 % in the spring.

Number of inseminations

Question arises as to the number of inseminations necessary to settle the ewes in the same estrous (Table 5.4). Benefical results have been observed by various workers and they claim an increase in conception rate from 4 to 10 percent by doing two insemination per estrous period. It is considered advisable to do second insemination on the day following first insemination of ewe still in estrous. Similar results have been obtained by Osikowski (1965) in Merino ewes at 24 hours interval following natural service.

Volume of semen and number of spermatozoa per insemination

Kunznetsov (1934) recommended that at least 50 million sperms per insemination are necessary for conception. Experience shows that dose as low as 0.1 ml of neat semen is satisfactory for higher conception rate (Sinclair, 1957; Keast and Morley, 1949). A volume of 0.05 to 0.2 ml is found satisfactory which is expected to contain 50 to 100 million sperms. A fairly good quality of ram semen should contain on an average about 150 million sperms per 0.05 ml.

Higher conception rates have been claimed in ewes by using neat semen as compared to the rates where diluted semen is used (Mies Filho and De Almeida Ramos, 1955b). In general, it is observed that conception rate of about 55 to 60% can be claimed to a single insemination if a dose of 0.5 to 1.5 × 10^8 spermatozoa in 0.05 to 0.1 ml of semen, diluted or undiluted, is used for cervical insemination.

Site of insemination

Dun (1955) observed that the external appearance of the cervix varied a good deal and difficulties may be experienced in locating the opening to the cervix. Because of the peculiar structure of the cervix of the ewe, it is not possible to deposit semen in uterus. It may be placed (i) in the opening to the cervix, (ii) on the cervix as it projects into the anterior end of the vagina or (iii) in the anterior end of the vagina.

Insemination

When the ewe is presented for insemination the lips of the vulva should be cleaned with cotton wool soaked in normal saline. A speculum lubricated with white vaseline or liquid parafin is slipped into the vagina with a rotary motion and the cervix is located and imprisoned in the opening of the speculum. The semen is drawn into the insemination pipette and the syringe and the tip of the insemination pipette is inserted gently into the opening of the cervix. About 0.1 ml of semen is deposited into the opening of the cervix and then the insemination pipette and the speculum are withdrawn.

Heterospermic inseminations

Favourable results have been claimed and higher conception rates obtained by use of pooled semen samples from different rams (Aslanjan and Camuha, 1950; Mamedjarv, 1952; Stojanovskaja and Mesjacev, 1957). However, similar results could not be reproduced in Australia (Dunn, 1956).

Detection of estrous in ewes

There are two ways of detection of estrous in ewes. One is by hand teasing in yards with a vasectomised ram. A loose piece of soft thick cloth is tied to the girth and kept hanging in front of preputial orifice, to avoid natural service. In the second method, vasectomised rams with breast painted with red clay (Geru under

Indian conditions) is allowed to run free in the flock. The fluid paint is renewed daily. Ewes which are marked with paint are separated for insemination and next morning they are teased again to confirm if the heat is over or else a second inseminaton is done to those which are still in estrous.

Ewes remain in estrous for about 30 hours on an average. In exceptional cases, it may extend over 2-3 days (Table 5.8). The ovulation may vary considerably in relation to the onset of heat. Mckenzie and Terrill (1937) observed that ovulations in the ewes generally occur at the end of heat period. Insemination done before the end of heat period results in higher conception rates than those at or after the end of the heat period.

Anestrous ewes can be brought in heat by 1/M injections of progesterone for 10 days and PMSG 400 i.u. on the last day of progesterone injections (Antoine *et al.*, 1992).

Faulty techniques resulting in low fertility

Some obvious things that lead to failures are (1) Fear of not mustering ewes quickly enough so that they are kept in small paddocks and often underfed (2) Faulty mustering ewes in estrous missed and not presented till 24 hours too late (3) Failure to look after rams prior to the collection time (4) Too much time spent in drafting sheep (5) Failure to recognise vaginal mucus changes, thus leading to faulty testing (6) Faulty management of donor rams causing loss of libido (7) Faulty management of rams e.g. mixed age groups together leading to fighting, mounting and a wastage of semen (8) Bad handling and unnecessary handling of ewes after insemination and (9) Failure to have strong and fit teasers.

Failure in A.I. technique

(A) Failure to control temperature of semen

(B) Failure to pick up and position cervix quickly. Glass speculum with a lip is used in order to position the cervix and get good lighting.

(C) Using too much air in the syringe as an air pump. This blows the semen out of the cervix entirely.

(D) Not withdrawing speculum before depositing the semen. If this is not done, a negative pressure is set up and the semen is withdrawn (at times completely from vagina).

(E) Lack of speed.

Table 5.8 : Reproductive life of females

Attributes	Cow	Baffalo	Camel	Mare	Sheep	Goat	Sow	Bitch
Age at puberty (Months)	24-31	36-41	44-48	10-24	10-12	10-12	5-8	6-8
Age at maturity (Months)	28-35	37-42	45-49	30-36	12-15	8-15	8-12	12-18
Type of estrous cycle	Poly estrus	Poly estrus	Poly estrus	Seasonally poly estrus	Poly estrus	Poly estrus	Poly estrus	Monestrus
Length of estrus cycle (days)	18-21	20-24	10-21	19-25	16-18	15-24	19-20	—
Duration of estrus	10-24 hr (16)	10-32hr (30-36 hr)	1-7 days (3-4 days)	2-11 days (6 days)	30-60 hr (36-48 hr)	30-60 hr (36-48 hr)	1-3 days (2 days)	4-13 days (7 days)
Time of ovulation	14 hr after onset of heat.	18-24 hr after onset of heat.		24-28 hr before end of heat.	12 hr after commencement of heat	Last day of heat	30-36 hr from the commencement of heat.	24-36 hr from the commencement of heat.
Gestation period (days)	278-290 (284)	305-342 (310)	330-340 406)	329-346 (336)	143-159 (148)	147-155 (148)	121-116 (114)	62
Proper time of insemination	Middle to the late of estrous	Later half of estrous.		every alternate day begining second day of estrous.	18-24 hr after onset of estrous.	24-26 hr after onset of estrous.	12-30 hr after onset of estrous.	2-3 days after onset of true, estrous.
Breeding season	Through out year.	Through out year	Nov.-Feb.	Mar.-June	Through out the year except during May	Through out the year except during May and June.	Through out the year	Aug.-Sept. Feb.-March

Estrous during pregnancy

Estrous in the pregnant ewes is associated with follicular growth, but ovulation, breeding and conception may occur during an existing pregnancy which would lead to superfoetation. Estrous has been observed in pregnant ewes (Matter, 1974; Younis and Afifi 1979; Kandasamy and Pant 1980; Sinha *et al.,* 1980). It has been observed in first month of pregnancy. It is not usually accompanied by ovulation (Williams *et al.,* 1956). There is a fear among sheep farmers if mating takes place abortion may result (Sinha *et al.,* 1980).

Gestation period

The duration of gestation in native and cross bred ewes has been reviewed extensively by Honmode (1970) and Kaushish (1971). The mean gestation length varies from 147 to 153 days. The average gestation period varied from 151.2 to 152.2 days in Nali, Lohi and crosses of former with Mandia and Nellore (Kaushish and Arora, 1974a). Ramamurti (1963), Singh *et al.,* (1977) and Kaushish *et al.,* (1986) reported duration of gestation in Native, Nilgiri and Bikaneri and Soviet Merino, respectively (Table 5.9).

The gestation period is shorter for exotic than indigenous ewes (Sahani and Pant, 1978), for ewes with exotic crossbred lambs, than those with indigenous lambs (Rao *et al.,* 1978), for lambs with lighter birth weights (Kishore *et al.,* 1980) and for female than male lambs (Narayanaswamy *et al.,* 1975). The gestation is positively related with weight at service (Kaushish, 1989).

The effect of breed, birth weight, sex of lamb, birth weight of dam at service, pH of amniotic fluid and post service gain in weight on gestation length has been studied by Ostgard, (1957) and Kaushish and Arora (1973, 1974 a). Gestation has been reported to have positive correlation with pH of amniotic fluid and negative with post service gain in weight and weight at service (Kaushish and Arora, 1974 a).

A study on post partum estrous interval has been carried out by Rawal *et al.,* (1986b) to determine the effects of season and sex of lamb on this trait in Muzaffarnagri sheep. Following single normal lambing, twin lambing and abortions, gestation period averaged 137.4 ± 4.82; 168.1 ± 23.63 and ± 172.8 + 13.22 days, respectively (Table 5.12). The sex of lamb did not have significant effect

PARTURITION CALANDER

Table 5.9 : The Table gives the probable date of parturition, when the females are served on the 1st or 16th day of the month. In case the date of service is after these dates, add the number to parturition date given in the table.

Date of service		Cow		Buffalo		Sheep & Goat		Mare		Camel		Bitch		Cat	
		Probable date of parturition													
Jan.	1	Oct.	11	Nov.	7	May	31	Dec.	4	Feb.	11	Mar.	4	Mar.	1
Jan.	16	Oct.	26	Nov.	22	June	15	Dec.	19	Feb.	26	Mar.	19	Mar.	16
Feb.	1	Nov.	11	Dec.	7	June	1	Jan.	4	Mar.	14	Apr.	4	Apr.	1
Feb.	16	Nov.	26	Dec.	23	July	16	Jan.	19	Mar.	29	Apr.	19	Apr.	16
Mar.	1	Dec.	9	Jan.	5	July	29	Feb.	1	Apr.	11	May	3	Apr.	29
Mar.	16	Dec.	24	Jan.	20	Aug.	13	Feb.	16	Apr.	26	May	18	May	14
April	1	Jan.	9	Feb.	5	Aug.	29	Mar.	4	May	12	June	3	May	30
April	16	Jan.	24	Feb.	20	Sept.	13	Mar.	19	May	27	June	18	June	14
May	1	Feb.	8	Mar.	7	Sept.	28	Apr.	3	June	11	July	3	June	29
May	16	Feb.	23	Mar.	22	Oct.	13	Apr.	18	June	26	July	18	July	14
June	1	Mar.	11	Apr.	7	Oct.	29	May	4	July	12	Aug.	3	July	30
June	16	Mar.	26	Apr.	22	Nov.	13	May	19	July	27	Aug.	18	Aug.	14

Table 5.9 continued .....

Date of service		Probable date of parturition													
		Cow		Buffalo		Sheep & Goat		Mare		Camel		Bitch		Cat	
July	1	Apr.	10	May	7	Nov.	29	June	3	Aug.	11	Sept.	2	Aug.	29
July	16	Apr.	25	May	22	Dec.	13	June	18	Aug.	26	Sept.	17	Sept.	13
Aug.	1	May	11	June	7	Dec.	29	July	4	Sept.	11	Oct.	3	Sept.	29
Aug.	16	May	26	June	22	Jan.	13	July	19	Sept.	26	Oct.	18	Oct.	14
Sept.	1	June	11	July	8	Jan.	29	Aug.	4	Oct.	12	Nov.	3	Oct.	30
Sept.	16	June	26	July	23	Feb.	13	Aug.	19	Oct.	27	Nov.	18	Nov.	14
Oct.	1	July	11	Aug.	7	Feb.	28	Sept.	3	Nov.	11	Dec.	3	Nov.	29
Oct.	16	July	26	Aug.	22	Mar.	15	Sept.	18	Nov.	26	Dec.	18	Dec.	14
Nov.	1	Aug.	11	Sept.	7	Mar.	31	Oct.	4	Dec.	12	Jan.	3	Dec.	30
Nov.	16	Aug.	26	Sept.	22	Apr.	15	Oct.	19	Dec.	27	Jan.	18	Jan.	14
Dec.	1	Sept.	10	Oct.	7	Apr.	30	Nov.	3	Jan.	11	Feb.	2	Jan.	29
Dec.	16	Sept.	25	Oct.	22	May	15	Nov.	18	Jan.	26	Feb.	17	Feb.	13

Table 5.10 : Duration of different stages noted in the process of parturition.

Genetic groups	Sex of the lamb	First stage (min)	Second stage (min)	Third stage (min)
		Mean ± SE	Mean ± SE	Mean ± SE
Nali	M	343.5 ± 19.50 (24)	33.6 ± 2.14 (27)	154.9 ± 8.40 (29)
	F	396.0 ± 14.37 (15)	33.8 ± 2.04 (30)	135.5 ± 14.69 (30)
	Both sexes	367.7 ± 17.51 (39)	33.7 ± 2.09 (57)	145.1 ± 11.60 (59)
Lohi	M	259.0 ± 38.07 (10)	33.1 ± 1.61 (14)	131.7 ± 16.16 (14)
	F	421.3 ± 28.41 (9)	32.3 ± 1.81 (13)	131.6 ± 7.11 (12)
	Both sexes	288.5 ± 33.25 (19)	32.8 ± 1.73 (27)	131.6 ± 10.50 (26)
Nellore × Nali	M	327.8 ± 29.11 (7)	33.8 ± 0.80 (7)	161.4 ± 13.17 (7)
	F	330.6 ± 18.72 (8)	33.8 ± 0.88 (10)	135.5 ± 13.19 (10)
	Both sesces	329.3 ± 24.41 (15)	33.8 ± 0.88 (17)	146.1 ± 13.1 (17)

Table 5.10 continued .....

Genetic groups	Sex of the lamb	First stage (min)	Second stage (min)	Third stage (min)
		Mean ± SE	Mean ± SE	Mean ± SE
Mandia × Nali	M	459.4 ± 96.69 (9)	37.8 ± 2.89 (9)	124.0 ± 12.40 (8)
	F	468.6 ± 74.44 (5)	41.8 ± 5.81 (5)	113.7 ± 17.25 (4)
	Both sexes	462.7 ± 83.65 (14)	39.3 ± 4.32 (14)	120.6 ± 14.92 (12)

on postpartum estrous interval in case of single births. Effect of season was also non significant. One percent ewes came in heat within 60 days. The average post partum estrous interval following twin lambings and after abortions was abnormally long. 28.1% ewes came in heat within 151-180 days after abortions.

Pregnancy diagnosis

There are several ways by which pregnancy can be diagnosed in sheep. Some methods are relatively simple. The failure of an ewe to return to estrous gives an indication that she is pregnant.

Physical examination of ewes in late pregnancy is simple and quick. The lamb can be felt by gently palpating the ewes abdomen. The udder becomes firm and enlarged in pregnant ewes.

A rectal probe can be used to palpate the uterus of ewe from about 70 days of pregnancy and is reliable when used by an experienced operator (Plant, 1980).

Richardson (1972) reviewed sophisticated methods of pregnancy diagnosis and she found over 80% accuracies by vaginal biopsy, ultrasonic foetal pulse detection and radiography. The foetal pulse detection is more practicable and is reviewed by Thwaites (1981) and Wani (1981). This method can accurately predict whether or not an ewe is pregnant if used after 60 days of pregnancy (Aswad *et al.*, 1976; Wani and Sahni, 1981) and the rate of lamb heart beat can be used to predict the date of lambing.

Trapp and Slyter (1983) reported that ultrasonic scanning instruments developed to measure carcass fatness, when used between days 69 and 112 of pregnancy, gave an accuracy of pregnancy diagnosis of about 90%.

By laparotomy, pregnancy can be detected by as early as 30 days by skilled workers (Pacho, 1973).

If the date of breeding is known, an analysis of plasma progesterone, 17 days later will determine whether ewe is pregnant or not (Robertson and Sarda, 1971). For lactating ewes, progesterone levels in the fore milk can be used to diagnose pregnancy (Ayalon and Shemesh, 1979).

Parturition

The act of parturition is termed as lambing in ewe. Normal parturition consists of expulsion of normal viable foetus from the uterus through maternal passage by natural forces alone at a stage when the lamb is capable of independent existence (Kaushish and Arora, 1974 b). It is difficult to predict the exact day of parturition in Nali sheep by studying the antepartum blood cholesterol content. (Kaushish and Arora, 1974 a).

Time of parturition

The time of parturition is one of the most critical periods in the life of a ewe. During this process, the ewe could possibly be injured and future reproduction efficiency be impaired temporarily or permanently. The distribution of births appears to vary with the breed of the ewe and the management system (Kaushish *et al.*, 1973; Younis and El-Gaboory, 1978; Tomar, 1979; Bhaik and Kohli, 1980). Highest number of lambings took place between 3 and 6 a.m. and 3 and 6 p.m. (Kaushish *et al.*, 1973) in some native breeds and their crosses whereas, maximum (40.0 percent) lambings took place between mid night and 6 a.m. in Russian Merino ewes (Kaushish and Sahni 1975) (Table 5.11).

Signs of parturition

The symptoms of the onset of parturition consist of changes in the birth canal, progressive extensibility to permit safe passage of foetus, and hypertrophy of the mammary glands. The vulval lips become resilient and oedematous, two to six times their normal size. The ewes usually lay on the right side streching the hind legs as far as possible, especially during contractions and frequently showed upward turning of lips. Frequent grinding of teeth were also observed. About 6 to 8 h before the expulsion of foetus the vulval labiae were about 0.5 cm apart and this distance increased to 1.0 cm about one hour before parturition. The respiration rate increased about half an hour before the appearance of water bag (Kaushish and Arora, 1974 b).

Process of parturition in sheep

The process of parturition has been studied in detail by Tiwari *et al.* (1969a) and Kaushish (1971) (Table 5.10). The time taken for the completion of parturition was longest in Mandia × Nali and shortest in Lohi. The post service gain in weight of dam was negatively correlated with the time taken for the second and third stages of the parturition (Kaushish and Arora, 1974b) (Table 5.10). The duration of sequence of events in the second stage of

Table 5.11 : The percent occurrence of births during different intervals in Russian Merino

Time intervals	Groups	Two year old			Three year old			Overall		
		Single	Twin	Combined	Single	Twin	Combined	Single	Twin	Combined
0.01-3.00	G1	13.41	0.00	13.41	11.76	10.29	22.06	12.38	6.67	19.05
		(5)	(–)	(5)	(8)	(7)	(15)	(13)	(7)	(20)
3.01-6.00	G2	24.34	2.70	27.04	8.82	8.82	17.65	14.29	6.67	20.50
		(9)	(1)	(10)	(6)	(6)	(12)	(15)	(7)	(22)
6.01-9.00	G3	2.70	0.00	2.70	8.82	4.41	13.23	6.67	2.86	9.50
		(1)	(–)	(1)	(6)	(3)	(9)	(7)	(3)	<10)
9.01-12.00	G4	10.81	0.00	10.81	1.48	2.98	4.41	4.77	1.90	6.67
		(4)	(–)	(4)	(1)	(2)	(3)	(5)	(2)	(7)
12.01-15.00	G5	16.22	2.70	18.92	8.82	5.89	14.85	11.45	4.76	16.19
		(6)	(1)	(7)	(6)	(4)	(10)	(12)	(5)	(17)
15.01-18.00	G6	18.92	0.00	18.92	2.98	4.41	7.42	8.57	2.86	12.38
		(7)	(–)	(7)	(2)	(3)	(5)	(9)	(3)	(12)
18.01-21.00	G7	5.40	0.00	5.40	10.29	1.48	11.76	8.57	0.95	9.50
		(2)	(–)	(2)	(7)	(1)	(8)	(9)	(1)	(10)
21.01-24.00	G8	2.70	0.00	2.70	8.82	0.00	8.82	6.67	0.00	6.67
		(1)	(–)	(1)	(6)	(–)	(6)	(7)	(-)	(7)

Figures in paranthesis denote the number of observations

* Zero hours start from midnight

parturition has also been studied in detail by them (Kaushish and Arora, 1974 a).

Relationship among weight of placenta, number of cotyledons and birth weight has been studied by Kaushish and Arora (1983). The average number of cotyledons varied from 66.6 to 74.1 between groups. The weight of placenta ranged from 0.195 to 0.259 kg. The sex of the lamb did not affect the placental weight and the number of cotyledons significantly. There was a positive relationship between placental weight, number of cotyledons and birth weight (Kaushish, 1971). Weight of placenta as percentage of lamb weight generally centered round about 8.5 percent in Nali, Lohi and the crosses of former with Mandia and Nellore (Kaushish and Arora, 1975). It was little lower in Bikaneri and Mandia (Tiwari *et al.*, 1969a). The effect of weight at service on time taken for expulsion of foetus and placenta has been studied by Kaushish and Arora (1973).

Post partum estrous :

Most ewes are seasonal breeders, consequently there are few attempt, for the ovaries to function until the breeding season occurs or that particular breed. In the breeds which have been selected to produce two crops of lambs each year, there is need for breeding soon after lactation ceases. As soon as weaning occurs, these breeds usually show a cyclic activity in the ovries accompanied by estrous and ovulation (Table 5.12).

Induction of parturition

Hormones like dexamethasone or flumethasone (Bosc, 1972) can be used to induce lambing in sheep. Hormones have been used to combat prolonged pregnancy of Karakul ewes in South Africa (Roux and Wyk, 1977). The interval between injection and parturition depends on the stage of pregnancy (Aswad *et al.*, 1974) and if the ewes are injected too early, the lambs die (Webster and Haresign, 1981). There is a great scope for this in tropics.

Interlambing interval

It is the interval between two successive lambings. Normally, it varies between 7 and 12 months, but may be as long as 20 months also. Values in the literture of lambing interval give flock means of 218 days for Morada Nova sheep (Teixeira *et al.*, 1980) and 408 days for Mandya sheep (Purushotam, 1978).

Table 5.12 : Averages (Mean ± S.E.) of post partum estrus interval following single normal lambings and abortions in Muzaffar-nagari ewes.

Interval (days)	Single births		Abortions	
	% of ewes	Mean ± S. E.	% of ewes	Mean ± S. E.
>60	19.1	35.5 ± 1.94	14.0	35.5 ± 6.20
61-90	16.1	77.6 ± 1.18	5.3	77.0 ± 1.0
91-120	12.3	103.9 ± 1.27	7.0	101.0 ± 3.10
121-150	12.3	137.1 ± 1.28	7.0	135.0 ± 4.77
151-180	12.9	163.8 ± 1.38	28.1	167.4 ± 1.90
181-210	9.1	196.0 ± 1.60	17.5	194.8 ± 2.95
211-240	7.1	229.5 ± 2.14	1.8	240.0 ± 0.00
241-270	2.3	254.1 ± 2.70	7.0	249.8 ± 3.11
271-300	3.9	285.2 ± 2.20	5.3	272.7 ± 1.53
301- and above	4.5	353.5 ± 7.03	7.0	426.7 ± 47.06

Long interlambing intervals may be due to long postpartum anestrous period, failure to conceive, death of all embryos or abortion. Ewes which do not conceive for two successive breeding seasons are culled under improved managemental conditions. Lambing interval decreases with parity upto 4, suggesting that young ewes which are still growing take longer to regain condition after lambing. Interlambing period is also affected by season, (Fall *et al.*, 1982).

Reducing age at puberty

Age at puberty depends upon the nutrition of lamb. Some breeders allow gimmers to mate immediately after they attain puberty. There is no evidence to suggest that early first lambing results in diminished reproductive capacity (Wilson and Durkin, 1983). The hormonal stimulation of lambs is of practical use in the tropics. The gonads of prepubertal sheep respond to exogenous hormones, but animals stimulated in this way do not commence spontaneous estrous cycle (Hunter, 1980).

Reducing interlambing interval

An 8 month of lambing interval is possible provided that nutrition and management are satisfactory (Naude and Grant, 1979). Hormonal treatment can be used to stimulate estrous where seasonal anestrous from photoperiod results (Fletcher *et al.*, 1980).

The period between lambing and conception (closed service period) of poorly fed ewes is about 180 days if no hormonal treatment is given. Intensively fed West African Dwarf ewes conceive at an average of 43 days after lambing (Berger and Ginisty, 1980). Estrous can be induced by using intravaginal sponges as early as 17 days after lambing (Nie Kerk, 1979). Honmode *et al.*, 1971a) induced heat in local Malpura ewes by inserting sponges 30 days after lambing. Early postpartum breeding of sheep in Ghana reduced lambing interval to 6 months (Brown *et al.*, 1972).

Sheep loose weight if bred frequently (Sahni and Tiwari, 1974b). A seasonal production on lambs results in high lamb mortality (Labban and Ghali, 1969) and poor growth rate (Ganesakale, 1975).

6

Embryo Technology

Under traditional sheep farming system low lamb output per ewe per year is a major factor limiting the energetic efficiency of sheep production. Over the past 50 years reproductive physiologists all over the world have tried to evolve total management package, including reproductive management to realize the full reproductive potential of the species commensurate with the potential biological ceiling of litter production and an economic lambing interval. To achieve this goal, the new reproductive biotechnological inovations have been successfully used so much so that these techniques are being presently used in several countries on a commercial scale.

The reproductive innovations for advancing sheep breeding season, more frequent lambing, induction of multiple births, synchronised breeding with fixed time artificial insemination, controlled and compact lambing are now practical realities. The possibility of embryo technology in sheep has been clearly demonstrated. Through hormonal manipulations and appropriate strategic breeding of ewe and lambs at 7 to 8 months of age is now a distinct possibility.

History

Heape (1890) was first to transfer eggs from the uterus of one Rabbit into another. Later on similar efforts were made by other workers in transfering eggs in other domestic animals. The use of the technique was mostly limited to laboratories only during earlier Stages, but now its use on commercial scale is being exploited. Normally, a sheep produced 6-7 lambs in whole of its life, but it can be increased several times in case of superior dams by the use of this technique. Many workers have done researches on embryo transplantation (Church and Shea, 1975; Hinkelman, 1979; Moore and Eppleston, 1979; Willardsen *et al.*, 1977; Dunn, 1980; Bradford and Kennedy, 1980) superovulation (Honmode and Tiwari, 1968, 1973; Honmode *et al.*, 1971a; Elsdon *et al.*, 1974; Wilmut and Sales, 1981) and embryo storage (Kardymowiez *et al.*, 1976; Willardsen *et al.*, 1977; Patil and Bhosrekar, 1978; Smorag *et al.*, 1978).

Increasing the lambing frequency beyond the traditional once a year crop, has been possible with use of appropriate reproductive technology, including hormonal treatment. It was possible to achieve two lambings every 13 months or 3 lambings every 24 months, lambing rate varying from 1.82 to 3.70 lambs per ewe per year in different countries. Integration of the various reproductive technological innovations into a single package of practice have been tried with the objective of producing lambing all the year round at intervals of 4 months. Hormonal induction of multiple lambings in flocks that show an inherently low twining percentage has been demonstrated to be practically successful even under commercial farming systems when the question of perinatal lamb mortality is adequately tackled through appropriate lamb management. More recently, the multiple lambing programme has been linked with estrous synchronisation, the set time A.I. and induced lambing for better economic management.

The ovulation rate can be increased if the ovaries are stimulated with exogenous gonadotrophins. Superovulation used in conjunction with embryo transplantation constitutes a method of accelerating and improving the reproductive performance of selected ewes. These techniques have been of considerable interest to shepherds and persons concerned with improving reproduction.

Synchronization

The success of embryo transplantation depends upon the efficient heat detection in prospective embryo donors and recepients so that various operations are carried out on specific days. Vasectomised or aparoned rams can be used for heat detection. The use of progestagens and prostaglandins facilitate the planning of subsequent operations. Daily injections of progesterone in oil, progesterone impregnated intravaginal sponges or ear implants are all effective as are various other progesterone treatments (Lamond, 1964; Falkenberg *et al.*, 1964; Karihalo and Singh, 1967; Shelton and Moore, 1967; Honmode, 1969, 1971; Honmode *et al.*, 1971 b).

The most commonly used method of estrous synchronisation in sheep is a vaginal sponge, impregnated with progestagen, withdrawal a single dose of PMSG is given (Britt, 1979; Combellas *et al.*, 1980). Using sub-cutaneous implants gives similar results although the interval from removal of progestagen to estrous and duration of estrous are both shorter with implants than sponges (Faure *et al.*, 1980). It is also possible by daily injections of progesterone (Dhanda and Arora, 1977) and by feeding melengestrol acetate (Tripathi, 1977). Synchronisation can be done more conveniently by giving two injections of prostaglandin at an interval of 10-12 days (Haresign, 1973). the schadule of giving two injections gives better results than that of giving single injection. Age, breed, season and time during estrous cycle affect the response to a particular treatment. PMSG has been used most extensively in super ovulation. Relatively small proportion of follicles present in the ovaries are responsive to PMSG (Hay and Moore, 1978). The dose of the PMSG should be adjusted in such a way that it does not cause ovulation of immature follicles. In most cases a dose of 20-25 i.u. of PMSG per kg, body weight has been used.

PMSG can also be used in conjunction with progestagens (Moore, 1970). Progestagen priming is necessary if anestrous ewes are to be used as donors (Lamond, 1964) and is also essential if prepuberal ewes are to be superovulated (Trounson and Willadsen, 1977). The sheep should be kept on progestagen for atleast ten days and injected with PMSG about 3 days prior to planned onset of estrous. Any of the following methods can be used for superovulating ewes:

(i) Synchronisation with subcutaneous (S/C) progesterone implants (350 mg progesterone, Abbot labs) for 10 days or two entra muscular (I/M) injections of 125/Ug Cloprostenol (Estrumate ICI Ltd.) 10-12 days apart. Almost all ewes come in heat 24-48 h following progesterone withdrawal or second injection of PGF2.

(ii) On day 11-12 of the following cycle an 1/M injection of 1250 i.u. PMSG (Folligon, intervet labs).

(iii) 48 h later an 1/M injection of 125 μg Cloprostenol after which the ewes are penned without access to food and water and checked regularly for onset of estrous. The majority of the ewes come in heat 24 h later practically all by 36 h.

Super ovulation

Factors controlling early embryo development are poorly understood and consequently the ability to culture embryos *in vitro* is limited. Oviduct specific glycoproteins have been reported in several species including the sheep (Sutton *et al.*, 1984; Gandolfi *et al.*, 1989). These proteins, together with those from serum transudates, are thought to contribute towards the establishment of an unique oviductal environment. It was found by Walker *et al.*, (1988) that about 90% of the sheep embryos cultured for 3 days from the zygote stage developed to Day 14 of pregnancy in synthetic oviduct fluid medium (SOFM7) supplemented with heat inactivated human serum.

The ovaries may be stimulated with exogenous gonadotrophins to increase ovulation rate (Hunt *et a.l, 1971).* Various workers (Pandey *et al.*, 1972; Sengupta *et al.*, 1978) have attempted superovulation with variable degree of success in India. The lambs born in multiple births have less chances of survival than those born in single births (Seth *et al.*, 1972). Sharma *et al.*, (1992) observed increase in the number of corpus leutium with the increase of age and body weight of dam 6 days post estrous in superovulated sheep.

Breeding

If natural service is done, it should be done on several occasions during heat, but failure of fertilization occurs in superovulated ewes. The failure is due to decreased transportation of sperms, through the cervix. Fertilization rate can be increased by doing insemination in the uterus of superovulated ewes (Killeen and Moore, 1971; Trounson and Moore, 1974). The effect of PMSG on the lambing performance has, also been studied (Sengupta *et al.*, 1976).

Collection of embryos

Eggs are normally collected when they enter the fallopian tubes. The collection of fertilised ova is done 3, 2, 4 and 6 days after fertilization in case of sheep, cow and pig and mare, respectively. Normally 40-80 percent fertilized eggs can be recovered by surgical means. Surgical method of embryo collection has been adopted in laboratory (Daniel, 1971) and domestic animals (Dzillk, 1971; Murray, 1978). Fertilized egg moves to uterus from fallopian tubes in 3 to 4 days. The number of embryos collected from fallopian tubes exceeds those collected from the uterus, but the embryos collected from the latter are more viable (Newcomb and Rowson, 1975) and can be frozen more successfully. Use of medium 199 is more popular for the embryo collection (Rowson *et al.,* 1969). Different mediums used for embryo collection are given in Table 6.1. Streptomycin and penicillin are mixed at the rate of 50 mg and 1 lac i.u. per litre. Bovine Serum Albumen must be added to this medium at the rate of 0.3 to 1.0 percent. The pH of the medium is kept between 7 and 8 but a pH between 7.2 and 7.6 gives best results. Double distilled water should be used for making the medium. Detailed information has been given by Gwatkin (1972) for making medium.

Collection of eggs from superovulated ewes within first few hours of ovulation results in low recovery rates. The best results are obtained if no tube is inserted into oviduct. For collecting embryos on day 12, 13 and 14, a curved large bore glass tube is inserted into the uterine horn and 15-20 ml of medium is flushed from the tube of the uterine horn or from the oviduct to be collected in a glass cup. The incision is closed with an absorbable suture. Premature ovulation and embryo collection in ewes has also been tried by Heard and Walker (1992).

Embryo transfer

Embryo transfer has been reviewed extensively by Trounson and Rowson, (1976) and Willardson and Polge, (1980). It allows for rapid multiplication of valuable genotypes. This technique does not yet have a high rate of success in sheep (Armstrong and Evans, 1983) although it has been used to multiply wool producing genotypes in India (Zanwar, 1981).

Table 6.1 : Mediums used for embryo collection

Quantity in mg/l.	Improved Delbeco (Gwatkin, 1972)	Hemice mixture GIBCO	Improved fallopian tube fluid (Tervit *et al.*, 1972)	Tissue culture medium 199 GIBCO	Vitan medium (Gwatkin, 1972)
Sodium chloride	8000	7400	6300	8000	5140
Potassium chloride	200	285	533	400	350
Calcium chloride	100	33	190	140	–
Magnesium chloride	100	–	100	–	–
Magnesium sulphate	–	153	–	200	294
Sodium bicarbonate	–	1200	2106	350	1900
Disodium Phasphate	1150	154	–	48	–
Potassium di hydrogen phosphate	200	83	162	60	162
Glucose	1000	1100	270	1000	1000
Sodium Pyruvate	36	110	36	–	36
Sodium lactate	–	–	370	–	2416
Calcium lactate	–	–	–	–	527
Ribose	–	–	–	0.5	–

Table 6.1 continued ....

Quantity in mg/l.	Improved Delbeco (Gwatkin, 1972)	Hemice mixture GIBCO	Improved fallopian tube fluid (Tervit *et al.*, 1972)	Tissue culture medium 199 GIBCO	Vitan medium (Gwatkin, 1972)
Deoxiribose	–	–	–	0.5	–
Vitamins	–	Yes	–	Yes	–
Nuclie acid	–	Yes	–	Yes	–
Trace elements	–	Yes	–	Yes	–
Bovine Serum Albumen	–	–	–	–	3000
Cholesterol	–	–	–	0.2	–
Sodium acetate	–	–	–	50	–
Lipoic acid	–	0.2	–	–	–
Tween 20	–	–	–	–	–
Glucothion	–	–	–	0.5	–
Vitamin E	–	–	–	0.01	–

The embryo to be transferred, is picked up with a pasture pipette in 0.04 ml of medium. The tip of the pipette is inserted 2-3 cm into the uterine lumen through the wall from uterotubal junction and the fluid containing embryo gently expelled.

Sheep embryos for transplantation may be collected from a few hours after ovulation upto 14 h after onset of estrous. In ewes which are mated naturally, collection rate may be as high as 75% of the eggs ovulated. Maximum egg recovery rates are obtained if collections are made while the eggs are still in the oviduct. A ewe may be used as a donor several times provided sufficient time is allowed for the establishment of normal cycle (Jainudeen and Hafez, 1974).

Care at embryo transplantation

Developmental stages ranging between one cell and elongated gastrula (day 14) have been successfully transferred in the sheep. If the main interest is to maximize the number of lambs produced, embryos ranging between eight cells (day 3-4) and early blastocysts (day 7) are most convenient to work. with. The embryonic survival rates have varied, but under optimal conditions some 60% or more of the embryos transferred can be expected to give rise to lambs (Moore & Shelton 1964 a,b, Rowson & Moor 1966); Moore & Trounson 1977). Differences with respect to breed between donor and recipient are of no consequence as long as the recipient is sufficiently large to carry and give birth to the type of lamb produced (Hunter *et al.*, 1955; Lowson & Rowson 1972). Healthy, mature ewes which have previously lambed should be chosen as recipients.

In the planning of embryo transplantations, one should aim at precise synchrony of estrous in donor and recipient. Although acceptable pregnancy rates are still obtainable if there is an asynchrony of ± 1 or even ± 2 days (Rowson & Moor 1966), the best results are probably obtained when donor and recipient are at precisely the same stage of the estrous cycle, (Moore & Shelton 1964b). Recipients that came into estrus slightly later than the donor are preferable to recipients that came into oestrus earlier. Day 13 and day 14 embryos should be transferred to day 12 recipients, since luteal regression is initiated about day 13 in the nonpregnant ewe (Rowson & Moore 1966). Ideally, the developmental stage of the individual embryo, rather than the stage of estrous cycle of the donor should govern the choice of recipient, e.g. a morule of some 30 cells should preferably be transferred to a day 5 or 6 recipient even if it was collected from a day 7 donor. Similarly, the stage of development rather than the age of the embryo should decide the choice of recipient for an embryo which has been cultured or stored *in vitro.*

If two embryos are to be transferred per recipient one should be transferred to each side of the genital tract; a single embryo should be transferred to the side ipsilateral to the ovary carrying the *corpus luteum.* The overall embryonic survival rate is about the same whether one or two embryos are transferred per recipient. Embryos less advanced than eight cells should be transferred to the oviduct, more advanced embryos to the uterus.

One cannot escape the conclusion that although relatively efficient methods of superovulation and embryo transplantation have been developed in the sheep, the practical application of these techniques would only be economically feasible in exceptional circumstances. It is equally evident that they are potentially extremely valuable to investigators of basic mechanisms and relationships in reproductive physiology, genetics and developmental biology.

Embryo storage

For practical use in egg transfer experiments, it is desirable to maintain the egg *in vitro* in a more or less dormant state for various periods of time allowing accurate synchronization of egg development. Storage of eggs also makes possible long distance transportation of embryos (Chang and Marden, 1954; Adams, 1961; Hunter *et al.,* 1955). Storage of embryo is needed between collection and its transplantation. Normally embryo can be stored at 37°C upto 24 hours. Embryo mortality rate increases if it is transplanted after 24 hours (Davis and Day, 1978). The storage of the embryo, at four cell stage is easier in sheep (Tervit and Glood, 1978). Embryo can be stored between 15 to 25°C from few hours to one day, but it can be stored for some days between 0 and 10°C (Boland *et al.,* 1978).

Cryopreservation of embryos

Cryopreservation harms the embryos. If dimethyl sulphoxide is added to the medium, the freezing occurs much below zero degree celsius. The embryo is rapidly cooled upto 0°C, after that the temperature is lowered upto - 7°C at the rate of one degree per minute. Thereafter temperature is lowered @ 0.3°C per minute upto - 33°C, then the embryo is lowered in liquid nitrogen.

Difficulties experienced in superovulation

Effect of nutrition on embryo has been studied by Dunn (1980). Factors affecting superovulation have been studied in detail by Armstrong and Evans, (1983). Some of them are :

1. Response varies from animal to animal.
2. Eggs pass rapidly through the fallopian tubes
3. Repeated collection of eggs damages egg productability of the sheep.
4. If this method is followed in young sheep the fertility of the eggs is reduced.
5. Sometimes the ova instead of entering the fallopian tubes enter the abdominal cavity.
6. Age and plane of nutrition affects it.
7. It also depends upon the number of times collection has been done earlier.

In Vitro Fertilization and embryo Transfer

Mammalian eggs can be fertilized *in vitro* by a relatively simple culture technique. Ovulated eggs or oocytes matured *in vitro* or oocytes collected from the ovary just before ovulation could be used for fertilization *in vitro*.

Eggs are incubated with capacitated spermatozoa in a suitable culture medium for 2-5 hrs. Capacitated spermatozoa could be obtained either from female reproductive tract after artificial insemination or spermatozoa capacitated *in vitro* could be used. For *in vitro* capacitation of spermatozoa, sperms are incubated in a medium containing tubal fluid, follicular fluid or heat treated serum for 2-6 hr, depending upon the species. A drop of capacitated spermatozoa is then mixed with eggs under paraffin oil which was previously equilibrated with the same medium. The petri dish is incubated in a humidified atmosphere containing 95% air and 5% CO_2. After 2-5 hr the eggs are recovered, washed with fresh medium and cultured for a further period of 3-5 days. The eggs are transferred to receipient females when they are at the stage of morula or early blastocyst. Only females with their reproductive cycle synchronised with the age of the embryo can successfully accept the embryo.

7

Wool

Unlike other ruminating animals wool continues to grow in case of sheep. The sheep are shorn twice or thrice in a year. Some southern breeds like Nellore, Mandya and Tenguri are hairy. The wool fiber is tensile and hygroscopic in nature. It does not catch fire so easily. Good quality wool fibers are thin, soft and fine thread can be prepared from these. They can absorb colour very well. The hair are not that tensile and can not take colour so easily.

Normally the fiber diameter of wool varies from 12-70 micron (μ). The fiber diameter of superior wool varies from 15-25 micron. The wool produced from Rambouillet and Merino sheep is of this quality. The number of crimps are more in the wool. The fiber diameter of native wool varies from 30-70 micron.

Fiber length, number of crimps and diameter plays important role in the determination of wool quality. The fineness of the fiber is known by the number given after making thread. Bundles are made of 256 yard long thread from ordinary wool or 560 yards long from combed wool. The number of such bundles per pound is known as count. If 10 such bundles weigh one pound, the count of such wool is ten. The merino wool is of 60-70's count. The wool quality of different breeds in presented in Table 7.1.

Chemical composition of wool

Wool is made up of protein, sterols and fats. It has 99 percent fat and one percent other materials. Wool has about 50 percent, keratin, 8 percent carbon, 16.5 percent hydrogen, 3.5 percent nitrogen and sulphur and oxygen together constitute 22 percent.

Table 7.1 : Mean values of wool parameters of some sheep breeds.

Breed	Half yearly wool yield (kg)	Fiber diameter (μ)	Medullation percentage	Staple length (cm)	Remarks	References
1	2	3	4	5	6	7
Arabi	1.81	26.2	10.9	–	Annual clip	Ashmawy and Al–Azawi, 1982 a,b
Barki	–	33.9	14.0	11.4	Annualclip	Guirgis, 1973.
Barki	1.90	–	–	9.8	Annual clip	Marai, 1975.
Barki	0.93	–	–	–	Annual clip	Aboul–Naga/1977.
Barki	3.51	13.4	–	–	Annual clip	El–Oksh *et al.*, 1979.
Bellary	0.33	59.0	43.4	–		Nagarcenkar and Bohra, 1983.
Bellary	1.23	50.0	43.4	–	Annual clip	Acharya, 1978.
Chhota nagpuri	0.18	51.5	83.6	–		Nagarcenkar and Bohra, 1983.
Chokla	0.94	29.8	23.0	6.5		Arora *et al.*, 1975 b.
Chokla	1.20	26.9	18.0	4.8		Mahajan *et al.*, 1980 ab
Chokla	–	23.9	12.5	4.2		Pant *et al.*, 1980
Coimbatore	0.41	–	–	–		Mishra and Singh, 1979.
Coimbatore	0.37	41.1	58.2	5.8		Nagarcenkar and Bohra, 1983.
Coimbatore	0.27	39.3	66.0	6.9		Singh and Rathore, 1987
Corriedale × Coimbatore (50%)	1.14	–	–	–	Annualclip	Mishra and Singh, 1979.
Corriedale × Coimbatore (50%)	0.39	29.4	28.0	6.1		Singh and Rathore, 1987.
Corriedale × Coimbatore (62.5%)	0.47	27.8	22.3	4.5		Singh and Rathore, 1987.

1	2	3	4	5	6	7
Corriedale × Coimbatore (75%)	0.48	25.2	13.7	4.6		Singh and Rathore, 1987.
Deccani	–	35.1	35.1	5.9		Daflapurkar *et al.*, 1979
Deccani	–	38.8	65.5	8.0		Ghanekar, 1983.
Deccani	0.74	34.9	42.2	7.0		Acharya, 1978.
Deccani	0.36	52.4	73.8	8.2		Nagarcenkar and Bohra, 1983.
Deccani	0.74	37.4	34.1	6.9	Annual clip	Ghanekar, 1983.
Deccani	0.34	–	–	–		AICRP on sheep, Ann. Rept. 1984.
Deccani	–	15.8	36.6	4.6		Deshpande, 1987.
Deccani	0.39	43.2	–	6.6		Charyulu, 1988
Corriedale × Deccani	–	14.5	28.8	5.2		Deshpande and Sakhre, 1987.
Mandya × Deccani	–	19.3	40.9	8.3		Deshpande and Sakhre, 1987.
Dorset × Deccani	0.66	30.9	–	8.4		Charyulu, 1988.
Doreset × Deccani	0.51	–	–	–		AICRP on sheep Ann. Rept. 1984.
Merino × Deccani	0.74	24.9	–	10.4		Charyulu, 1988.
Merino × Deccani	0.61	25.3	–	–		AICRP on sheep Annual Report, 1984.
Patanwadi	0.45	29.6	19.1	5.7		Daflapurkar et al., 1979.
Patanwadi × Deccani (50%)	0.38	25.3	33.2	6.1		Daflapurkar et al., 1979.
Patanwadi × Deccani (50%)	–	35.7	38.1	7.9		Ghanekar, 1983
Patanwadi × Deccani (75%)	–	29.3	27.2	7.2		Ghanekar, 1983

1	2	3	4	5	6	7
Hamdani	2.00	30.0	94	–	Annual clip	Ashmawy and Al–Azawi, 1982 a,b.
Jaisalmeri	–	36.5	–	–		Chatterjee and Arora, 1974.
Karadi	–	48.1	19.4	19.3	Annual clip	Ghoneim *et al.*, 1973.
Karadi	1.6	28.8	9.6	–	Annual clip	Ashmawy and Al–Azawi, 1982 a,b.
Libyan Barbary	2.7	32.5	30.0	13.1	Annual clip	Labban, 1973.
Magra	–	41.7	–	–		Chatterjee and Arora, 1974
Magra	1.4	–	–	–		Mittal and Pande, 1975.
Magra	–	30.8	37.0	–		Sattar *et al.*, 1978.
Malpura	–	47.9	–	–		Chatterjee and Arora, 1974
Malpura	0.5	39.0	79.7	6.7		Mahajan *et al.*, 1980 a, b.
Malpura	0.6	40.6	89.7	5.0		Kishore *et al.*, 1978.
Mandya	–	23.0	57.0	4.4		Deshpande and Sakhre, 1987.
Mandya × Deccani	–	19.3	40.1	8.3		Deshpande and Sakhre, 1987.
Najdi	2.2	66.5	87.8	–	Annual clip	Pritchard *et al.*, 1975.
Nali	1.0	—	–	6.9		Ram *et al.*, 1978.
Nali	–	30.9	37.8	6.2		Singh *et al.*, 1981a
Nali	–	37.0	80.0	4.5		Pant *et al.*, 1980
Nilgiri	0.8	–	–	–		Bhuvanakumar and Subramanian, 1981
Ossimi	2.1	–	–	11.6	Annual clip	Marai, 1975.
Patanwadi	0.45	29.6	19.1	5.7		Daflapurkar *et al.*, 1979.

Following amino acids are found in wool:

1.	Allanine	2.	Arginine	3.	Aspartic acid
4.	Cistine	5.	Glutamic acid	6.	Glycine
7.	Histidine	8.	Hydroxilin	9.	Leucine
10.	Lysine	11.	Methionine	12.	Tyrosine
13.	Phenylalanine	14.	Tryptophan	15.	Threonine
16.	Velinamile nitrogen	17.	Cystine		

Physical properties of wool

The quality of the wool is known from its thinness. It is expessed as counts. Normally thick wool fibers are longer. The length of fine wool, medium wool and coarse wool fibers is 4-12, 6-17 and 10-30 cm, respectively. The number of crimps are less in coarse wool and more in the fine wool. The friction decreases if fingers are moved from the root to tip of the wool fiber. It increases if the fingers are moved in reverse direction. Wool fibers can absorb more water as compared to other fibers. It can absorb water upto one third of its weight. The diagram of wool follicle is depicted in Fig. 7.1.

Chemical properties of wool

It dissolves in acid and alkali solutions. Acids break peptide chain bonds. Caustic soda solution dissolves wool fiber. Most oxidizing agents are used for decolourisation of fiber and preventing it from shrinkage. Wool becomes yellow if kept in warm humid places. It starts changing colour at 100°C.

Mechanical properties of wool

Tensile strength is very high in wool fibre. Its length can increase by 70 percent when pulled slowly in water. Wool fibers behave differently at high temperature. If a fiber is stretched in boiling water for 30 minutes, it does not return to its original length, when kept in cold water. If it is kept in boiling water for 15 minutes, it decreases in length. Tensile strength depends upon temperature and humidity. It could be 30-40 percent under arid conditions and 50-70 percent in humid climate. Felting is observed in wool.

Factors affecting wool production

Level of wool production and fleece structure may be affected by various climatic factors and seasonal changes which

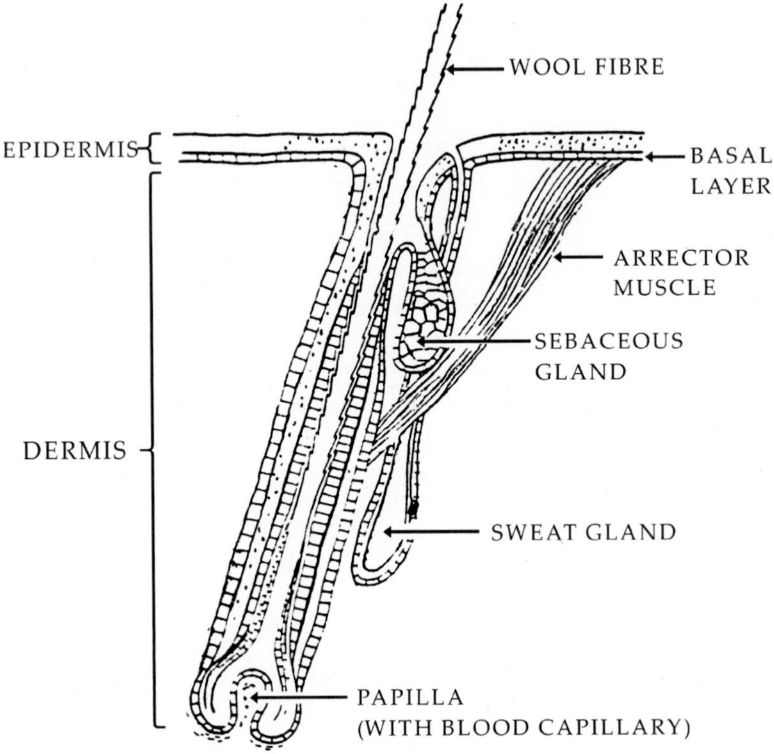

Fig. 7.1 : Diagram of a Wool follicle

can influence the shedding and regeneration of follicles, fiber diameter and fiber length growth rate (Bhasin and Desai, 1965).

In cold countries sheep is shorn annually. In India it is normally shorn twice a year but in hot states they are shorn thrice a year. Shearing even in warm climate results in change in the microclimate of sheep. Shearing in a warm environment might stimulate wool growth or in cold conditions might bring about a stress response from adrenal cortex. Injections of adrenocortico-trophic hormone (ACTH), cortisone and the application of cortisone ointments depress wool growth.

Humidity and Atmospheric temperature

It has been found that humidity differences of 25-30 percent and temperature differences of upto 30°C had no effect.

On the other hand there is a marked correlation between the rate of wool growth and mean atmospheric temperature amongst maiden Corriedale ewes fed a uniform maintenance diet throughtout the year.

Seasonal rhythm

The seasonal rhythm persists amongst Corridale ewes fed a ration of constant composition at rates which maintained the effective body weights of the ewes constant. Marston (1955) found five fold differences in the rate of wool production of fine wool sheep between lush pastures of spring and the poor pasture of winter. The fundamental causes of the seasonal rhythm of wool growth remains obscure.

Photoperiodicity

The ewes produced lesser wool in winter than in summer. Nutrition, pregnancy and lactation can override or magnify the seasonal effect, but it can be isolated and measured, provided these factors, are adequately controlled.

The photoperiodicity rhythm may be one factor and it is probable that the seasonal rhythm in temperature may be another.

Hormones

Speculation of the mechanisms involved here is perhaps unwarranted at this stage though the probable connection between hormone output and temperature and photoperiod rhythms may be mentioned. Atmospheric temperature is known to affect the output of the thyroid. It has been shown that thyroxin administration and deprivation influences wool growth.

The thyroid gland has a permissive effect on wool follicle development and wool growth. Thyroidectomy reduces wool growth to about 40-50% of normal. Administration of thyroxine to normal sheep stimulates wool growth. Dry ewes given a single 60-100 mg implant responded by an increase in wool production by 15%.

Oral administratione of L-thyroxine to Suffolk sheep brought about an increase in wool production per unit area of skin due to an increase in fiber length growth rate and also caused an increase in milk secretion during the second to eighth week of lactation in Black Leicester × Scotch Black face ewes.

Cortisol

The main glucocorticoid secreted by the adrenal cortex of the sheep is cortisol and the plasma level of cortisol in the sheep is exceptionally low compared to other species. Administration of cortisol or ACTH to normal sheep suppresses wool growth and complete inhibition can be produced even though the plasma cortisol level remains below that of other species for which the plasma levels are known. Raised levels of plasma cortisol cause "break" and "tenderness" of fleece. Stress conditions such as injury, disease, starvation and extremes of environmental temperature stimulates adrenal cortical secretion and a break in the fleece following such conditions is probably medicated via the adrenal cortex.

Growth hormone

Which has a direct action, does not appear to have a permissive effect on wool growth in that it does not permit growth in hypophysectomized sheep more over it does supplement the response to thyroid hormone in these sheep.

Other pituitary hormones such as prolactin, gonadotrophic hormones or oxytocin do not appear to have any significant permissive or regulatory effect on wool.

Physiological state

It has been found to have significant effect on wool growth. In general entire males produce more wool than castrated males which produce more than females. There is no real evidence, however, to suggest that these differences are determined by anything other than differences in size and nutritional state.

The reproductive cycle in females can reduce wool growth significantly. During late pregnancy, depression of wool growth rate in the range of 20-40 percent has been found and even in breeds which exhibit a marked seasonal depression, pregnancy can depress the rate still further.

In lactating ewes, wool growth is also reduced by upto 30 percent or more, despite the accepted increase in voluntary intake associated with lactation.

The average reduction in clean fleece weight between handicapped animals (twins and progeny of young ewes) and unhandicapped (singles and the progeny of adults) animals born in the same year was 150 g. The period most sensitive to nutrition deficit is from 30 days before to 35 days after birth (Fraser, 1954).

Age

Rate of wool production in most breeds increases to a maximum between four and five years and then declines, often rapidly.

Nutrition

The most important. factor affecting adult wool growth however, is the current level of nutrition (Fraser, 1954; Kapoor *et al.*, 1972). Increase in energy intake, except at very low levels of protein content in the diet, can have an immediate effect on wool growth detectable within a few days. It is generally accepted that wool growth rate is dependent on energy rather than protein intake. Wool protein contains a high proportion of the high sulphur amino acid, cystine, and it has been shown that variation in the availability of the sulphur amino acids to the follicle can affect both fiber growth rate and fiber composition.

It is not possible to estimate the actual amount of energy which produces wool. The amount of metabolizable energy which is drawn upon for wool production is likely to exceed materialy the combustible energy for wool fiber and of the secretions which accompany it which is 3% of the metabolizable energy derived from rations (Marston, 1955). Deficiency of minerals such as calcium and vitamin D are associated with depressed wool growth. These act indirectly possibly by affecting health and appetite.

Carbohydrate supplementation to sheep increases body weight and fleece weight (Fraser, 1954). The addition of carbohydrate to diet converts a negative nitrogen balance into a positive one (Krishnan, 1939). Carbohydrate is necessary for mitosis to take place and there is importance of glycogen in the follicle for wool growth.

Breeds

In five wool breeds, winter growth rate has been found to be around 85% of that in summer, whereas in British hill breeds the winter rate may be as low as 30% of that in summer. The wool quality of Deccani and their crosses with Dorset and Merinos, Patanwadi and Deccani crosses has been studied by Daflapurkar *et al.*, (1979) and Deshpande and Sakhre (1987).

The importance of differences between strains and probably breeds of sheep is indicated by recent work with merinos indicating small but significant differences in efficiency of conversion of feed to wool.

The behavioural differences between individual sheep subjected to competition for limited feed, are another source of variation. Feeding groups of sheep once per week gave higher mean body weights and fleece weights than feeding them equal total amounts on daily basis. Difference of social order within a flock or group of sheep also determines wool production.

Wool shearing

The process of removal of wool from sheep or harvesting of one of the most valuable products of sheep i.e. wool is known as shearing. A poor job, lowers the quality and thus the market value of the wool. To avoid this, the shearing has become a technique rather than a process.

Time of shearing

Wool acts as a protective cover in sheep against heat and cold and it may therefore be shorn at a time when the animal needs it to be done for its comfort. In India, the hot weather begins from March/April and if wool is not shorn then, the animal would not be able to keep its body cool. The sheep are therefore shorn in February/March. Shearing during these months has another advantage that as mating is done after shearing, crutching of ewes and removal of wool around the testicles in rams i.e. preparation of ewes and rams for breeding is not done as a separate operation.

As the shorter wool dries up quicker than longer wool, shearing is done before the onset of winter i.e. in September/October. Shearing done in September/October has another advantage i.e. the wool is not too short when the winter begins and it acts as an insulator as well. Incidentally, wool itself is not a particularly good insulator but air is entangled among the fine wool fibers that most effectively serves the purpose.

Therefore, the shearing is generally done twice a year i.e. in February/March and September/October to provide a protective cover to sheep from heat and cold and also as a cash crop to the shepherd twice a year.

Methods of shearing

Shearing can be done by any of the following methods.

1. Hand shearing
2. Machine shearing
3. Chemical shearing

1. *Hand shearing* : Hand shearing is the most common method of shearing under Indian conditions. It is an intersting sight to see an Indian shepherd shearing his sheep. He ensures that the animal is comfortable during shearing as well as clips as close to the skin as possible and maintains an adequate speed in his operation. An experienced shepherd can shear as many as 40-50 sheep in a day and there are not many cuts on the body of sheep. In advanced countries like Australia, sheep breeder prefers to shear his costly animals by hand shearing to avoid cuts by machine shearing. Shears with sharpened blades are essential to obtain uniform cuts and effect quick shearing. Now shears with a better springing capacity is the handy one available, which saves a good deal of energy and as a result shepherd does not get fatigued and can shear more sheep in a day.

The main disadvantage of hand shearing is that it is time consuming and fleece can not be cut evenly but it has advantages also. By hand shearing, the fleece can be graded on the basis of individual cuts and can be done at any place.

2. *Machine shearing* : Machine shearing in India has been introduced only two decades ago. At present machine shearing is done in the states by respective, Animal Husbandry departments.

The shearing machine is a metal comb over which a cutter is driven. A system of flexible gut, encased and broken by various joints, makes it possible to reach the various parts of the sheep. These machines are either electric or diesel operated.

The main advantage of machine shearing is that it is less time consuming and also works very close to the body and as such economical. But little negligience during shearing can cause cuts or wounds.

3. *Chemical shearing* : This technique was developed at USDA Station, U.S.A. by feeding 24 mg/kg body weight of cyclophosphamide, which causes a break in the wool. The wool is removed

during a 3 day period starting from 12th day of administration of the drug. But it has been observed that the removal of wool is not uniform. Due to the doubtful effect in the meat causing baldness of the consumer, this method is not used commercially.

Prepration of animals for shearing

The animals must be washed 4-5 days prior to shearing in clear water to remove the soiled materials. This improves the shearing efficiency, reduces skin cuts and reduces stress on animals. This gives a good market appeal to the wool. The animals should not be either over fed or underfed before shearing. They should be shorn on a clean floor. They should be shorn only two times in a year. The climate should be neither very hot nor very cold at shearing. To prevent canary colouration of wool, they should be shorn in February and July in semi arid areas. Wool shorn from abdomen, and legs should be kept separately. Sheep should be protected from intense heat after shearing.

Care of sheep after shearing

After shearing sheep may be kept in shade and provided with quality pasture and sufficient drinking water. The feed and water intake increases by 13 to 23% and 11 to 12%, respectively after shearing. They may be put to routine management practices when they attain about 1" wool on the body surface which is sufficient to protect the animals from solar radiations.

Dipping

Shortly after shearing, an important operation in sheep husbandry is dipping for the control of ectoparasites. The main ectoparasites are tick, body louce and itch mites.

Synthetic insecticides which have greater residual effect on the fleece e.g. D.D.T., B.H.C. or aldrin are generally used for the destruction of ectoparasites. (Table 7. 2) Organophosphorus insecticides such as melathion are used due to their high quality insecticidal properties.

Time of dipping

The time when dipping is given to sheep in respect of health of sheep, life cycle of the ectoparasites and the type of insecticide is very important. While shearing some cuts on the body are expected, therefore dipping should be done 12 to 15 days after shearing so that the wounds or cuts caused during shearing are either healed or scabbed over.

Table 7. 2: Medicines used for dipping

S. No.	Chemical	Concentration in dipping solution	Remarks
1.	Arsenic sulphate	0. 2%	It contains 0. 13% soluble arsenic
2.	D.D.T.	0. 5%	To make 0. 5% solution dissolve 20 kg of 25% wetable DDT in 1000 lit water
3.	Nicotin (Tobacco)	0. 1%	15 kg tobaco leaves moistened in 500 l water can be used.
4.	Phenyl	0. 76%	It has 0. 36% carbolic acid
5.	Lindane	0. 031%	Gamma isomer concentrate solution is used for lambs
		0. 05%	This solution is used for adult sheep.

Sheep dip

Sheep dip is a long, narrow swimming bath, 0. 75-1. 00 m in width, 1. 50 m deep and 0. 60-1. 00 m long. At the enterance there is a collection chamber where sheep are collected for dipping followed by a forcing pen from where sheep are driven for dip. At the exit, there is a draining pen. The length of the dip can be increased as checking of sheep in the bath is not good. The floor of the bath should be made to slope gradually, it should slope backwards. These dips can be made of cement, concrete, or of steel. A fairly big outlet pipe at the bottom is needed to syphon out the liquid and the sediments collected at the bottom. Therefore the dip should be constructed at a higher level. A plug is filled in the pipe. If the outlet is not provided to the dip, then it has to be bucketed out or pumped out to clear it. The dips can be filled either by a pipe from the tap or separate draining pens can be constructed near the dip.

Time in bath:

The time, the sheep should remain in the bath depends on the wool they are carrying, when they are dipped. Half a minute is sufficent for sheep carrying one month's growth of wool. The object is thorough saturation of the fleece to the skin.

Precautions in the dipping

1. The dipping bath should be thoroughly, cleaned out some days before the dipping.
2. The liquid capacity of the bath should be accurately gauged.
3. The directions given for the use of medicines should be strictly followed.
4. The bath should be thoroughly stirred from the bottom throughout its length just before the dipping begins, during the dipping and each time new dip and water are added.
5. All scum should be skimmed off as it forms.
6. Start the dipping early in the morning and continue only so long as to allow all sheep to be dry before evening.
7. Sheep should not be thirsty as they are then apt to drink the wash or lick the floor or draining pens. There should be provision of water in the collection chamber.
8. Dip the rams first, then the rest of the adult sheep and the lambs last.
9. Dipping should not be carried out in cold or wet weather.
10. Care should be taken to see that the sheep are not over crowded in the draining pens. After draining, sheep should be turned out into shed, yards or where there is shelter.
11. All the cartoons or packages of dip material should be disposed before they reach in any body's hands.

Skirting and Primary classing of Wool

Skirting and classing of wool is done on the shearing floor in most of the agriculturally advanced countries. However, there is no such practice in India. As early as 1939-40, the Directorate of Marketing and Inspection suggested simple method of washing the sheep before shearing and skirting coloured and soiled wool from rest of the main fleece. The fleece of the adult sheep and lambs were also suggested to be kept separately, from rest of the main fleece. The fleece of the adult washing and skirting was tried in Jodhpur state on experimental basis and the shepherds obtained 5.7 to 33.2% premium in sale value through these procedures. Though a large number of shepherds were attracted to take advantage of the scheme but due to shortage of supervisory staff and other financial difficulties the scope of the scheme remained limited.

Development of Grading Systems

To improve upon the traditional wool marketing system and to get higher returns to the sheep breeder and at the same time to prepare the wool clip according to the requirement of wool industry, grading systems were explored and introduced. It was first started under export of raw wool. The grading is based on colour, clean wool yield and type. The general characters described ensure that the consignment is properly cleaned and is free from admixture with other fibers and extraneous matter.

Under UNDP assisted project large scale sheep shearing, wool grading and marketing were taken up in the year 1963 in Rajasthan. Wool was purchased by Government, through extension centres and pooled at the wool grading and marketing centres. The wool thus pooled was graded into 90 types as detailed below.

Fineness	Length	Colour	Burr
Super fine A	2″ up long	White	L B upto 3%
58^S up	2″ down short		
Fine A 54^S /56^S	2½″ up long	Light yellow	M B-3 to 6%
Medium B			
46^S /50^S			
Strong C 40^S/44^S			
Coarse D 40^S down			

In addition types of black, greasy, pieces, stainy wool etc. were also kept separate.

Under the above system wool is graded by trained wool graders under the supervision of bin inspectors and grading officers in the types detailed above. The grading and marketing programme was to be taken up in 8 states and actually it was operated in five states viz. Rajasthan, Jammu & Kashmir, U. P., Punjab and Haryana. A nearly similar wool grading system has been evolved at I.S.I. (I.S.S. 2900: 1970) as follows:

Fineness		Length	Colour	Burr
Below 34.4	A	Above 75 mm	White	LB-upto 3%
34.4 37.0	B	Below 75 mm	T.W.	
			Light yellow	MB-3 to 5%
37.1 to 40	C		Heavy yellow	HB-above 5%
40 and above	D		Other colour	

However, no licence under IS 2900:1970 now revised to IS 2900:1979, is being operated in the country. The specifications under above grading are as under:

Style	Code No.	Description
Good	1	Good colour, yield and sorted Burr 0-3% good quality & lengths.
Good average	2	Yellow wool, nicely sorted, Burr 0.3%, equal in quality and length.
Average	3	Average yield, nicely sorted, 3-6% Burr.
Inferior	4	Poor length, burr 6% and more.

The wools are described as per examples given below :

AA – Good colour, 48^{s} quality burr, 0-3% nicely sorted, length above 50 mm.

B – Yellow wool, $40\text{-}48^{s}$ quality, nicely sorted, burr 3-6%, length above 50 mm.

CC – Average, $30\text{-}40^{s}$ quality, nicely sorted, burr 3-6%, length above 50 mm.

DD – Dirty wool, quality below 30^{s}, burr above 6%, length above 50 mm.

Where

A, B, C & D stands for wool quality single letter stands for length below 50 mm and double letter denotes length above 50 mm.

Appraisal of present Grading System :

Various research workers have critically evaluated the grading system introduced under UNDP programme. The details of the results on laboratory evaluation indicate that with respect to colour burr content and length, visual grading is reasonably correct. However, with respect to fiber fineness there was a very large overlapping and only 34 to 57 percent of the wool lots could be graded correctly. Visual grading factor, the product of average fiber diameter and Medullation percentage bears a higher correlation with the visual grades than the average fiber diameter alone.

Investigations carried out at Central Sheep and Wool Research Institute, Avikanagar (Raj.) also revealed a very serious overlapping of graded wool according to fiber diameter. On the basis of the average fiber diameter when the different grades were put into quality grades, it was found that there was serious down grading. Some of the lots which were graded as A or B or C should have actually been graded as 'A'. There were significant differences among visually graded lots in respect of medullation percentage between the wool lots of same grade, and also between the different breed wools of the same grade, further reflecting on ineffectiveness of visual grading systems. A very large number of grades i.e. 90 besides lower lines appears to have added to this confusion.

The Australian wool corporation is now recommending objective clip preparation (OCP) which does not require subjecting the wool to visual classing based on quality, length, soundness, style etc. It has been seen that within a mob (flock of animals run together and of same age group and sex) there is serious overlapping in the quality of lots assessed subjectively when subjected to laboratory tests and the spinning performance of clip perpared objectively and on traditional systems does not seriously differ. It has been further observed that in Merinos, 2/3rd of the total variation among sheep in the average fiber diameter is accounted for by the variation between fibers within a staple.

Wool Packaging

Wool packaging has to ensure adequate protection to the new material from the influence of environment and also to prevent pilferage. The procedure should be such that it preserves the quality of wool and is economical. The packs should occupy minimum space and should be of convenient weight for manual handling, which is very common in this country. The common practice in vogue is to pack wools of separate qualities in small jute bags and into bigger jute bags depending upon the bulk and quality of wools to be packed. Packing in the bags is done under pressure to accommodate more wool in gunny bags.

The suppliers of raw materials for the packaging industry must be educated and made to improve the quality of the indigenously manufactured materials. The need for constant flow of technological information from developed countries must be fully appreciated.

Storage

The storage of raw wool is an important operation in the process of wool grading and marketing. The storage can be of wool freshly shorn in the field, the graded wool in the godown of the grading centre and the wool stocked by the manufacturing units. The very purpose of storage will be defeated if during storage, the quality of the wool undergoes any deterioration due to the effect of varied temperature, humidities and light intensities which cause disintegration of the wool fibers and loss in strength. Wool stored in a warm place in moist conditions without proper air circulation is damaged and the wool fibers are considerably weakened. The storage of wool is being practised in the country in ambient conditions without paying requisite attention to the influencing factors which cause huge damage to wool quality reducing its value in terms of end use suitablity for manufacture of varied products.

Possible damages during storage

Damage to wool can occur due to microbiological attack; attack by insects, exposure to moisture etc.

By contamination with micro-organisms, wool is prone to deterioration of multiplying rapidly under favourable conditions like temperature and humidity which would cause the wool fibers to be stained with mildew or to lose strength. It has been reported that disintegration of wool fibers can take place by action of enzymes, which affect the scales of the fiber. If wool bales come in contact with water, fungus develops.

It has been reported that mildew development caused by fungi may be increased by the presence of alkaline content in various soaps, wool grease and majority of conditioning fluids. In hygroscopic soaps, moth growth can occur over a wide range of humidity. Besides, this, it can occur over a wide range of pH since both acid and alkaline conditions favour the same. The greenish yellow stains commonly found on wool are due to the action of the soaps. Mineral oils have a protective effect. Lower fatty acids retard moth growth, where as more complex fatty acids favour growth.

If wool is stored in a warm place, in a moist alkaline conditions without any air circulation, bacteria start to grow and produce enzymes which can rapidly attack the fiber. Some research workers have found that wool which have neither been injured mechanically nor modified chemically, is completely resistant to attack by enzymes. When the wool scale are damaged by mechanical means the wool becomes susceptible to enzymes. Even though, the very small portion of the wool is digested, the fibers are considerably weakened and their structure partly destroyed. Due to bacterial action large amount of heat is generated and temperatures as high as 60°C have been measured inside bacterially attacked bales.

Wool like fur and feathers is vulnerable to attack by moth larvae in certain conditions which are favourable for the growth of moths. It has been estimated that the progeny from a single female moth might under favourable conditions for reproduction, eat about 92 pound of wool in a year. In wool industry even in more advanced countries no control on the considerable damage caused in the warehouses has been practicable. An ideal moth proofing agent is that which provides permanent protection to the wool fibers from the moth attack. The moth proofing of wool and other animal fibers is effectively achieved by the impregnation of the fibers with chemicals that make unuseable as food by moth larvae. The widely published D.D.T., type of insecticides, however, have proved very effective against moth. Concentration of 0.1 to 0.15% are effective on wool and have been used in the mills and in dry cleaning in spray form.

The new insecticide "Dieldrin" has shown itself to be an effective moth proofing agent. As little quantity as one part in one thousand by weight on the wool has proved a sufficient protection. The moth-proofing agent should not be toxic to human being. For this reason arsenic derivatives are not used.

Apart from the consideration of the ideal conditions for storing wool, the matter of specifying some quick test for assessing chemical damage due to bacterial or moth attack is also essential to serve as quality control during wool storage.

Transportation

The wool is usually transported in trucks and rail wagons. No special trucks or wagons for transportation of wool packs (bales) have been made in the country. The present size of the wool packs is such that two compressed bales and three saggy bales could be placed side by side in a row in a truck, while the compressed bales can be placed only in "twos" leaving spaces in between the bales tend to slip. The trucks available are not suited for compressed bales.

Preparation of clip for sale

Sheep with its multifacet utility producing meat, wool, skins and manure, plays an important role in the national economy. This is especially important in arid and semi arid regions with marginal and submarginal land unfit for agricultural production even using modern sophisticated methodology for water and soil utilisation. The arid areas in India cover more than 3 lakh square kilometres. Sheep is the most important livestock for using the sparse vegetation available in arid and semiarid regions of the country.

India has about 48 million sheep producing nearly 47 million kilograms of wool. Out of this production about 15 percent is of finer quality and could be suitably used in combing process of manufacture and rest of the wool is either of carpet quality or suitable for woollen manufacture only.

The state of Rajasthan occupies a unique position in respect of wool production in the country. It has sheep population of 13.68 million according to the livestock census of 1982 and the estimated wool production is about 13.6 million kg. (Table 7.3). This quantity represents about 40 percent of the total wool produced in the country. Because of no organised channels in which to orderly market their products, the state is not receiving fully economic advantage, it would derive if established markets were set-up.

In order to effectively prepare the Indian wool clip for market, which could meet the requirements of trade and at the same time to ensure fair price to the sheep breeders, large scale sheep shearing, wool grading and marketing programme was started in the year 1964-65 in Rajasthan, Punjab, Haryana, Uttar Pradesh, Jammu & Kashmir, Gujarat, Madhya Pradesh and Maharashtra.

The UNDP programme was started in Rajasthan in 1964 and one Project Manager along with a team of 4 Australian experts remained in the state for a period of 5 years. Later this team extended its operations in other states mentioned above in a phased programme. Prior to the introduction of wool grading programme under UNDP, wool was classed and graded under

Ag-mark scheme sponsored by the Directorate of Marketing and Inspection, Ministry of Food and Agriculture, Govt of India and it was regulated under the Agriculture Produce (Grading and Marketing) Act 1937. The actual wool grading and marketing Rules 1961 were confirmed and it stipulated the grading of wool on the following lines. This grading was taken up predominently for export of wool from India to the continent and the USSR.

1. Indian Clipped wool

Wool is graded irrespective of length and designated as Indian clipped wool having a yield range of 70-90 percent. It includes shorn fleece wools from live animals and the colour includes white, tinged white, pale yellow, yellow and coloured wool (Kaushish, 1993).

2. Indian Pulled wool.

The other classifications given are for pulled wool which are obtained from the skins of dead animals without having treated with lime process. This wool indicated a yield range of 70-90 per cent and colour designations are the same as for freshly clipped wool.

3. Indian Tennery Wool (Limed)

In this category the classification given are for wools obtained from the skins of dead sheep by lime treatment. The yield range is 70-90 percent. The colour designations are the same as for other wools.

Other designations which are incorporated under Ag-mark scheme included Indian-Mixed wool, clipped graded, clpped cagrrded, clipped pulled, hill wool clipped, and Indian ginned wool.

An important feature of the wools graded under Ag-mark is that these wools have a maximum vegetable content of 30 percent for all wools except for coloured wool, where in the maximum vegetable content permissible is 4 percent. Export of wool having vegetable contents over these limits, is prohibited :

It may be mentioned here that the emphasis which has been solely given to Ag-mark wools is the yield, without giving any special reference to its quality and staple length.

The wool produced in the state of Rajasthan ranges qualitatively from medium to coarse types. The fleeces obtained from sheep, though vary greatly in length and fiber fineness, are most suited for carpet manufacturing. About 15 percent of the total production of the state may be used for the manufacture of apparels and also better type of fabrics may be produced by blending this type of wool with imported wool.

There is a vast scope for the development of wool industries in small scale as well as large/medium scale sectors. But the enterpreneurs may face difficulties in regard to supply of standardised wool, which may only be overcome through effective preparation of wool clip for market to enable the manufacturers to get required type of wool and at the same time the producer getting fair prices for the produce.

There are eight breeds of sheep in Rajasthan-namely, Chokla, Magra, Marwari, Jaisalmeri, Pugal, Malpura, Nali and Sonadi. Some of the Marwari and Magra sheep are shorn thrice a year, otherwise all the sheep are shorn twice a year. The sheep which are shorn twice a year normally give long staple wool to the extent of 70 percent and the remaining 30 percent may be classified as short staple wool, pieces, black, stains and dogs etc.

Under the existing conditions, the wool is sold to the merchants in a mixed form and no grading is done at the shearing point. The shepherd is more or less under the financial control of the local merchants because he depends upon them for his very existence. The shepherds are usualy poor and illiterate, knowing little of the outside world and because of no organised market sell their produce, they had to look to the local merchants to take over their commodities.

In the stores of the wool merchants in market, they employ the labourers for simply separating the white wool from yellow and for removing whatever black wool they see. Heavy burr is also . removed from the wool by clipping with scissors. The wool is then pressed and is sold to the local manufacturers.

The authorities have no control over the markets whereby the shepherds, could offer his produce through organised channels and get better price for his wool. The unfortunate shepherd has to take whatever price is offered to him by the village merchant. One can easily imagine that price at which it is first bought from the breeder is well within the profit range for the merchant. Under this order the nation is not receiving full economic returns from its wool industry as profit is going in the hands of a few.

In Rajasthan wool has been included as agriculture produce and regulated under the Krishi Upaj Mandi commodities but it has not been operative in big *Mandi* like Bikaner so far.

Under the large scale sheep shearing, wool grading and marketing project taken up by Rajasthan and eight other States, the wool is purchased by the department of Sheep & Wool through its extension centres. To some extent, the sheep are shorn by an organised method of mobile machine shearing units of the department.

This wool is then packed in big jute bags and despatched to wool Grading Centres, where it is graded by the trained wool graders in standard wool types.

These standards are based on the Australian system which conform with world wool requirements. When the grading of wool is completed, it is pressed up in standard wool bales keeping each quality separate. It is then catalogued in order of merit showing the number of bales, the qualities, length of staple, the colour and burr contents of the wool in each lot.

The separted show bales which are representative bales of the entrire lot, are shown on a show floor in order in which it is printed in the catalogue so as to make convenient and easy for the buyers to inspect and value each lot independently.

After the inspection of the wool is completed, the wool is offered in separate lots by public auction and sold to the highest bidder.

It is very essential that grading programme which was taken up over the last decade is properly streamlined and grading is done looking to the demand of the manufacturers. If sufficent attention is not paid to keep the various wool grades uniform, the buyers hesitate to pay correct price for it.

Under the large scale sheep shearing, wool grading and marketing programme started in many states of India, wool are graded mainly in 5 basic qualities i.e. superfine, fine, medium strong and coarse. They are further classed according to length, colour and vegetable contents. Greasy wools are graded separately and lower line (wools of belly, legs, stained wool) are kept separately and graded in various types. Thus in all about 90 types are made out of Indian wools.

Wool lines in vogue at present are as under:

Washed wools

	Super fine AALB	Fine BBLB	Medium CCLB
AASLB	Super fine AAMB	Fine BBMB	Medium CCMB
AASMB	Super fine AARB	Fine BBMB	Medium CCRB
	Strong DDLB	Coarse	
	Strong DDMB		
	Strong DDHB		

Colour

White, light, yellow and heavy yellow burr contents.

LB 0 to 3 percent
MB 3 to 6 percent
HB over 6 percent

Colour (BRANDS)

Black
Pcs.,
Stns.,
Cotts.

Greasy wools

AAGSY fine and medium
BBGSY Strong and coarse
Burr contents
LB
MB
HB

All the above types are also made in short lines and indicated by one alphabet i.e.

As- Superior fine short
A Fine short
B Medium short
C Strong short
D Coarse short.

Purchasing and selling of wool is essentially based on its yield contents. It means that if 100 kg of wool is purchased and when scoured it would give a clean wool content which may range according to the type of wool scoured. During washing the impurities like grease, dust, sand, suint etc. are washed away and clean wool remains. Normally under our conditions the sheep are washed 2-3 days before shearing. This wool is called washed wool. The other category of wool is from those sheep which are not washed. This wool is termed as greasy or unwashed wool. In the washed wools, normally we get a yield of 70-80 percent and this is also influenced by length, quality, colour, amount of grease, vegetable matter, dust or sand, density and to some extent the type of tip.

Finer the wool, the lesser the yield and coarser the wool, higher the yield. Our wools are heterogenous in nature i.e. a fleece of wool obtained from sheep has 3 to 4 qualities in it and therefore, the yield assessment has to be based primarily on the main quality of wool found in that particular fleece.

Other factors which influence the yield contents of wool are suint, vegetable matter and dust which is directly related to the type of washing practiced and the presence of inferior grades of wool in a particular quality of wool.

All these types will be packed in *boras* (bigger gunny bags), and despatched to the Wool Marketing unit of the federation where, required grading will be done and wool despatched off.

4. Utilisation of wool

There is a wide variation in the quality of wool produced by sheep reared in different agro-climatic conditions of the country. The quality of wool varies from 40's to 60's spinning quality in the wools grown in the Himalayan and dry Northern regions, while the wools produced by the sheep of the Southern regions are coarse and coloured. Annually about 9 to 10 million kg is exported as carpet wool earning valuable foreign exchange, while the country imports finer types of wools to the extent of Rs. 120 million. Due to restrictions on import, larger quantities of indigenous wools are being utilised in the various sectors of the textile industry. The estimated utilisation being 7 million kg in the organised mill sector, 7 million kg in the blanket manufacturing industries. Considerable difficulties are, however, experienced in the utilisation of the these wools for the manufacture of quality products.

In order to make larger quantities of better quality wools available for the indigenous industries and for export, a wool grading programme is being undertaken in Rajasthan and other wool producing states. Under this programme, the wools are graded on the basis of quality, the quality being assessed by subjective methods such as colour, etc. The ultimate utility of these grades for the manufacture of various end products is not yet known. The sheep improvement programmes undertaken in the country envisage crossbreeding of local coarse wooled sheep with exotic fine wool type to produce fine type of wools. The manufacturing trials on these wools will elucidate the correct utilisation of wools already produced and the new type of cross bred wools and also the extent to which blending of different types is desirable to ensure the required properties in various end products.

This would result in less dependence on finer type now being imported, in increased exports simultaneously bringing more financial returns to the products of quality wools.

Marketing of Wool

The North temperate zone produces 3.09 million kg, North-Western dry zone 23.0 million kg Southern and coastal zone, 7.22 million kg and East zone in 1.4 million kg wool. Southern sheep are mutton type and their wool is coarse.

Table 73 : Wool marketing in principal wool markets from 1977-78 to 1985-86 (Wool in quintal).

Year	Bikaner	Beawar	Kekari	Jodhpur	Wool marketing/ year
1977-78	77, 234	36,017	9,445	12,000	135,595
1978-79	61,549	34,554	10,974	13,000	120,077
1979-80	76,863	31,216	10,195	10,195	112,374
1980-81	84,448	37,294	14,483	12,000	148,225
1981-82	81,432	31,977	12,700	15,000	141,159
1982-83	92,055	36,279	13,759	18,000	154,884
1983-84	76,348	32,282	14,048	14,000	136,778
1984-85	75,296	34,518	16,279	—	126,193
1985-86	75,107	35,480	—	—	—
Av./Yr	**78,947**	**34,160**	**12,771**	**13,114**	**136,681**

The shephard sells his wool to local village trader or takes the wool to marketing centre or sells to the federation. Most of the wool is sold through wool commission agents @ 1.5% commission. The Krishi Upaj Mandi tax, on raw wool is 4% and 2% on woollen yarn. The wool price has increased from Rs. 1450/- per quintal in

78-79 to Rs. 2500/- per quintal in 1985-86 in Beawar. The amount of wool sold in different wool markets of Rajasthan is presented in Table 7.3. There is demand of Indian carpets overseas. Approximately 5.0 million square metre carpet is being produced each year. The export of woollen carpet in 1979-80 was Rs. 1.00 billion and in 1984-85 Rs. 1.57 billion. The demand of carpet wool is acute and the shortage of wool is due to higher demand of the carpet wool in the country. India imported 2.5 million kg New Zealand wool in 1981- 82. The carpet wool import duty was reduced from 45% to 20% in 1985-86.

Canary colouration of wool

Incidences of canary colouration increase with the increase in atmospheric temperature and humidity (Nagarcenkar, 1961; Narayan and Sharma, 1970). Different regions of sheep for observing canary coloration are given in Fig 7.2. This problem is more in Gujarat, Haryana, Punjab, Uttar Pradesh and Rajasthan. Sheep of this region produce 20 million kg wool according to 1966 animal census, which is 57 percent of the total wool produced in India. Canary coloration varies from 80-100 percent in autumn clip (Narayan, 1969). Canary colouration reduces the income from wool to sheep farmers. Such wools do not take up colours properly. Yellowing of wool starts with the onset of rainy season. It is more in breeds which produce more wool.

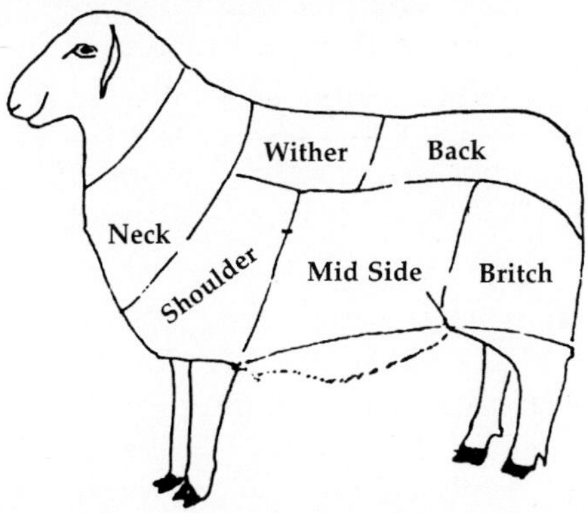

Fig. 7.2 : Different sites of Sheep for observing yellowing of wool

Yellowing is more in growing lambs as compared to adult sheep (Narayan and Sharma, 1970). Canary colouration starts decreasing after three years of age (Narayan, 1969). It is more in rams than in ewes (Narayan and Sharma, 1970). The quantity of canary colouration varies from region to region of the body. It starts from legs and spreads to other parts of the body.

There are more incidences of canary coloration of wool in Romney marsh than in Corriedale sheep in New Zealand (Hender son, 1965). It is more in Marwari breed than in Malpura breed (Narayan, 1969). It is less in coarse and less crimpy wool (Kapoor *et al.*, 1972). Canary colouration is of two types (Serra and deMatos, 1951) :-

1. Simple canary colouration : It goes with the washing of wool. It does not harm wool.
2. Special canary colouration : It does not go with simple washing. Such wool does not take different colours permanently.

Causes

1) Light : Sun light does not cause yellowing of wool (Nagarcenkar, 1964). Less yellowing is observed in Chokla sheep which are grazed between 8 PM and 5 AM as compared to those grazed during day (Anon, 1972a). White wool becomes yellow on keeping in sun light. It also becomes yellow if kept in ultra violet rays.

2) Climate : It is well known fact that yellowing of wool starts with the onset of rainy season when the atmospheric temperature and humidity are high. It starts from the month of June when the atmospheric temperature is 43.1° celsius and relative humidity 57 percent (Anon, 1972a). Wool becomes yellow if kept at pH 3.0 at 38° celsius for 1-2 weeks (Serra and De Matos, 1951). It becomes yellow if kept in alkaline solution at 150° celsius (Norton and Nicholls, 1960).

3) Hereditary : Yellowing is found more in some breeds than in others. It is found in Chokla, Hissar dale, Sonadi, Malpura, Jaisalmeri, Marwari, Magra in decreasing order (Narayan and Sharma, 1970). It has been found in decreasing order in Magra, Chokla, Rambouillet × Chokla (50%) (Anon, 1972b). No difference has been observed between Bellary and Bikaneri in this respect (Nagarcenkar, 1963, 1964; Nagarcenkar and Bohra, 1983).

4) Wool wax : It is more in wool in colder months and is more in fine wool breeds (Sutton, 1935). Table 7.4. If the quantity of wool wax is more in a wool, it does not become yellow (Chippalkatti *et al.,* 1965). The amount of wool wax differs with the breed (Anon, 1972a).

Table 7.4 : Amount of wool wax found in different breeds of sheep.

Breed	Percent of wool	Breed	Percent of wool
Sonadi	4.37	R × M (50%)	10.73
Malpura (M)	4.74	R × M (75%)	11.43
Jaisalmeri (J)	5.60	R × C (50%)	13.38
Pugal	6.05	R × C	14.05
Chokla (C)	8.71	R × C (75%)	18.10
Nali	9.63	Russian Merino	19.10
Magra	7.25	Rambouillet (R)	22.32

Wool wax was found more in spring clip of Nali breed (Chippalkatti *et al.,* 1965; Narayan and Sharma, 1970).

5) Micro organisms : Colour producing micro organisms develop in hot humid climate (Fraser and Mulcock, 1956).

6) Green fodder : Yellowing of wool develops in season when there is more green fodder available. Green colour of green fodder is not absorbed through intestine so there is less probability of yellowing due to it.

Remedies

Different measures have been adopted at times to reduce yellowing of wool. Some of them are :-

1) Changing the time of shearing : Normally the sheep are shorn in March and September. If they are shorn at the time when the yellowing starts, white wool can be prevented from yellowing (Narayan, 1969). So they should be shorn in the beginning of rainy season.

2) Washing of wool : Weekly washing of Chokla sheep reduces canary colouration to negligible amount (Narayan and Sharma, 1970).

3) Cross breeding : Yellowing of wool is found to be reduced in cross bred sheep.

4) Energy : Sheep should be given dry fodder during summer season to reduce sweating.

8

Pelt Production

Sheep in India have traditionally been kept for wool, mutton, skins, manure and to some extent for milk. There has been little emphasis on lamb pelt production, although it is known that the khatik community in Rajasthan does produce lamb pelts from Malpura breed on a small scale. An attempt is now being made to add this new dimension to sheep products industry in India.

Fur apparel is considered as a status symbol in many countries. Besides being an item of luxury, wearing of fur garments and other outfits made of furskins/furlings make the living during winter comfortable. Naturally, fairly large concentration of the industry in India is in the Kashmir valley and to certain extent, in the Rajasthan and other Northern regions. In addition to meeting local needs, properly dressed finished furs have great export potential and the value added to raw furs through such processing is manyfold. The industry in India, however, has so long remained traditional and primitive possibly due to insufficient availability of raw skins and lack of modernisation of the processing.

Importance of lamb pelt production

Lamb pelt production is an important facet of sheep industry in a number of countries, especially, U.S.S.R., Afghanistan and South Africa. Karakul is an important breed for producing lamb pelts. It is a fat tailed, coarse carpet wool breed, very well adapted to extreme climatic conditions and poor vegetation resources.

Karakul sheep population in Asia and Africa is around 31.5 million which is around 7.5 percent of the total sheep population in this region. The Soviet Union, with an annual production of ten million pelts, is still the leading producer, followed by South Africa, with about 5.5. million pelts per annum. The pelts are produced either through the slaughter of lambs within 24-48 hours of birth or through killing of unborn lamb removed through slaughter or through abortion of the mother around 130-140 days of pregnancy. The latter type of pelts known as fat tailed persian, are twice more valuable than the best types produced from slaughter after birth as they have better ornament, are lighter and more lustrous. The lamb pelts sell between $ 15 to 20 per pelt in international market depending upon the size and quality.

The quality of pelt is generally determined by the type of curls, their size and tightness and lustre. The size and weight of pelt are also important factors in deciding the quality. The curls are usually termed as long when the length of hair is more than 20 mm. Present trend is to get light pelts with short hairs. The size of curl is completed during foetal life itself. The pelts are classified into various types such as Jacket, Ribbed, Flat and Caucassian.

Fertility of ewes

The Karakul ewes in the farms of central Asia and Kazakhistan as a rule give one lamb in one parturition. Percentage of twins depends upon nutritional and better management conditions and varies between 2 to 10%. In some farms in Ukraine and Astrakhan, the twining percentage goes upto 40%. However, Karakul ewe has a potential for producing more lambs through superovulation caused by the use of pregnant mare serum gonadotropin (PMSG). PMSG is used abundantly for augmenting Karakul pelt production in the Soviet Union and about 4 million sheep are treated with PMSG every year.

Because the Karakul lambs are slaughtered in the early stages of their life, superovulation by the use of PMSG becomes interesting proposal. In U.S.S.R., 150 to 170 lambs per 100 ewes are produced in some farms by this method.

The early slaughter of lamb eliminates mortality in weakling lambs. The mothers whose lambs have been slaughtered may be rebred and thus can produce lambs more often. Karakul sheep produces milk also, particularly because the mother is without her lamb immediately after lambing, they are milked in many farms, where flocks are in well fed condition. There they give about 25-30 kg milk in one lactation with fat percentage from 5 to 7%. Their milk can be used as foster mothers for twin lambs or which have lost their mothers or their mothers have very poor milk production. The further elimination of a large number of lambs at birth will reduce the pressure on feed/fodder resources as for meat/wool they must be reared for a longer period.

The wool, mutton, skins, manure and milk will also be available from the flocks being reared for pelt production, thus pelt production will replace only the disposal of extra male lambs which would normally have been marketed for meat around 9 months to one year of age.

The Karakul sheep have shown adaptation to hot arid conditions. The greasy fleece weight of six monthly fleece has been observed to be 1.23 and 0.90 kg in case of rams and ewes, respectively. Among the native breeds, Malpura produced pelts of best quality followed in descending order by Sonadi, Nali and Chokla (Table 8.1). The ornamentation in native breeds is restricted to the part of skin whereas in case of Karakul, it is spread over whole of the skin. In case of Chokla and Nali and majority of crosses, for evolving fine wool breeds, largest percentage are of ribbed type. About 80% Karakul in the U.S.S.R. are black. The remaining 20% could be grey, brick sur, or rose colours. These coloured Karakuls are less adaptive to harsh climate and nutrition (Stoyanovskaya, 1966). The pelts produced by crossing native sheep may be black, brown, black on white or white on brown in colour.

Quality evaluation of pelts

The generalisations and broad principles deducted by Ivanov and Yudin (1932) and Dhillon, (1980) after investigations, a great mass of material provided the basis of development of systematic evaluation methods. The composition and properties of a Karakul pelt determine, its commercial value. As the most important component of a Karakul pelt is its specific curls, the quality of the pelt, by and large, depends upon the quality of its curls. The absence cf curls or curls of low quality may diminish the value of the pelt. Besides, there are a number of other properties which influence the quality of the pelt. The chief of these attributes are enlisted below :-

1. Curls kind, size and patterns.

Table 8 : Distribution of Karakul pelt types in Malpura, Sonadi and their crosses

Breed & breed crosses	Number	Jacket % age	Ribbed % age	Flat % age	Caucassian % age	No class % age
Malpura	39	10.00 (4)	52.50 (20)	27.50 (11)	5.00 (2)	5.00 (2)
Karakul × Malpura (halfbreds)	11	36.36 (4)	45.45 (5)	18.18(2)	Nil	Nil
Sonadi	50	22.00 (1)	22.00(11)	38.00 (15)	8.00 (4)	27.00 (18)
Karakul × Sonadi (halfbreds)	24	41.67(10)	33.33 (8)	20.83 (5)	4.17(1)	Nil

(Figers within parentheses indicate the number of observations)

2. The colouration of the hair coverings.
3. The quality of hair, silkiness, gloss, density, tightness, elasticity.
4. The size of pelt.
5. The weight of pelt.
6. The thickness of the skin.

Classification of curls

Karakul curls are classified into different types according to their size and shape as detailed below :

i) Pipes and Beans : These are the most valuable curls characterising the First grade pelts of Jacket and Ribbed types.

ii) Walnut curls : The pelts belonging to the first and second grades of caucasion group show the predominance of walnut curls.

iii) Feather curls : The second grade pelts of flat type are characterised by the presence of these curls. Feather curls are less valuable.

iv) Corkscrews, Rings etc : These are faulty and defective curls and hence undesirable.

Advantages of pelt production

Good quality carpets can be made from the sheep produced by crossing native sheep with Karakul rams. As the lambs are used for pelt production immediately after birth, so lamb mortality is reduced to zero. The sheep whose lambs have been used for pelt production, produce two lambs in a year or three lambs in two years, can be used for milk production. The expenditure on fodder, concentrate, health, shelter for such lambs is eliminated and meat of these can be used as protein supplement in feeds.

Making pelt

To make profits from pelt production, the pelt must be removed within 24 hours of birth. This should be allowed to cool on wooden board. Then meat, if any, sticking to pelt, should be removed and finely ground salt applied. It is then dried in shade. In the end excess salt is removed with a soft brush. Such pelt can be preserved upto six months.

9

Mutton Production

Most of the breeds of sheep of tropical regions, respond poorly to better nutrition, but Iranian sheep respond quite well to improved nutrition (Demiruren *et al.*, 1971). Sheep and goat have made great contribution to meat production in India, the reason being religious taboo against eating cow and pig meat. In India sheep are mainly reared for wool and mutton production. The sheep rearing is in the hands of landless and poor people. Most of the meat sold in the market is from old sheep or male lambs. The average, market weight of native male lambs varies from 13-16 kg (Kaushish, 1980). The mutton from old sheep contains large proportion of bones. Average mutton production per sheep is much less in India as compared to some western breeds. This is because our sheep farmers are illiterate, orthodox and economically backward and day by day decrease in pasture area for grazing sheep.

Mutton Production

The consumption and production of meat in India are given in Table 9.1 and 9.2. According to latest census, the estimated production of meat is 582,000 tonnes as against an estimated requirement of 10 million tonnes calculated on the basis of 50.7 g per head per day.

According to 1983 census there are 48 million sheep in India. Despite the fact that India stands sixth with respect to sheep population in the world, there was a steady decline in its population figures between 1966 and 1977. This might be due to high percentage of slaughter of sheep for mutton production (26.2 per-cent) as sheep alone contributes to 13.64 per cent of the total meat produced in the country and there is a steady increase in mutton production during the last few years.

Table 9.1 : Meat consumption per head in source countries (F.A.0. 1969)

Countries	g/day
Argentina (1963-65)	256
Australia	290
Burma (1964-66)	12
Canada (1967)	5
Ceylon (1968)	250
China (1964-66)	47
Denmark	166
Germany (F.R.)	200
Hong Kong	111
India	4
Indonesia (1963-65)	11
Japan	37
New Zealand	302
Pakistan	11
Philippines	43
Sweden	145
U.K.	205
U.S.A.	299
U.S.S.R. (1964-66)	106
Yugoslavia (1967)	85

FAO Production year book Vol. 23, 1969

Table 9.2: Production of meat in India (in'000 tonnes)

	1964	1965	1966	1967	1968	1972-76
Mutton	360	363	366	367	368	384
Beef	160	162	162	163	164	186
Pork	47	49	50	50	50	98

FAO Production year Book Vol. 23, 1969.

Contribution of sheep

It is a well known fact that sheep are being reared in the country for two main purposes namely mutton and wool. As per the estimated targets by 2000 AD, we propose to raise the wool production from 36.81 million kg to 56 million kg and meat production from 1.3 million tonnes to 2.1 million tonnes, almost a 51.9 per cent and 61.5 percent increase from the present level in wool and mutton production, respectively. Besides this the contribution of sheep to the livestock economy can be easily evaluated from figures indicated in Table 9.3.

Table 9.3 : Contribution of sheep to livestock economy in relation to world production.

Items (in million tonnes)	India	Per cent of world total
Meat	0.125	2.10
Wool	0.036	1.30
Skin	0.034	3.00

The number of well defined sheep breeds available in the country is around 40 and out of these Nellore and Mandya deserve a mention with respect to meat quality. These breeds attained weight of 10.94 ± 0.11 kg and 9.84 ± 0.12 kg by three months time.

Mutton Crossbred Lambs

An average adult sheep weighs 26.0 kg and the dressing percentage is 40. About 28 percent sheep are slaughtered every year and they produce about 4.45 million kg mutton. Average mature live weight of a number of Indian breeds are given in Table 9.5. Few Indian breeds attain mature live weight of more than 40 kg. Average mature weight of Sudan Desert sheep varies from 25 to

50 kg (Osman *et al.*, 1970). Mature Awassi and Arabi weigh 42 kg and 38 kg, respectively (Asker and El Khalisi, 1966). The sheep of many tropical breeds dress out poorly. Hejazi sheep dress out at 37 to 40 percent and Osman and El Shafie (1967) stated that the dressing out percentage of Sudan Desert sheep varied between 36.4 and 52.1. However, Demiruren *et al.* (1971) stated that the dressing out percentages of range fed ewes and rams in Iran were 56.5 and 55.3, respectively.

Table 9.4 : Proximate composition and energy value of fresh muscle mutton.

Cut of meat	Protein %	Fat%	Water %	Ash%	Cal/100 g
Breast	12.8	37	48	-	380
Leg	18.0	18	64	0.9	240
Loin	18.6	16	65	_	220
Rib chop	14.9	32	52	0.8	360
Shoulder	15.6	25	58	0.8	300

Ref. The Science of Meat and Meat Products. p 187 American Meat Insttitute Foundation, W.H. Freeman & Co., San Francisco. London.

Table 9.5 : Average mature live weight of some breeds of Indian sub continent.

S.No.	Breed	Average mature live weight (in kg)	Reference
1.	Baluchi	33-34	Lall, 1947
2.	Bellary	32-54	Kaura, 1941
3.	Bhadarwah	32	Kaura, 1942
4.	Bibrik	27-36	Kaura, 1942
5.	Bonpala	32-59	Katiyar *et al.*, 1981
6.	Chokla	23-35	Mathur, 1977
7.	Damani	23	Lall 1947
8.	Dumeri	27-32	Kaura, 1942
9.	Deccani	20-32	Kaura, 1941
10.	Hassan	23	Kaura, 1942
11.	Jaisalmeri	33-36	Acharya, 1982
12.	Jalauni	24-32	Lall, 1947
13.	Kaghani	23	Lall, 1947
14.	Kathiawari	27-45	Kaura, 1942
15.	Lohi	36-69	Kaura, 1942

16.	Magra	24-27	Acharya, 1982
17.	Malpura	24-41	Arora *et al.*, 1975a
18.	Mandya	29	Kaira, 1942
19.	Marwari	30	Acharya, 1982
20.	Nali	24-35	Arora *et al.*, 1975 c
21.	Nellore	38-41	Kaura, 1941
22.	Pugal	27-32	Acharya, 1982
23.	Sonadi	21-39	Arora *et al.*, 1977
24.	Thal	18-27	Lall, 1947
25.	Waziri	23	Lall, 1947

Means for increasing meat production

The dressing percentage of some of our breeds like Mandia and Nellore are comparable with some of the best mutton breeds. It varies from 36.8 to 51.8 percent. Dressing percentage is better in Malpura and Sonadi as compared to some other breeds found in India. The only solution to increase mutton production is by crossing our native breeds with exotic mutton breeds. Care was never given for mutton production in India. It was only in 1971, an All India Coordinated Research Project on sheep for mutton production was started. Meat production can be increased by following means:

1. Selective Breeding of sheep

The sheep found in Northern India are heavier than those found in South India. Magra breed is one of the heavier breeds. Madras Red is heavier than the other breeds of the area. Nali and Lohi have poor conversion of concentrate into mutton. By crossing heavier rams of a breed with heavier ewes of the same breed we can increase mutton production, but the gains will be slower.

2. By crossing Native breeds

This type of work was done in Andhra Pradesh and Tamil Nadu by crossing Bikaneri rams with local sheep. This did not give encouraging results because Bikaneri rams could not adapt themselves to hot humid climate. It was observed in Bihar that crossing of native sheep with Bikaneri rams produced heavier cross bred lambs. The ability to convert concentrate mixture into mutton varies from 14.0 to 19.8 percent. It is highest in Nali sheep. The dressing percentage varies from 44.8 to 48.3 percent.

3. By crossing Native sheep with Exotic rams

Crossing of Mandya sheep with Somali and South Down rams and crossing of Nellore and Bellari sheep with Corriedale rams gave encouraging results.

It has been stated in Israel recently that Awassi × Dorset Horn crossbreds grow at a 20-30 percent faster rate than pure bred Awassi during the first 90 days of life. Ramamurti (1964) showed that both live weight and wool yield of Rommey Marsh × Nilgiri crossbreds were significantly higher than those of the indigenous Nilgiri over a four year period. Lambs produced by crossing East African Blackhead and Dorset Horn were heavier at birth and had higher growth rate than the natives (Trail and Sacker, 1966).

Marketing terminology

The marketing terminology for sheep of varying age groups is different. A sheep from birth to 12 months is termed as 'lamb', a male castrated animal between three and six months age as 'Hog' and castrated afterwards as 'Tup'. A female which has not yet given birth to a lamb is known as 'Gimmer' and one which has given birth to a lamb as 'Ewe'.

Judging of sheep for mutton

At times it becomes difficult to judge a sheep based on the general appearance, the most deceiving factor being the covering of wool. To overcome it, we should view the sheep from a distance of at least twenty to thirty feet to assess the general breed characteristics, size, type of muscling, symmetry and style including the fleece growth, before handling the animal. Usually a lamb should be viewed while judging it for mutton purposes from four directions-front, top, side and rear. To be more precise ideal viewing from above directions would mean close observation about certain specific points.

Front view

i) The lamb should have a bold muscular shoulders including arms and forearms.
ii) The animal should be free from excessive skin in throat area.
iii) Chest should be wide and clean.
iv) Head should be reasonably fine (not too large) true to the breed type.
v) Legs should be set out wide on the corners of the body.

Side view

i) Legs should be of average length matching with the growth, age and age at maturity.

ii) Feet and legs should be squarely placed.

iii) The animal should be neat and clean both at the rear and front.

Rear view

i) The lamb should be long from the hocks to the pins.

ii) Long in the rump area and the back should be reasonably full and levelled out over the dock.

iii) Rear legs should appear muscular, firm and full and set well apart to give a "Meat type" appearance.

Top view

i) The animal should give "Coke bottle view" with a reasonably clean neck blending well and should have more spring of the rear ribs when compared to the front ones, with the . extra thickness spread towards the loin.

ii) The top line should be long and levelled.

After assessing the animal based on the above observations the animal should be handled to determine the degree of finish. Usually lamb buyers handle parts like backbone, ribs, edge of rack, back, hips, joints and docked tail hardness and firmness over the back and loin is a good indicator of condition of the lamb. While handling an animal it is always advisable to keep the fingers together and handle it firmly. Depth of the animal is judged by placing one hand on the top of the shoulder and another below the chest. You can run your hand on the shoulder to estimate smoothness, meatiness including muscling and degree and finish. The fullness of the shoulder is estimated just behind it. Smoothness, firmness and amount of finish can be estimated by moving the hand on the top of the shoulder, edge of the rack, back and loin including the dock. The width of the fore and rear ribs are estimated by using both the hands simultaneously. Following it, firmness and degree of finish of ribs can be evaluated. By grasping well the legs with both hands muscling of the leg can be assessed. Depth of the twist is judged by placing one hand on the top of the rump and other at the bottom of the twist.

In addition, if laboratory facilities are available one can also estimate the finish by estimating the rib-eye area and fat cover over the twelfth rib. According to consumer standards we consider 0.15 to 0.25 inch of fat cover and over 2.5 to 3.0 sq inch of rib eye areas as most desirable for a 25 kg weight carcass. Cameron (1992) correlated physiological responses for carcass lean content in sheep.

Chemical composition of carcass

The water contents of the carcass are highest in the lambs at birth, it goes on decreasing with the age and the amount of fat goes on increasing. The ash contents of the carcass decrease because of the ceasing of growth of the lamb with age. The amount of crude protein on dry matter basis decreases with age. This is because of increase in body fat.

Nutritional value of meat

The chemical composition of meat varies considerably, depending on the particular cut, the breed of the animal and the. degree to which the animal has been fattened and nourished. The fattening increases the fat content of mutton. The average composition of mutton is given in Table 9.4.

Protein

Mutton is a good source of protein. It contains essential amino acids such as lysine and threonine. (Table 9.6). It can be used to supplement cereal proteins which are deficient in lysine and threonine.

Table 9.6 : Essential amino-acid composition of beef, pork and mutton

Amino acids	Beef	Pork	Mutton
Isoleucine	5.1	4.9	4.8
Leucine	8.4	7.5	7.4
Lysine	8.4	7.8	7.6
Methionine	2.3	2.5	2.3
Phentlalanine	4.0	4.1	3.9
Threonine	4.0	5.1	4.9
Tryptophan	1.1	1.4	1.3
Valine	5.7	5.0	5.0

Ref. : The science of Meat and Meat Products, p 187. American Meat Institute Foundation, W.H. Freeman & Co., San Francisco, and London.

B Vitamins

It is a good source of folic acid, niacin, pantothenic acid and vitamin B-12 and fair source of riboflavin. Liver is a good source of niacine, folic acid, Vit. B-12 and Vit - A.

Minerals

Muscles are rich source of phosphorus and potassium, and fair source of iron, copper, cobalt, manganese and zinc. It is a poor source of calcium.

Carcass quality characteristics

The main characteristics influencing carcass value are as follows :

i) Carcass weight

ii) Carcass lean, fat and bone contents.

iii) Carcass confirmation and muscle thickness.

iv) The distribution of the weights of lean and bone between the different parts of the carcass.

v) Meat quality.

Characteristics of Meat

The value of mutton depends upon its nutritive value digestive value and consumer acceptance. The palatability of mutton depends upon colour, odour, flavour, juiciness, tenderness and texture. The evaluation of the eating qualities has been attempted by measuring the physical and chemical qualities of meat and correlating them with eating qualities. At present the taste- panel method is relied upon.

Colour

The colour of the freshly cut mutton is due to reduced myoglobin which is converted into oxymyoglobulin, which on further exposure is converted into metmyoglobulin leading to dark brown colour.

Aroma

Fresh raw mutton has a very slight aroma or odour. The mutton from younger animals has less odour than that from old ones of the same species. The feed fed also affects the aroma. The odours are dependent on the conditions of storage. Putrid, sour or rancid odours are because of various chemical and microbiological changes taking place in mutton held under unfavourable conditions.

Flavour

It arises from muscle fibers on cooking. It is dependent on the age of the animal, the type of food and condition of storage after slaughtering.

Juiciness

Good quality mutton is juicier than poor quality meat, the difference being due to the high content of intra muscular fat. Tender mutton is juicier. Mutton becomes less juicy with the increasing cooking losses of meat fluid.

Tenderness and texture

These are the most important characteristics of the eating quality of mutton. The texture depends upon the size of muscle bundles and the amount of connective tissue present in the muscle portion. Coarse grained muscles have large bundles, whereas fine grained muscles have small bundles. The tenderness depends upon the water holding capacity.

Factors affecting growth of lambs

Body weights : Body growth affects the composition of the lamb. It has been observed by many research workers (Reid *et al.*, 1968) that the 63 percent, 95 percent and 88 percent of variation in ash, protein and fat, respectively are due to changes in body weight at slaughter.

Breed : The rate of growth varies from breed to breed, even if given same type of fodder and concentrate to all the breeds. The growth rate is affected by the adult body weight (Jackson, 1974).

Sex : The sex hormones of males and females affect their growth rate. The ewe differ from the ram in that it has slower growth rate, a more early maturing carcass and reaches a lower mature size due to the effect of oestrogen in restricting the growth of long bones of the body.

Castration : The castration of male decreases their growth rate and they mature at an early age. The appreciation of the central role of the sex hormones in growth and development as manifested in the consequences of castration have led to extensive study of the effects of the administration of natural or synthetic hormones to meat animals in order to influence performance (Hafs *et al.*, 1971).

These days there is an attempt to improve our promising indigenous mutton breeds like Mandya and Nellore through proper selection and cross breeding with superior mutton breeds like Suffolk, Dorset and South Down. This has become necessary to reach the dressing yield of sheep, which is hardly 9.0 kg per animal to at least 12.0 kg per animal, a figure close to the world average. Parallel to it, there is a development of quality consciousness among the consumers in the country and to meet the demand and regulate marketing of live sheep, selection of a market lamb would play a vital role in future.

10

Milk

Milk is of commercial importance in some countries, but is used for home consumption in most of the countries. Some Indian communities and nomadic sheep farmers during migration of flock use sheep milk for kneeding flour for making *chapaties.* Such *chapaties* do not dry up. Ghee is prepared from sheep milk in some countries. They are normally milked in the morning. Awassi is the most productive milk breed. It can yield upto 266 kg of milk in one lactation.

Milk producing breeds

Some of the Indian breeds of sheep can produce relatively large quantities of milk. A Lohi sheep can yield upto 3.6 kg milk in a day (Kaura, 1941), whereas, Kuka sheep can produce 1.8 to 3.6 kg of milk per day (Kaura, 1942). Sonadi breed can produce 1.0 to 1.5 kg milk daily (Anon, 1953). Malpura, Rambouillet x Malpura, Chokla, and Rambouillet x Chokla can produce 45.0, 54.0, 41.4 and 45.9 kg milk, respectively in a lactation of 90 days (Table 10.1). Maximum milk yield per day has been observed to be 1.0 kg (Sahani *et al.,* 1975). Mean lactation for Bharat Merino has been reported to be 45.3 ± 1.7 kg.

Milk production can be affected by the weight of sheep (Castellanos and Valancia, 1982, Acharya and Bawa, 1971), nutrition and season (Sahni *et al.,* 1975; Acharya and Bawa, 1971), birth weight of lamb (Acharya and Bawa, 1971) and breed, age and sex of lamb (Wallace, 1948; Acharya and Bawa 1971). Maximum milk yield has been observed in fourth lactation. Sheep milk contains 6 to 8 percent fat. Milk production does not affect wool yield.

Lactation can effect the length of the rebreeding interval, suckling by lamb hastens the involulation of uterus. It has been observed that ewes that lamb early but which lose their lambs, are more likely to become pregnant again during the same breeding season, although Hunter (1968) has pointed out that this is an unsatisfactory basis for comparison.

Anatomy of udder

The right and left halves of the udder are separated by inter mammary groove. There is no communication between the two halves. This prevents the spread of infection from one part to the other. Each half is composed of secretory and some supporting connective tissue. The secretory tissue consists of numerous *alveoli* or tiny chambers lined with many secretory cells. Each alveolus is supplied with tiny capillaries which lie outside the secretory cells. Small muscle fibers surround each alveolus, which cause contraction of alveoli and produces "let down" of milk. Each alveolus is drained by terminal duct. The cluster of alveoli and terminal ducts

Table 10.1 : Milk yields in sheep

Country	Breed	Yield/ day (kg)	Period measured	Source
Brazil	Santa Ines	0.18	92	Bellaver *et al.*, 1980 a
Brazil	Morada Nova	0.21	92	Bellaver *et al.*, 1980 a
Cyprus	Chios	1.27	209	Louca, 1972.
Cyprus	Chios	1.28	90	Mavrogenis, 1982.
Egypt	Barki	0.71	84	Aboul-Naga *et al.*, 1981
Egypt	Merino	0.85	126	El-Sherbiny *et al.*, 1981a
Egypt	Ossimi	0.75	126	El-Sherbiny *et al.*, 1981a
Egypt	Ossimi	0.78	84	Aboul-Naga *et al.*, 1981
Egypt	Rahmani	0.77	84	Aboul-Naga *et al.*, 1981
India	Avikaline	0.47	92	Kishore *et al.*, 1983
India	Avivastra	0.45	92	Kishore *et al.*, 1983
India	Chokla	0.46	90	Sahni and Tiwari, 1975 b.
India	Jaisalmeri	0.40	90	Sahni and Tiwari, 1975 b.
India	Malpura	0.50	90	Sahni and Tiwari, 1975 b.

India	Merino	0.70	112	Mallikeswaran *et al.*, 1980
India	Nilgiri	0.65	112	Mallikeswaran *et al.*, 1980
India	Merino × Nilgiri	0.67	112	Mallikeswaran *et al.*, 1980
India	Rambouil let × Chokla	0.51	90	Sahni and Tiwari, 1975 b.
India	Rambouil let × Malpura	0.60	90	Sahni and Tiwari, 1975 b.
Iran	Awassi	1.16	207	Kaschanian, 1973.
Italy	Sardinian	0.86	44	Dattilo and Congiu, 1979.
Isreal	Assaf	1.34	156	Eyal *et al.*, 1978.
Kenya	Corriedale	1.36	70	Bush, 1965.
Lebanon	Awassi	0.68	150	Bhattacharya and Harb, 1973.
Morocco	Middle Atler	1.90	28	Kabbali, 1977.
Maxico	Pelibue	0.53	84	Castellanos Ruelas and Valencia, Z.M. 1982.
Nigeria	West African Dwarf	0.42	70	Adu and Ngere, 1979.
Phili-ppines	Shropshire	0.06	73	Villegas and Cruz, 1960
Saudi Arabia	Najdi	1.35	84	Ramadan *et al.*, 1977
Trinidad	Blackhead Persian	0.63	84	Butterworth *et al.*, 1968.
Tunisia	Tunisian Barbary	1.06	35	Sarson, 1972.
Vene-zuela	West African	0.60.	56	Combellas, 1980.

form a lobe. Terminal ducts form *interalobular* ducts, which in turn join to form interlobular ducts. These further join to form galactophores which empty into the milk cistern. This is continuous, immediatly below with the teat cistern. The teat cistern is joined with *streak canal,* a narrow tube that opens at the lower end of the teat. The streak canal is surrounded by a sphincter. which remains constricted and prevent leakage of milk until milking commences.

Milk Secretion

Hormones play part in the development of mammary tissue, initiation and maintenance of milk secretion and ejection of milk into the cistern and teats. The ovarian steroids are involved in the development of mammary tissue and rate and extent of ducts and alveoli development. The hormonal mechanisms that control the initiation of milk secretion at lambing are complex and not completely understood (Demiruren *et al.*, 1971) but the changes in the ovarian steroids are involved particularly the increase in oestrogen in relation to progesterone, the considerable release of oxytocin at lambing and the increase in the secretion of adrenal steroids.

It starts after the birth of a lamb. The blood which nourishes the foetus in the uterus, starts flowing in milk vein after birth, which enlarges and starts supplying the udder. Before the birth the ovarian hormones suppress the lactogenic hormones, but after birth, the secretions of ovarian hormones are suppressed. Excess production of lactogenic hormones results in the secretion of milk.

In addition to hormones, the size of ducts also regulates the blood flow which in turn affects the milk secretion. All essential components of milk are brought to udder by blood, which in turn are converted into milk.

Table 10.2 : Composition of plasma and milk

S.No.	Attributes	Blood plasma	Milk
1.	Water	91.00	87.00
2.	Glucose	0.05	4.80 (Lactose)
3.	Serum albumen	3.20	0.52 (Lactalbumen)
4.	Serum globulin	4.40	0.05 (Globulen)
5.	Amino Acids	0.03	2.90 (Casein)
6.	Lipids	0.09	3.80
7.	Phospholipids	0.20	0.04
8.	Cholesterole	0.17	Traces
9.	Calcium	0.09	0.12
10.	Phosphorus	0.11	0.10
11.	Sodium	0.34	0.15
12.	Potassium	0.03	0.15
13.	Chlorine	0.35	0.11
14.	Citric acid	Traces	0.20

Table 10.3 : Composition of colostrum (g/dl).

S.No.	Components	Species		
		Cow	Goat	Sheep
1.	Water	73.3	81.2	58.8
2.	Protein	17.6	5.7	20.1
3.	Fat	5.1	8.2	17.7
4.	Carbohydrate	2.2	5.7	2.2
5.	Ash	1.0	0.9	1.0

It is clear from Table 10.2 that as compared to blood milk contains 90 times more sugar, 9 times lipids, 5 times potassium, 13 times calcium and 10 times phosphorus.

Chemical composition of colostrum

The onset of lactation occurs with parturition. Initially the secretion is of colostrum which is high in fat, protein and immunoglobulins and low in lactose (Table 10.3). During first 2-3 days the composition of the secretion changes from that of colostrum to that of normal milk.

Chemical composition of milk

The concentration of all constituents is much greater than in cows milk resulting in the energy content being approximately 65 percent greater. The ewe suckling two lambs generally produces 40 percent more milk than an ewe at the same level of nutrition but suckling a single lamb. The number of lambs suckled has a greater effect on milk yield than a wide range of nutrition in both pregnancy (Wallace, 1948) and in lactation (Gardner and Hogue, 1964).

The composition of milk depends upon the production level of the ewe and the number of lactations. The composition of milk changes with the advancement of lactations. The concentration of fat, protein, solid not fat and ash is highest in colostrum. The concentration of these decrease with the advancement of lactation and then again increases at the end of the lactation. The concentration of lactose first increases and then decreases (Ashton *et al.*, 1964). The concentration of calcium increases in the beginning of lactation and then becomes stable. The concentration of magnesium remains stable throughout. The concentration of sodium is directly related with lactation length and inversely with potassium (Ashton and Yousef, 1961).

Table 10.4 : Chemical composition of milk obtained from cow, baffalo, sheep and goat.

S.No.	Component	Unit/dl	Cow	Buffalo	Goat	Sheep
1.	Water	g	87.3	87.0	85.6	78.9
2.	Solid not fat	g	8.5	9.5	9.0	10.3
3.	Total solids	g	13.2	10.5	17.0	–
4.	Ash contents	g	0.8	0.8	0.9	1.1
5.	Lipids	g	4.5	7.0	4.9	8.9
6.	Carbohydrates	g	4.6	5.0	4.7	5.0
7.	Proteins	g	3.5	3.8	4.1	6.2
8.	Lactose	g	4.8	4.4	4.7	4.6
9.	Calcium	mg	125.0	–	130.0	193.0
10.	Chloride	mg	103.0	–	159.0	54.0
11.	Phosphorus	mg	12.0	–	16.0	–
12.	Potassium	mg	138.0	–	181.0	190
13.	Sodium	mg	58	–	41	–
14.	Magnesium	mg	12	–	16	–
15.	Sulphur	mg	30	–	16	31

Table 10.5 : Average composition of Milk of domestic animals.

Species	Composition (%)				
	Total solids	Fat	Protein	Lactose	Ash.
Sheep (*Ovis aries*)	19.3	7.4	5.5	4.8	1.0
Cattle (*Bos taurus*)	12.7	3.7	3.4	4.8	0.7
Cattle (*B. indicus*)	13.5	4.7	3.2	4.9	0.7
Buffalo (*Bubalus bubalis*)	17.2	7.4	3.8	4.8	0.8
Goat (*Capra hirca*)	13.2	4.5	2.9	4.1	0.8
Horse (*Equus caballus*)	11.2	1.9	2.5	6.2	0.5
Donkey (*E. asinus*)	11.7	1.4	2.0	7.4	0.5
Bacterian camel (*Camelus bactrianus*)	15.0	5.4	3.9	5.1	0.7
Dromedary (*C. dromedarius*)	13.6	4.5	3.6	5.0	0.7

Source: Jenness, 1974.

Milk consists of water, lipids, proteins, sugar, minerals, vitamins, enzymes and some cellular material. There is a great deal of variation in milk composition among breeds and individuals. Milk produced by cow, buffalo, sheep and goat appears to be similar, but they differ in chemical composition (Table 10.4 and 10.5).

Carbohydrates

Lactose is the main carbohydrate and is synthesised in alveoli, it does not occur anywhere in mammalian tissues.

Protein

The chief type of proteins found in milk are casein, lactalbumin and lactoglobulin. Casein and lactoglobulin comprise about 90 percent of the protein of milk. These two and lactalbumin are synthesised in gland from amino acids.

Fat

The lipids of milk consist of true fat, phospholipids, cholesterols, pigments and fat soluble vitamins. The true fat forms the large part of milk lipids, other components comprise only about one percent of total lipids.

Minerals and Vitamins

Calcium and phosphorus found in milk are directly derived from blood. The same is also true for vitamins.

The concentration of milk proteins, fat, carbohydrates, ash and solids not fats are higher than that found in cow, buffalo and goat milk. The water contents of ewe milk are the lowest. The composition depends upon fodder, climate, atmospheric temperature etc.

Milk production

The amount of milk produced by an ewe affects the growth rate of lambs. The milk yield of the ewe can be estimated by two methods. The difference in body weight of lambs before and after feeding milk gives the amount milk suckled. This is done several times in a day. We can calculate the milk yield from this (Owen, 1957).

In second method sheep is given an injection of oxytocin and then she is milked 2 to 4 times in a day. The total amount of milk thus collected gives the yield of ewe for the day (Coombe *et al.,* 1960).

Factors affecting milk production

The use of sheep for milk production is not very popular in India. Thus the studies of the factors affecting milk production in ewes are rare. However, regardless whether the sheep are used for milk production or not the morphological traits have an important influence on lamb rearing (Sagi and Morag 1974; Kirk *et al.,* 1980). The good udder formation, particularly position and size of teats ensures proper lamb rearing. Abnormalities of udder and teat morphology make milking difficult. This may be responsible for higher mortality in lambs and inflamation of udder. Gonzalo *et al.,* (1985) observed existence of close relationship between morphological traits of udder and susceptibility to mastitis. Milk yield of ewe is affected by the morphology of the udder (Gootwine *et al.,* 1980; Labussiere *et al.,* 1981, Mavrogenis *et al.,* 1985; Woolny and Nizrdkowaski, 1985). Correlations between length, width and depth of udder and milk yield have high positive correlations (Milkus 1968). Some of the factors have been disscussed below :-

1) Breed : The milk yield of the ewe differs from breed to breed (Salevno and Malossini, 1968; Ricordean and Flament, 1969).

Lohi, Malpura and Sonadi are good milkers, whereas Chokla is a poor milker.

2) *Age* : A ewe in good health improves in her milk yield upto her fourth lactation. The milk of a young ewe is much richer in fats and solids than that of aged ewes. The amount of fat and protein increases with the increase in lactation length. (Ricordean and Flament, 1969).

3) *Length of lactation* : The peak in milk yield is reached in first 2 or 3 weeks. It starts declining after that.

4) *Period of year* : In spring and early rainy season when the grass is young, ewes produce more milk than they do in dry, hot weather.

5) *Estrous* : In majority of the cases, estrous reduces the milk yield.

6) *Water supply* : Milk is reduced if the ewes are given restricted supply of water.

7) *Weight of ewe* : Milk production to large extent depends upon the size of the ewe.

8) *Method of suckling* : Milk production increases by 40 percent if a ewe suckles twins (Peart *et al.,* 1972). Merino ewes suckling crossbred lambs produce more milk than when suckling pure bred lambs (Moore, 1966).

9) *Nutrition* : Milk production is directly related with nutrition. Concentration of calcium and phosphorus increase if the ewes are given concentrate mixture (Ashton *et al.,* 1964).

10) *Fear and excitation* : They have ill effect on milk production. They cause the release of adrenaline, which in turn reduces the milk production.

11) *Health* : The milk from healthy ewes is nutritious. The milk yield decreases in mastitis. It also results in increase in chloride concentration and decrease in lactose. Amount of fat also decreases in disease.

12) *Number of lactations* : The amount of solid not fat and fat increases with the number of lactations.

13) *Dry period* : If dry period is longer, ewe will produce more milk.

14) *Hereditary* : It affects the milk production.

15) Hormonal imbalance: Udder development and milk synthesis are affected by the release of estrogen, oxytocin, and prolactin hormones. Oxytocin helps in the release of milk.

In addition to these milk production is also affected by change of place, weather, constitution of udder, time of milking etc.

11

Health Management

Productivity and the financial success of sheep management depends on balancing the cost of production with the return from sales. In this crude equation the contribution by the ewe, which is most likely to vary between seasons, is the value of lambs available for sale. Diseases cause great losses in case of sheep. Diseases affect through mortality and morbidity. Change of climate decreases the adaptability of sheep to diseases. Incidences of diseases increase if the sheep are fed succulent grasses in place of dry grasses, given stagnant dirty water to drink and allowed to sit on moist floor at night. Grazing sheep in large numbers in limited area not only depletes the grass but also increases the incidences of diseases due to parasites. They can also develop cough, pneumonia, diarrhoea as a result of this.

Unlike other animals, it becomes difficult to observe symptoms in sheep. They continue to move with the flock even if their health is very bad. A sheep lags behind in the flock only if it is sick. Sheep farmer should observe carefully the behaviour of his sheep from the time they move out for grazing till they return from grazing. He should have a round of the flock in the morning before they move out for grazing. Cough, sneezing and unthriftyness are the signs of sickness.

Disinfection

The disinfection of a building implies the elimination from a house of all microorganisms that are capable of causing disease. A disinfectant is the agent that is capable of achieving this and in livestock farming it is usually a chemical agent. The process of disinfection of a building can take place by the use of natural disinfectants and by artificial disinfectants.

Sunshine, heat and some chemical agents kill bacteria. Sunshine is the most powerful disinfectant. Its efficiency is entirely due to the ultra violet range of wavelength and its most effective range is between 2800 and 2400 Å. Dessication by wind and fresh air also contributes to the disinfection. The temperature of air and high temperatures accelerate the destruction of exposed micro-organism.

Sheep sheds can be disinfected by using steam, boiling water and by synging. Some important disinfectants are given in Table 11.1 All the bacteria are not killed by one disinfectant. Humidity reduces the killing power of these disinfectants, therefore, cleaning is very essential.

The disinfection of farm buildings is generally carried out by using chemical disinfectants. Chemical disinfectants act by coagulating or denaturation of the protein of the bacteria. They are phenols, alcohols, acids and salts of heavy metals. Metallic salts prevent the development of the cell because they are protein precipetants. Mercury and silver salts are the most potent. Phenols also act as protein denaturants.

Before disinfection operation is taken up, dung sticking to walls should be removed. They should be scrubbed thoroughly. If the floor of the shed is not made up of cement concrete, 8-10 cm soil should be removed and fresh soil mixed with slacked lime should be put. The cement flooring can be disinfected with 4 per cent sodium carbonate solution. The shed should not be used for tying animals within 24 hours of disinfection.

It becomes difficult to diagnose diseases in sheep. Recording of following attributes helps in diagnosis :

1. *Body temperature*: Veterinary clinical thermometer is used for recording rectal temperature. It is inserted deep into rectum and kept there for one minute. The temperature recorded in the morning is considered to be normal. The temperature recorded in the after noon is little higher and it returns to normal in the evening.

2. *Pulse rate*: The pulse is examined by catching the tail at the base between two fingers. The number of pulse perminute gives the pulse rate.

Table 11.1 : Common disinfectants and their usages

Name of disinfectant	Utility	Concentration	Mode of application	Remarks
Bleaching powder	It is a powerful disinfectant, rapidly inactivated by organic matter, can be used for disninfection of sheep sheds when a contageous disease has occured.	It should not contain less than 30 percent chlorine.	Dusting	Protect it from air, light and moisture.
Boric acid	Weak antiseptic can be used for washing of infected eyes.	2% aqueous solution	Splashing	Good for minor infections
Caustic soda	It has powerful germicidal action. Destroys virus of sheep pox and Foot and Mouth Disease. It can also be used for the disinfection of sheep sheds.	2% aqueous solution for disinfection of shed and 5% for killing anthrax and B.Q. spores.	Spray	While spraying protect eyes, hands and skin, use gloves.
Cresol	Its solution is prepared by dissolving in soap solution. It is effective against Tuberculosis and Johnes disease bacteria but not against viruses and spores, can be used for the disinfection of shed.	2-3 percent solution	Spraying	Prepare in soft water as soap precipitates in hard water.

3. *Respiration rate* : Inhalation and exhalation of air in the lungs is known as respiration. These can be counted by keeping a hand infront of nostrils or by counting the expansions and contractions of chest per minute. The normal values for rectal temperature, pulse and respiration rates are given in Table 11.2. (Kaushish *et al.*, 1976).

Depending upon the aetiology, sheep diseases can be divided into following groups:

A. Bacterial
B. Viral
C. Protozoan
D. Parasitic and
E. Miscellaneous.

Bacterial Diseases

Bacteria are present everywhere in the world. They are present in fresh and saline water, inside and outside the body and in hot and cold water. They differ in shape and size. They may be round or rod like. Some of the bacterial diseases found in sheep are described below.

Haemorrhagic septicaemia:

It is a contagious disease which occurs in rainy season. It is caused by a bacteria named *Haemorrhagic septicaemia.* This disease develops in animals which are not vaccinated.

Symptoms : The body temperature rises to 40-42° celsius. Animal feels pain in throat region and has difficulty in respiration. It becomes lazy. The tongue protrudes and there is salivation. If the animal lives for a day or two it develops pain in belly.

Treatment : Animal can be saved if given intravenous injection of 33 percent sulphamezathine at the outset of the disease.

Prevention : The disease can be prevented if sheep are given vaccination against this disease.

Black Quarter (Black Leg)

This is also known by the names of Fud, Suja, Chirchira, Budta, Takli, Ghatia Tav etc. Mostly sheep in good health are affected. It is caused by *Clostridium chauvei.* (Minnett : 1948). This bacteria lives in soil and develops in the intestine. The bacteria

Table 11.2 : Rectal temperature, pulse and respiration rate of sheep during different seasons (Kaushish *et al.*, 1976)

Attributes	Time	Summer season	Rainy season	Autumn season	Spring season
Rectal Temp. (°C)	Morning	37.8 ± 0.05	37.2 ± 0.02	37.5 ± 0.08	37.6 ± 0.27
	Evening	38.5 ± 0 06	38.0 ± 0.03	38.8 ± 0.05	38.8 ± 0.06
Pulse Rate (P.M.)	Morning	82.6 ± 2.14	82.1 ± 1.90	69.4 ± 1.77	82.6 ± 2.41
	Evening	99.3 ± 2.88	120.5 ± 3.36	94.8 ± 2.16	94.6 ± 2.40
Respiration	Morning	31.7 ± 2.02	56.9 ± 4.39	39.2 ± 2.50	26.1 ± 1.10
Rate (P.M.)	Evening	56.3 ± 2.97	127.2 ± 5.35	41.5 ± 1.80	43.9 ± 2.30

enters through scratches in skin or injuries. This disease occurs in rainy season.

Symptoms : Temperature rises to 39.5-41.5° celsius. The sheep becomes lame and stops eating and rumination. The skin becomes purplish black, swollen, oedematous and leathery over the lesion. Crepitation which is common in cattle usually not seen much in sheep. The swelling is hot and painful. The disease may cause sudden death or the animal may show stiffness and straddling gait. The pulse rate varies between 90 and 100 per minute.

Treatment : Antiserum prepared is helpful. Injections of terramycin are also helpful.

Prevention : Sheep should be vaccinated in the beginning of rainy season. Dead animals should be burried and skin should not be removed. Lime should be put on dead body. Isolate the suspected animals. Disinfect the utensils used for affected animals. Burn the fodder used by dead animals.

Enterotoxaemia

The disease develops in adult as well as in lambs. It is caused by *Clostridium welchi* Type D. (Gardner, 1973). This disease develops due to supplementation or sudden change in diet. This disease develops when the sheep are grazed in gram harvested fields or at the start of rainy season. This disease develops in lambs after heavy milk feeding.

Symptoms : Sheep circles in one direction. It starts smelling earth and pretends as if it is grazing. Rise in body temperature is also observed occasionally. There may be sudden death or the sheep may jump into the air, fall and go into convultions. If these live longer, head may be thrown back on one side. The temperature may rise by 2-3 degree celsius. Presence of sugar in urine ((2-6 percent) is an important feature. Sheep may show nervous symptoms. Mortality rate is 90 to 100 percent.

Treatment : It was thought that there is no treatment for this dreaded disease. The author and one of his colleagues while working at Central Sheep and Wool Research Institute, Avikanagar (Rajasthan) have developed a treatment for this disease. They could save 91.6 percent of the sheep in which the symptoms of this disease developed by drenching 50 mg of potassium permanganate in 100 ml of water. Such sheep started grazing after 2-3 days. (Dube and Kaushish, 1981).

Prevention : To prevent this disease, sheep should be vaccinated. Careful feeding programme of younger stock helps in control of infection.

Foot rot

This disease develops during rainy season and continues till winter. If diseased animal is not treated in time, the animal becomes weak and may die. The disease is caused by *Sporophorus necrophorus.*

Symptoms : Affected sheep becomes lame and lags behind. Bad smell is emitted from the affected feet. The hooves may fall off. Sheep becomes unable to graze, grows weak and may die in 12-15 days.

Prevention : Sheep should not be allowed to graze in pastures where the ground is wet. Sheep shed should have aluminium or cement concrete foot bath which should have 20 percent copper sulphate or 6 percent formalin solution. All sheep they return from grazing, should be passed through it. Affected sheep should be retained in the shed, till they become alright. They should be vaccinated (Egerton and Roberts, 1969).

Lamb Dysentery

This is a dangerous disease of lambs. It is of bacterial origin. *Clostridium perfringens* Type B causes severe dysentery in young lambs. In affected group of lambs morbidity may reach as high as 20 to 30 percent. It is prevalent in cold weather and in over crowded farms.

Symptoms : The syndrome of lamb dysentery usually occurs in lambs less than two weeks old and is manifested by sudden death without premonitory signs in peracute cases. In more common acute form, there is severe abdominal pain, recumbency, failure to suckle milk, the passing of brown fluid faeces some times containing blood. Death usually occurs after a period of coma and within 24 hrs of the onset of illness.

Treatment : Contact veterinary doctor for treatment. Wool around anus should be cut to prevent flies to lay eggs.

Prevention : If there is disease at the time of birth of the lamb, change the enclosure 2-3 times. Vacciation, preferably with type specific toxoid or bacterin is the only preventive measure available.

Lamb arthritis

The incidences of this disease are more in the lambs born in dirty sheds. It could be suppurative or nonsuppurative in nature.

a) Non suppurative : It is caused by bacteria. The lamb becomes lame from one or two feet. Lamb feels pain in joints when pressed. Stops suckling. Pus is formed in the swollen joint. Lamb may automatically become alright.

b) Suppurative : Caused by *Streptococci, Corinal* or *Coli* bacterium. Joints become swollen and painful. Mobility of joints decreases. Temperature rises 40.0 to 41.0° celsius. Bad smelling pus comes out of swollen joints. Disease spreads from this pus.

Treatment : Get treatment from veterinary doctor. Economic aspect should also be kept in mind while treating the affected lamb. Naval cord must be painted with tincture iodine at birth. Keep lambs in clean sheds.

Inflammation of Naval cord

It is a common disease of lambs. It may also result in death of lamb due to tetanus. Attack of *Streptococci* or *Corinea bacterium* occurs due to the soiling of naval cord and there may be pus formation. Sometimes maggot formation results (Newson and Thorpe, 1938).

Symptoms : Lamb becomes restless and feels pain. There may be fever and deaths also.

Prevention and treatment : Naval cord should be cut at a distance of 2-3 inch and tincture iodine applied. Medicine should be applied till it becomes dry and falls down.

Listeriosis

It is an acute infectious disease of sheep caused by *Listeria monocytogenes.*

Symptoms : The affected sheep becomes sluggish with conjunctivitis, nasal discharge from one or both nostrils. One ear is dropped indicating paralysis, head is drawn to one side and sheep may move in circles. The temperature may rise to 41.0° celsius in early stages but may fall to 39.0°, celsius. Death may occur in 1-3 days. Recovery is rare. It is a communicable disease of sheep. (Kaushish, 1980).

Prevention : Isolate sick animals. Observe sanitary practices. Rodents spread this disease, so they must be killed. Most cases are observed due to silage feeding so this point must be considered. No commercial vaccine is available.

Tetanus

Tetanus is a highly fatal and infectious disease of sheep caused by toxins of *Clostriduim tetani,* which is found in soil in large numbers. (Haberman and Wellhorner, 1972). It can affect animals of all the ages. High incidence of tetanus may occur in young lambs following tail docking. Docking by the use of elastic band ligatures appears to be especially hazardous (Mahoney 1949). Bacteria enter the body through wounds and produce toxins under anaerobic conditions.

Symptoms : Cases appear 3-10 days after docking and shearing in sheep and lamb. There is rise in temperature and a general increase in muscle stiffness is observed. There is trismus with restriction of jaw movements, prolapse of third eye lid, stiffness of hind legs, causing straddling gait and animal becomes sensitive to light and sound. The animal dies within a few hours of the development of symptoms. The temperature and pulse are within normal range in early stages but may rise later when muscular tone and activity are further increased.

Prevention : The treatment of this disease is very costly. The response to treatment is poor. Lambs should be put on clean place immediately after birth. Change shed if the spores of this bacteria are present there. Give tetanus toxoid injection within 15 days of birth. Tetanus antitoxin is usually administered but is of little value once symptoms appeared.

Pneumonia

The disease occurs due to hot, cold, dust particles, bacteria, viruses and fungi in young and adult sheep. Lambs may suffer, if not protected, from cold. Mishra *et al.* (1970) did bacteriological examination of morbid materials from cases of sheep pneumonia. Verminous penumonia has been reported from Bengal by Mohan, (1948).

Symptoms : There is copious discharge from nose which becomes thick with the passage of time. Cough and rapid respiration also occur. Lambs stop suckling, become weak and may die.

Treatment : Animals should be kept in comfortable place. They should be given antibiotics and antipyretics.

Tuberculosis

It is caused by *Tuberculus bascili.* The disease can be contacted through drinking raw milk. It can spread from sheep to man (Kaushish, 1992b).

Symptoms : This causes tonsils. This may affect intestine and lungs.

Prevention and treatment : Sheep should be examined for the disease and given specific treatment. Affected animals should be separated from healthy animals.

Brucellosis

This disease is neither as prevalent in sheep nor as severe as vibriosis. Once a sheep is affected the infection continues through out the life of sheep. The disease affects the rams. *Brucella ovis* causes epidedymitis in males and may cause abortions in ewes. It is caused by *Brucella meletences, B. abortus* or *B. ovis* (Meyer 1974). The bacteria of this disease enter body through injuries, eyes or through skin.

Symptoms : The latent period of this disease varies from 15-20 days. At first, there is rise in temperature which falls and again rises. There is lot of sweating, pain in knee joints, anorexia followed by weakness. There occurs mucopurulent discharge from vagina within few days of service. The abortion occurs near term.

Prevention : Wear hand gloves while handling aborted foetus or placenta of such sheep. Aborted foetus should be burned. Get rid of the infected rams. Vaccination of rams gives good results. General hygienic measures be practised.

Vibriosis

Abortion is caused by *Campylobacter foetus* which was formerly called *Vibrio foetus,* during last 6 weeks of pregnancy. Diagnosis is from the history and isolation of causative agent from foetal stomach, visceral organs and brain.

Symptoms : The disease is suspected in flocks in which there is an increase in the number of services per conception. Abortion may be the only symptom but in most cases there may be brown mucopurulent discharge for a few days. The foetus born is oedematous.

Prevention : General hygienic measures and disposal of foetal discharge and membranes will help a long way in checking the infection. An ovine vaccine is effective in preventing abortion when used prior to breeding or in early gestation (Bryner, 1976).

Leptospirosis

This disease is caused by *Leptospira pomona,* or *L. grippotyphosa.* These bacteria develop in blood after entering through skin, mouth, genital organs, eyes or through respiratory tract. They leave the body through urine. There may be acute haemolytic anaemia with haemoglobin in urine.

Symptoms : There is sudden rise in body temperature, followed by anorexia and abortion. Sheep may also die in 3 to 4 days.

Prevention : Aborted foetus should be burned or buried. The vaccination of this disease is not yet available (Weisman, 1967).

Anthrax

This disease is caused by *Bacilus anthracis.* This is a very serious disease.

Symptoms : It is a percute disease characterized by septicaemia and sudden death (Kaushish, 1992b). In acute cases sheep may die in a short period, but fever, muscle tremor, dyspnoea and congestion of mucosa may be observed. Blood may come out of natural openings such as nostrils, mouth, anus, urethral orifice, which does not coagulate. On postmortem examination, blood spots are observed. There may be enlargement of spleen.

Treatment and prevention : General hygienic measures should be taken. The dead body should be buried after putting lime on it. Sheep should be vaccinated (Lincoln *et al.*, 1964).

Oxytetracycline @ 2-5 mg/kg body weight when given parenterally has proved to be useful (Johnson and Percival, 1955).

Viral Diseases

Some of the important viral diseases of sheep are described below:

Blue tongue

Blue tongue was first reported in India in sheep and goat in Maharashtra in 1961 and subsequently report on the occurrence of this disease from other parts of the country has also been made. In Rajasthan the prevalence of disease was first noticed by Lonkar *et al.*, (1983).

General outbreak of disease is mostly seen after rainy season mainly between September and November depending on onset of rain. The autumn season is the main breeding season of sheep and by the time this disease occurs, sheep are in early stage of pregnancy i.e. foeti are around 60-90 days. There may be loss in birth weight and congenital abnormalities. Blindness, parrot mouth, breachygnathia, buldged head, micromosia and curved neck. Rams showed aspermia for as long as sixty days.

Rinder pest

Though this disease is not a problem in Rajasthan but it has become a reemerging disease in the coastal areas of Andhra Pradesh.

Symptoms : Sheep occasionally suffer from it. It is characterised by rise in temperature and rapid respiration. The erosions are in the mouth and there is congestion and inflammation of abomasum. The faeces become loose.

Prevention : General hygienic measures should be taken. Fresh purchases should be quarantined. Vaccination should be taken up.

Conjunctivitis

This may affect one or both the eyes of lambs as well as adult sheep. This disease spreads very fast, if there is some injury of eye (Kaushish, 1992b).

Symptoms : Eyes become red, there is swelling of eyes and lot of lecrymation. Animal may not be able to open its eyes. The outer layer of eye ball becomes white, tears start becoming thick and pus may also appear in tears at later stage. The eyesight is affected. Sheep becomes alright in 5 weeks.

Treatment : The affected sheep should be kept in darkness. Ten percent solution of Zinc sulphate if put twice a day in affected eye, gives good results. Puttinga mixture of sulphacetamide powder (1 part) and boric acid (2 parts) in affected eye, twice a day after cleaning gave encouraging results.

Ecthyma

This is also known as jadia. It is caused by contagious pustular dermatitis virus. The virus can survive for long under adverse conditions. This is a contagious disease of lambs. It spreads rapidly in humid climate. It can also spread to teats from suckling lambs.

Symptoms : The commisures first become red, then there is pus formation. These become dry and scabs are formed and fall off in 10-15 days.

Prevention : Lorexan, Himax or penicillin ointment should be applied after removing scabs. Separate the affected animals. Animals can be vaccinated on the underside of thighs by scarification. The vaccine can be prepared by grinding 1% scabs in glycerine saline (50 percent) solution.

Foot and Mouth Disease (F.M.D.) :

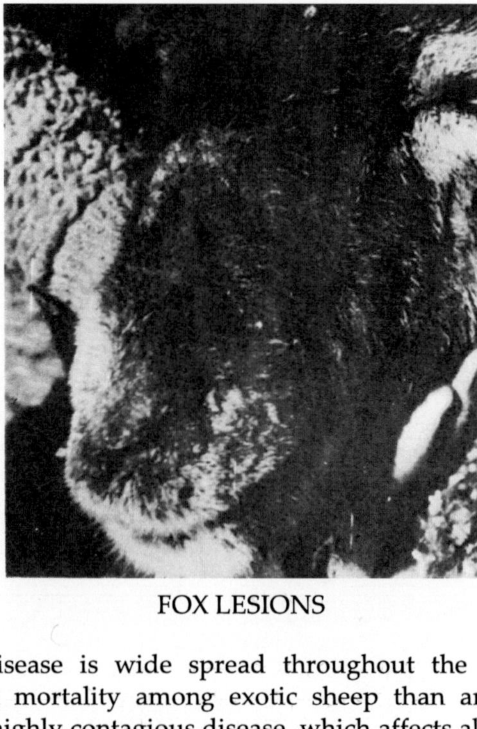

FOX LESIONS

This disease is wide spread throughout the world. This causes high mortality among exotic sheep than among native sheep. It is highly contagious disease, which affects all the groups. The disease can occur at any time of the year.

Symptoms : The lesions may develop on the upper side of tongue, gums, udder, interdigital spaces etc. Suckling lambs and new born may show high mortality due to mayocardial involvement. It is followed by salivation, lameness, if not treated there may be maggot formation. The disease may continue from 2-3 weeks to a few months.

Treatment : Affected feet and mouth should be washed with potassium permanganate solution. Boroglycerine should then be applied in mouth and phenyl on affected feet. Affected sheep should be given soft green nutritious grass and given rest.

Prevention : The sheep should be vaccinated every year. The protection against disease is 100 percent if vaccination is done in time (Bachrach, 1968). Hygienic measures should be taken and sick animals should be isolated.

Sheep pox

It is very contagious disease of sheep. It affects sheep of all the ages. The mortality rate is higher in lambs than in adults. It deteriorates the quality of hides. The morbidity is high. It spreads by air and by contact.

Symptoms : The disease is characterised by rise in temperature (41.1°C), eye lids are swollen, mucous discharge from nostrils. There are erruptions on non wooly areas such as nostrils, eyelids, lips, chest, abdomen and mammary glands. There are eruptions in the mouth causing excessive salivation. Lungs may become pneumonic The lambs may die within a few days of appearence of symptoms (Kaushish, 1992b).

Treatment : Isolate affected animals. Apply antiseptic cream on pustules and give crushed, boiled barley or maize mixed with whey or milk. Bury dead lambs after putting lime.

Prevention : Vaccinate sheep in February/March. They develop immunity in 2-3 weeks.

Rabies

This disease occurs due to a virus which is present in the saliva of the affected animal. The virus enters the body because of biting by the affected animals. The virus affects the nervous system. The syptoms appear within 14-90 days of biting. This time depends upon the distances of place of biting from brain. Two types of symptoms are observed.

Symptoms : i) *Silent form* : The animal remains depressed and dies in 1-4 days.

ii) *Excited form* : Sudden change in behaviour occurs. It becomes excited, runs to bite. There occurs change in voice, lower jaw hangs downward and there is salivation with protruding tongue. This is followed by paralysis. Animal is unable to drink water and dies in 2-4 days.

Treatment : There is no treatment for this disease.

Prevention : If a dog bites, it should not be killed and keep a watch. If it does not die in 10 days, it is not suffering from rabies. Clean the wound with carbolic acid thoroughly. The sheep should be given anti rabies vaccine.

Fungal Diseases

Only three diseases from this group are important from disease point of view. They are :

(1) Mycotic dermatitis

It is caused by *Actinomyces dermationomus.* Smear from the exudate and scabs show branching mycelia and the broader filaments consists of coccoid bodies. It is more important in young lambs than in adults. Treat the animal after shearing with 1:500 copper sulphate 1:200 or Anabae a prepartaion containing quaternary ammonium compound after shearing.

2) Ring Worm

(Tricophyton spp.) Lesions are on the face, ears and head (rarely on the legs) as circular patches covered with dry grey yellow crusts. Treat cases after softening the crust with soap or oil and apply Tincture Iodine.

3) Coccidiomycosis

A deep fungus *(Coccidioides immitis)* causing suppurative lesions in the lymph nodes which histologically show granulomatous lesions consisting of double contoured body (may be having radiating clubs). Organisms from the lesions are isolated on artificial media as Cotton white mass having hyphae. It is not a very important condition in sheep as compared with cattle.

Protozoan Diseases

i) Red water Disease

This disease is spread by ticks. It is caused by thieleria and babesias. Blood appears in urine. There is high rise in temperature. Injections of Berenil, Bebesan or tripan blue have proved to be very effective. To prevent this disease, sheds should be kept clean and free from ticks.

ii) Coccidiosis

Out of the protozoan infections, coccidiosis is the most important. *Emeria arloingi* is the single most significant species causing coccidiosis in sheep.

Symptoms : Lose faeces often thin and watery mixed with blood. Lambs lose appetite and decline to move. The temperature may rise by 1-2 degree celsius. The animal becomes weak and dies.

Prevention : The lambs should be stall fed. As a rule dry feeds are less likely to cause problem. Treatment with sulphamezathine and nitrofurazone is advocated.

iii) Abortion due to Listeria

The abortion occurs in the last month of gestation. Lamb dies 2-3 days before abortion. Affected sheep and lambs show nervous symptoms. These protozoan are found in large numbers in placenta, naval cord and uterine discharge. They enter through mouth by eating infected material. This disease is not as severe as brucellosis. Not more than 7 percent sheep are affected by this disease.

Hygiene is very important to prevent the spread of the disease. Once a sheep has aborted it will not abort for next 2-3 years (Dennis, 1961). There is need to discover preventive measures for controlling listeriosis which has been recorded by Katiyar (1961 a).

iv) Toxoplasmosis

Ovine abortion is caused by *Toxoplasma gondii.* Abortion occurs during the last month of gestation and weak full time lambs may be observed. There may be lesions of focal necrosis on placenta. Diagnosis may be confirmed by observing white necrotic areas on the cotyledons.

Lamb gets infection directly from sheep. Resorption of foetus occurs. This disease affects sheep of young age. Nervous symptoms are noticed. Protozoa are present in placenta. *Prophylaxis*: There is no known method of control due to lack of proper epizootology. General, hygienic measures are advised.

Parasitic Diseases

Depending upon the place of their occurrence, they can be classified as external or internal.

A) External parasites

A number of external parasites are found in sheep. Some of them are harmful but some transmit diseases. The external parasites could be ticks, wingless flies, lices, mites etc. The first type of external parasites are those which come to suck blood and leave the host e.g. ticks and some types of flies. Second type of parasites are those who remain temporarily on the host e.g. lices and mangemites.

i) Blood sucking parasites

a) Ticks : It is an important parasite of sheep. They are dark brown in colour, hairy body with six legs (Soalsby, 1968). On an average, they are 6 mm in length. They suck blood and fall on the ground after engorgement, there they lay eggs. Nymphs hatch out of eggs in 1-3 months. They suck blood and again fall on the ground and become adult in 3-4 weeks time. Ticks are agents for the transmission of disease organisms. The wounds caused can allow invasion by disease organisms. These include looping ill caused by a virus. Tick borne fever caused by one of the *Rickettsia* species, causes abortions and tick pyrexia.

Prevention and treatment : It consists of spraying, dipping or dusting the entire flock at regular intervals. The sheep should be dipped in insecticide suspension for one minute. Normally it will itself come out by climbing the steps, incase it is not able to come out, it should be helped (Kaushish and Sharma, 1982). The housing area should also be treated. Except in heavily infected flocks, one or two treatments per year will control ticks. Best results are obtained if the animals are treated when the wool is short.

b) Flies : They are found throughout the world. Their bite is very painful and the affected sheep may stop grazing. They help in spreading some diseases such as blue tongue, rift valley fever and sheep ked. The examples of these flies are, Tebanid, Sand flies and some types of mosquitoes. Mellophegus is a blood sucking wingless fly. If present in large numbers it can cause anaemia and skin irritation.

ii) Temporary parasites

a) Nasal fly : The fly *Oestrus ovis* is found all over the world. The adult fly flies around the head and deposits larvae near the sheep's nostrils. The larvae cause irritation resulting in coughing and sneezing. In severe out breaks some control can be achieved by treating the sheep to eliminate the larvae in nostrils.

b) Maggots : Screw worm fly lays eggs around the edge of the wound. They stay in wound from stage I to stage III. The pupae fall on the ground. Ointments containing organophosphorus compounds are effective in these cases. Dipping in B.H.C. or D.D.T. solution gives good results.

c) Myiasis : It is caused by laying maggots by a particular type of meat eating fly. The repeated strikes by these flies are known as slow fly strike. Heavy losses occur during summer season. Flies of *Lucilia* species cause blow fly attack. Some of the *Calliphora* species can also cause primary attacks. Secondary attacks are caused by *Chrysomyia* species. The blow fly lays its eggs in damp, soiled areas of wool around the hind quarters of scouring sheep. The larvae burrow into skin and feed on living flesh. The attacks can be prevented by dipping sheep in long acting insecticides. The sheep will be able to come out of the dipping tank herself if she is unable to do so, she should be helped (Kaushish and Sharma, 1982).

iii) Permanent parasites

a) Lices : These could be of two types, biting and blood sucking. They are flat wingless insects. They lay one egg per day close to the skin. First stage nymph comes out in 10-15 days which moults into second and third stage and ultimately becomes adult, which starts laying eggs after sucking blood for 3-4 days. It completes its life cycle in 30-45 days. The biting lices are found very close to skin and feed on dandruf. Irritation occurs due to biting. Sheep rubs its body against walls and trees. These can be killed by dipping sheep in D.D.T., B.H.C. or organophosphorus solution. This disease spreads by coming in contact with other sheep. Treatment and prevention is same as for ticks.

b) Mange : These cause great losses of wool. The sheep affected with this parasite may die if not treated. These are very small in size. Female seroptic mites lay eggs deep into the skin she lays 3-5 eggs per day. The larvae mature in about 17 days. It completes its life cycle in 3-4 weeks. These mites resemble those found in human beings (Mellanby, 1972). Soroptic mange can be found in ears also on pole (Little John, 1968). These could be of following types :

Sorergetus mange : This affects all the age groups. The sheep rubs the affected parts with feet and teeth. The skin becomes thick and wool dry.

Chorioptic mange : It affects hind feet which may spread to scrotum and udder. The skin of the affected part becomes red and there may be dry scab formation also.

Soroptic mange : It affects the woollen parts of the body. Sheep bites the affected part with teeth.

Sinclair (1949) has discussed about the eradication of sheep scab. This can be controlled by dusting shed with gamexane or slacked lime. The shed can be kept vacant for 10-15 days. Dipping of sheep in lime sulphur solution with nicotine sulphate kills mangemites. Jawahar *et al.,* (1976) tried benzyl benzoate and tetraethyl thiuram monosulphide against sarcoptic mange.

B) Internal parasites

Internal parasites are found in all types of climate. Several species of trematodes, cestodes and nematodes are present in the digestive tract including liver and mesentric channels and other organs of sheep, consequently sheep is said to be a "museum of parasites". There is nothing more important than control of worms in sheep because these worms interfere in growth of lambs, produce unthrifiness, emaciation and anaemia, loss in production, pave ground for bacterial and viral infections, diminish resistance to other diseases in extreme cases, cause severe mortality in the flocks. Incidences of helminths have been reported from Bihar (Bali, 1973), Jammu and Kashmir (Bali 1972 a, b; Bali and Fotedar, 1972; Bali *et al.,* 1976). Losses from parasites have been observed more commonly in young sheep than in adults and in areas receiving more than 20" rainfall. They result in losses amounting to millions of rupees. (Anon, 1972 b). It has been reported by Morris *et al.,* 1977) that the prevention of internal parasites may increase profits by 1200 percent. Since these diseases are chronic in nature, they are particularly of economic importance as they cause tremendous waste in maintenance of debilitated animals (Dunn, 1948; Gordon, 1967; Georgi, 1969). Few most important helminthic diseases have been discussed below:

Leaf worms : (Liver fluke disease)

It is also known as liver fluke disease. It is one of the most serious diseases known to occur in endemic areas. The affected animals showed high rise of temperature with sudden deaths in most of the cases. No other concurrent infections in these animals were found. About 2 per cent sheep and goat slaughtered in Bhubaneswar have lesions in the lungs due to fasciolosis infection. About 50 percent of these animals did not show corresponding liver infection. Mathur and Dutt (1970) studied the variation of liver fluke egg counts within a day and incidence of infection in sheep and its intermediate snail host. In heavily infected areas this is one of the most serious diseases and is the major limiting factor in sheep development. Fascioliasis is also found to be mixed with bacterial infection such as *Clostridium oedimatiens* which may develop and the disease is known as "Black disease". *Fasciola gigentica* ranks foremost in undermining the health of our sheep.

Treatment : Carbon tetrachloride is recommended for the control of Fascioliasis. Mathur and Dutt (1970) studied the treatment of *F. gigentica* infection in sheep and conducted field trials in Himachal Pradesh for control of the disease. Lone *et al.*, (1972) conducted field trials with Trodex and Distodin against fascioliasis in naturally infected sheep.

Control : The control of this disease is important because of the fact that although by treatment there is a reduction in the incidence of infestation but reinfection is difficult to control. Do not allow water to stagnate. To prevent development of snails, spray such medicines which kill them in and around ponds. Allow sheep to graze grass grown in the viscinity of ponds only in the afternoon, (Boray 1969; Armour, 1972; Over and Dykistra, 1975).

Amphistomiasis

Large number of amphistome species named as parasites of sheep (Hiji 1935; D'Souza 1949) are from investigational work undertaken mostly on account of pathogenic significance. *Paramphistomum* sp., *Cotylophown* sp., *Calicophoron* sp., *Ceylonocotyle* sp.,

Gastrothylaz sp, *Ficshoderius* sp. and *Olvnria* sp. have been reported by Bhattacharyulu and Pande (1969).

Dharmendra Nath (1971) described the detailed incidence of severity of infection of immature amphistomiasis in sheep. The disease occurred from September to March and percentage of mortality was found to be 27.4%. Sharma Deorani (1965) studied the histopathological effects of immature amphistomiasis of sheep. Dharmendra Nath (1971) made observations on allergic skin test for diagnosis of immature amphistomiasis in sheep. Bali and Fotedar (1972) reported *Pammphistomum fischoedar* from sheep of Jammu and Kashmir. The above authors in the same year studied the morphology of *Paramphistomum* and *Ceylonocotyl* from sheep of Jammu and Kashmir. Bali (1972 a) reported the occurrence of *Calicophoron* in sheep of Jammu and Kashmir. Dharmendra Nath (1971) reported observations on the treatment of amphistomiasis of sheep. In the same year he also reported the effect of some anthelmetics against amphistomiasis in experimental infected sheep. Jain and Kamlapur (1971) reported the occurrence of *Calicoporon concorchis* in sheep in India.

Round worms

Several species of nematods and cestods are present in the stomach and intestines of sheep and goat. Consequently most clinical infections are designated correctly as parasitic gastro-enteritis rather than a distinct gestritis or entritis. There is relatively little information on the number of those parasites required to produce clinical parasitism death.

Dhar and Singh (1970) studied some effects of cortisone administration on the natural and acquired resistance in lamb to the nematode *Oesophagostomum columbianum.* Effect of experimental *O. columbianum* infection has been studied by Sarles (1944). Chellappa and Gopalakrishnan (1977) reported incidences of *Galgeria pachyscellis.* Sheep appeared to suffer more than goats from helminth parasites.

In 1937, 60 ml of copper sulphate solution was suggested as treatment dose for sheep. Number of anthelmetics were tried in the last two decades with varying results. Hiregoudar (1970) have reported outstanding efficacy of Nilverm (Tetrmesole) as an anthelmetic for gastro intestinal nematodiasis in sheep. Fenbendazole has been tried by Kalita *et al.,* (1978) and Uppal *et al.* (1977)

in sheep. Makkar *et al.* (1974) tried 4 cyano-6-nitro phenol against *H. controtus* worms in sheep. Bali *et al.,* (1976) and Bali and Singh (1977) tried a new anti cestodial drugs in sheep. Bali and Singh (1977) also compared different anthelmentics in sheep and goat nemodiasis. Resistance of *H. controtus* against phenothiazine and thiabendazole in sheep has also been reported.

Lung Worm Infection in Sheep

Lung worm disease is an enzootic bronchopneumonia due to the presence of the roundworm in the bronchi. The lung worm of sheep are fairly cosmopolitan in distribution. Commonly occuring species in sheep and goats are, *Dictyocaulus filaria; Protostrongylus rufescens; Cystocaulus ocreatus* and *Muilieruis capilleris.* Lungworm disease is world wide but it has been reported from the tropics and from countries where the winters are mild. Exact knowledge is lacking on the prevalence of this disease in different parts of the world but wherever, it is common as a clinical infection it is serious, weather often determines its prevalence in any locality and where it is sporadic a wet season may precipitate a general attack. Lack of systematic data on the lungworm disease in India makes it difficult to assess the magnitude of this problem. However, the work carried out so far indicates that the lungworm infection in sheep is very common. Contribution in this field has been made by Bhatia and Pande, 1950. A recent survey in Himachal Pradesh and Jammu and Kashmir has shown that the disease is wide spread and occurs in endemic form causing severe morbidity and high mortality. In a survey in the Garwal hills and in the neighbouring areas, more than 10,000 sheep and 4,400 goats were screened. The rate of infection was around 46% in young lambs and sheep. The heaviest infection has been recorded in animals between 6-12 months of age (Dhar *et al.,* 1972). Westerheim (1937) reported from Norway that in its humid atmospher lungworm disease occurs frequently in goats and almost every old animal either harbours parasites or shown injuries caused by them. Genus *Dictyocalus* is of most of economic significance (Poynterand Selway, 1966).

Treatment

i. Cyanacethydrazide :- @ 15 mg/kg body weight can be given by sub-cutaneous route. It can be given as drench @ 17.5 mg/kg, body weight.

ii. Diethylcabamazine :- can be given through intra muscular injections. The dose rate is 40 mg/kg body weight.

iii. Tetramisole :- @ 15 mg/body weight can be given as drench.

Control

Where clinical disease occurs, animals should be moved from the infected ground to clean pasture. Avoidance of grazing in wet or marshy fields is helpful. Nutrition of the animals appears to play an important role in the incidence and severity of infection. Animals should be well fed. Where lungworm infection in sheep is severe, strategic preventive medicine with cyanacethydrazide or diethylacarbamazine should be attempted.

Tape worms

Several types of tape worms are found in the intestine of sheep. Some of them like broad tape worms (*Moniezia* Sp.) which are lodged in small intestine and can cause blockage. Larval forms of tape worms, called bladder worms, may cause severe damage in sheep. The larval form of *Taenia multiceps* are lodged in the brain and spinal cord of sheep and causes gid. Sheep moves blindly in circles. This disease can be prevented by regular dosing of dogs against causative tape worms. Adult tape worm may reach a length of 2-4 metres. Their segments are shed with dung balls. Crivid mites found in pastures become infected by eating these eggs. Sheep picks up infection by eating these mites. They take 30-40 days to develop in sheep. They are found in large numbers in hoggets. Increase in numbers of tape worms affects the growth rate of lambs (Soulsby, 1965).

Table 11.3 : Common tape worms found in sheep.

Name	Common name	Place where found
Avitalina stripetata	Tape worm	Small intestine
Echinococus cyst	Hydrated worm	Liver and lungs
Helidometra giardi	Tape worm	Small intestine
Monezia expansa	Tape worm	Small intestine
Cysticercus ovis	Sheep muscles	Heart and muscles

Miscellaneous Diseases

Choking

It occurs due to the eating of mango seed or clothes. Sheep moves with head side ways, stops eating. If it drinks water, it comes out through nostrils. As soon as choking is observed sheep should be given water to drink, so that the choking material goes into rumen. If it does not succeed, operation of oesophagus is the only solution.

Constipation

It can occur due to over eating or diseases of intestine. The dung comes out in small quantity at a time. It may be dry covered with mucous. Sheep becomes lazy and stops grazing. Giving anaema with luke warm soap water relieves the symptoms.

Diarrhoea

This may occur due to indigestion or giving strong medicines. It may also occur due to some protozoon diseases. The hind quarters may be soiled with loose faeces and rumen may appear to be empty. Such sheep should be given barley water. Sheep can be given opium, Katechu or Kaoline but sheep should be given 750 ml linseed oil before giving medicine.

Injuries

Injuries can occur at any part of the body due to sharp objects. Efforts should be made to check haemorrhage. The wound should be cleaned with weak potassium permanganate solution or some dilute antiseptic solution such as detol or savelon can be used for this purpose. Then some antiseptic ointment can be applied.

Dystocia

It may be due to foetal or maternal causes. It may result due to abnormal presentation of lambs. Sometimes mating small sized ewes with bigger rams also results in dystocia, it may be due to oversized foetus. Maternal dystocia is more frequent in primiparous animals. A veterinarian may be consulted.

Mastitis

It occurs due to retention of milk, injury and inflammation of udder and teat. *Pasteurella haemolytica* and *Staphylococcus aureus* were recorded in 33.3 and 53.3 percent cases of mastites milk (Das Kornel, 1992). The udder should be emptied and hot water fomentation should be done. If there is injury it should be cleaned with weak antiseptic solution and some ointment be applied. If it is due to bacterial infection, then intramammary antibiotic can be given.

Retention of placenta :

Failure of placenta to be expelled during third stage of parturition is a common problem. It is associated with asynchrony of endocrine mechanism, dystocia, uterine inertia, twinning and induction of parturition with dexamethasone. It may occur in abortions also. There may be pus formation in the uterus. Yellow foul smelling discharge may come out of uterus. Normally placenta comes out itself in 6-12 hours. If is takes moretime, some antibiotic may have to be kept in the uterus.

Tympany

It occurs due to eating of rotten or fungus infected fodder or concentrate mixture. It may also occur due to eating of a lot of Berseem in the morning. It can be treated by making the sheep to run, by giving asphotida or terpentine oil. Weak solution of formaline can also be given. If there is no relief puncture can be made with trocar and cannula in left flank.

Intussuception

It is rather of a common occurence in sheep due to several reasons i.e. nodular worms, change in feed, local intestinal lesions. The animal looks dull, off feed, kicking at the belly with no rise in temperature and frequent straining with no defecation, an excitment with symptom of colic at later stage and the animal becomes recumbent, and dies. No rational treatment except surgery. Mild purgatives with astringants may be of little avail.

Table 11.4 : Common anthelminitics

Name of Anthelmintic	Dose	Method of giving	Remarks
Nicotine + Copper sulpate solution	30-40 ml/adult sheep	Drench	If poisoning symptoms appear give tea decoction
Dicestol		As drench or electuary tablet can be swallowed.	For tape worms
Carbon tetra chloride	1-3 ml/adult sheep	As drench mixed with whey or liquid parafin	Effective against liver flukes.
Hexachlorethane	0.2 g/kg body weight	Given through stomach tube.	Effective against flukes and stomach worms.
Hexachlorethane	0.015-0.02 g/kg body weight	Given as drench	Effective against round and tape worms.
Nilverm		Given as drench	It is a broad spectrum anthelmintic
Piperazine adepate	4-8 g/adult sheep depending upon weight	Mixed with feed, can be given as a drench	Most effective against ascaris and nodular worms
Phenothiazine	5-15 g	Given as drench	Effective against strongyles.
Thiabendol	0.05-0.1 g/kg body weight	Given as drench	Very effective against strongyles and other stomach worms.

Drenching pneumonia

It is one of the most common and important pathological conditions in sheep. It is characterised clinically by increased respiration, coughing and abdominal breathing. Occasionally one can come across "Aspiration" or "drenching" pneumonia during mass drenching operation. Avoid forced drenching and allow the animal to deglutinate the drench itself. If some fluid has entered in respiratory tract lower the animal's head immediately and slap few times on its head.

Acute Impaction of the Rumen

The ingestion of large quantities of highly fermentable, carbohydrate feeds cause an acute illness due to the excess production of lactic acid in the rumen. Clinically the disease is manifested by dehyration, blindness, recumbency, complete ruminal stasis and a high mortality rate. Antihistamine such as Avil, normal saline solution 500 ml. I/V and soda bicarbonate 4 g orally will help.

Poisoning caused by

The chlorinated Hydrocarbon insecticides : This group includes Dichlore dipheny trichloroethane (DDT), benzene hexachloride (BHC), its pure gamma isomer Lindane, aldrin, dieldrin, chlormane, toxaphane, methoychlor, isodrin and heptachlor etc. They are used extensively as insecticides to control external parasites on sheep or to control insect infestation ingestion, aspiration and percutaneous absorption are possible portals of entry into the body. Most of the insecticides accumulate in the fat depots, may be excreted in the milk in dangerous amounts. The toxic effects produced include increased excitability and irritability, followed by muscle tremors, weakness and paralysis and terminal convulsions in severe cases, salivation and teeth grinding is noticed. Complete anorexia occurs.

Organophosphorus insecticides : The group includes some of the most dangerous chemicals which are highly toxic, but are used and recommended for the control of insects attacking plants, available for control of internal parasites of sheep. Some of the compounds are Carbophenthion, Demton, Daznon, Dimethylaparation. As anthelmintics Rueleam (Asuntol), Coral (Bayer) and Neguvon (Bayer), have been used. The clinical signs of poisoning are profuse salivation, muscle stiffness, dyspnoea, with open mouth breathing, diarrohea and muscular tremors. Atropin sulphate 2.0 mg/ body weight; 1 /3 to be given I/V in 10 ml distilled water and rest S/C.

Snake bite

Venomnous snakes fall into 2 classes : the alpine snakes which include cobra, and the vilerino snakes which comprise russels viper. Alpine snakes have short fangs and tend to chew their victims. Their venom is neurotoxic and kills the victim by paralysing the respiratory centres. Viperine snakes attack once and then withdraw. Their venoms mainly haemolytic causing pronounced local damage. If antivenom are not available and the bite is located in an area where tourniquets cannot be applied, excision of an area of skin, 7 cm in diameter is done. No treatment is possible due to sudden bite in grazing areas.

Electrocution

One of the natural calamities where death ensues in few seconds. Short circuit in electric line, leakage of current in wire fencing are its causes. It should be very well differentiated from lightening stroke because there was no evidence of lightening fall and moreover there were no signs of scorching and singing of skin and hairs oh the body of the animals.

12

Nutrition

Sheep can eat many types of fodders, grains and cakes. It can also eat some types of shrubs and leaves of a variety of tree species. Samples of a variety of fodders eaten by sheep have been collected by Maheshwari and Talpatra (1967). They took samples from oesophagus and ruminal fistulae. A sheep eats about 800-1200 g of fodder on dry matter basis daily.

Sheep is able to meet throughout the year, all its requirements of nutrients from grasslands, except during winter season. Only phosphorus requirement cannot be met from the pastures (Joshi and Ludri, 1970). The digestible crude protein (DCP) and total digestible nutrients (TDN) of some fodders are given in Table 12.4.

Protein and energy are essential for weight gain and better production. The ratio between protein and energy contents are essential for their better utilization (Shukla and Pal, 1972). Researches have been conducted on protein requirements of rams and ewes by several workers (Singh and Mahadevan, 1968, 1969, 1970). Body weight of sheep increase if, they are given 250-450 g of concentrate mixture (Khot *et al.*, 1962). Fifty percent increase in body weights have been observed in Malpura sheep if they are given 400g of concentrate mixture in addition to grazing (Bhatia *et al.*, 1971; Patnayak *et al.*, 1971). Semen quality deteriorated when sheep were fed cowpea upto one and a half year (Honmode *et al.*, 1971d).

The major advantage of sheep in agricultural production system is their special ability to utilize, mostly natural grasses, crop stubbles, tree loppings and to some extent bushes to produce saleable products e.g. meat, wool, pelts, skins etc. Sheep are selective grazers preferring short and low set grasses, legumes and a wide variety of low forbes and thus do not compete with goats and cattle which mostly utilize shrubs and bushes and tall grasses, respectively. They can utilize very sparce and low set vegetation because of extremely close grazing nature. This is possible because of clefted upper lip, thus waste land or marginal and sub marginal land can be profitably utilized for sheep production. However, large areas which should receive attention for improvement in this manner are now being put under crop production. The density of livestock per unit grazing area has greatly increased due to an increase in the livestock population and shrinkage of grazing lands. Practically a little or no supplemental feeding is provided to sheep. The principal role of sheep is therefore as a grazing animal and the aim of efficient sheep production should be to maximise output of pasture and natural range lands with some supplementation with conserved fodders and agroindustrial byproducts, except in very highly specialized production systems e.g. fat lamb production, where certain percentage of feed would have to contain cereals, oil cakes, brans etc.

Nutrient Requirement of Sheep

Sheep are reared for two purposes-(i) For wool purpose and (ii) For mutton purpose. A good pasture may fulfil the maintenance requirement of sheep. The need for extra nutrient arises when pasture is poor. The word nutrient means the component of feed, fodder or feeding stuff which supports the animal's body. To maintain health and productivity of sheep "Balanced ration" is essential. In the balanced ration, all the essential nutrients are found in adequate quantity and proportion. The nutrients are protein, carbohydrate, fats, minerals, vitamins and water Table 12.6. Deficiency of any of these nutrients will result in deficiency disease. Protein, fat and carbohydrates are known as macronutrients while minerals and vitamins are known as micronutrients, because they are required by sheep in very small quantities. The energy i.e. Total Digestible Nutrient (TDN) requirement is 1-3 kg/100 kg body weight. The requirement of protein and energy increase during stress i.e. lactation, breeding etc.

Nutritional Requirements of Sheep

The nutritional requirements of sheep have to be met through the available feed resources by adopting scientific feeding practices that will largely overcome the common constraints Table 12.1. Work on nutrient requirements of sheep in this country is far from comprehensive and, therefore, separate feeding standards for sheep have not yet been developed.

Table 12.1 : Nutrient requirements of a sheep weighing 35 kg.

S. No.	Nutrient	Daily Requirement
1.	Total dry matter	900 g
2.	TDN	450 g
3.	DCP	45 g
4.	Calcium	2.4 g
5.	Manganese	1.0 g
6.	Phosphorus	1.2 g
7.	Common Salt	5.0 g
8.	Potassium	4.2 g
9.	Sulphur	1.5 g
10.	Copper	8.0 mg
11.	Cobalt	0.2 mg
12.	Iron	60.0 mg
13.	Iodine	30.0 mg
14.	Molybdenum	0.1 mg
15.	Carotene	1.5 mg
16.	VIT. A	750.0 i.u.
17.	VIT.D	200 i.u.

On dry matter basis

NRC standards for lambs, start from a body weight of 27 kg, whereas the adult body weight of most sheep breeds of our country is about 30 kg or less Table 12.7 and 12.8. Ranjhan (1980) has tabulated the nutrient requirements for lower body weights which are essentially based on extrapolation of ARC and NRC recommendations. There is an urgent need to work out the requirements of various nutrients under our conditions and to lay down uniform feeding standards.

Studies have shown that about 2 litres of water is required per sheep/day during winter and 3.5 to 4.0 litres during summer.

This requirement would further increase if the animals have to walk long distances. In general, an adult sheep should consume feed from 2.5 to 3.0% of its body weight for its maintenance, whereas a growing lamb should consume 4 to 5% of its body weight to show adequate growth. Energy is the most important nutrient on which the requirements for other nutrients depends. Joshi and Ludri (1970) found 1.3 kg/100kg body weight as the minimum amount of TDN required. Gill and Negi (1971) recommended 809 g of DM and 401 g of TDN per day for maintenance of rams averaging 30 kg liveweight. Bhatia *et al.,* (1978) found that the grazing rams consumed 343 to 766 g TDN in different months without loss of body weight. Singh *et al.,* (1979) recommended 30 g TDN and 5 g of DCP per kg metabolic weight for maintenance of the ewes and optimum wool production. Patnayak *et al.,* (1979) observed 397 g TDN and 34g DCP could maintain rams weighing about 32 kg. All these results indicate that for adult ewes approximately 10 g TDN are required per kg live weight for maintenance and wool production. It appears that weaner lambs can consume 400 to 650 g TDN/day and gain on an average 50 to 150g/day.For obtaining optimum growth in weaner lambs of about 150g/day, at least three times the maintenance requirement, about 30g TDN per kg live weight of lamb, is needed.

The optimum ratio between DCP and TDN energy consumption in adult sheep is 1:10; whereas for grazing lambs it is 1:8.5 to 9.0 (Maheshwari, 1977). Gill and Negi (1971) found that 23 g of DCP was required for a ram weighing 30 kg. But Singh and Mahadevan (1970) reported that the minimum DCP requirement for maintenance was 12.1g/day for a ram weighing 35kg. Recently, Pachauri and Negi (1979) working on Gaddi ewes reported that 23 g of DCP was required by a 20 kg ewe to maintain nitrogen balance. Grazing studies conducted at Central Sheep and Wool Research Institute, Arid Region Campus, Bikaner have shown very low level of DCP intake (7 to 9 g/day) by Marwari ewes for maintaining their body weight. Mathur (1973) found that upto 10 months of age 13 percent of crude protein in the diet was adequate and from 10 to 16 months of age, 10 percent crude protein in the diet was found adequate. The approximate requirement of DCP for maintanance appears to be 1 g or a little less per kilogram live weight. Mahadevan (1961) reported that 6.7 g of calcium and 6.0 g of phosphorus and 25 mg of carotenes are required per day for a sheep weighing 30 kg.

Nutritional requirements of sheep in relation to certain nutritional diseases have also been reported and it is indicated that vitamins and minerals particularly trace elements should be properly supplemented.

Feed Resources

Sheep will have to depend mainly on stubble grazing, crop residues, tree leaves and agro-industrial by products, because it may not be possible to spare any cultivable land for sheep grazing in times to come. The natural rangelands are dominated by shrubs and grasses with a few trees. Among the bushes, *Ziziphus* species is very common. These lands can be improved by protecting them from biotic factors and reseeding with perennial and high yielding grasses and legumes. Technologies have been evolved for establishing perennial grasses and legumes etc, introduction of fodder legumes as inter-crops in rain fed cereal crops and raising the fodder resources through silvipastoral practices. The availability of dry matter from tree leaves in the country has not been assessed but it is roughly estimated that 60% of the feed available to sheep and goats come from top feeds alone. The most common fodder trees available in the grazing lands in arid areas are Khejri *(Prosopis cineraria)*. Kankera *(Gymnospora spinosa),* Babool *(Acacia spp.)* and Neem *(Azadirachta indica)* etc.

Feeding management of top feed resources has to be geared not only to the increased nutrient availability and preservation of sources but also to the applied feasibility and economics. It will thus be necessary to workout the relative efficiencies of the various practices of harvesting and utilisation, in terms of their actual nutritive values coupled with cost benefit ratios.

Water

Water is more vital for the maintenance of the animal's life than any other feed component. It is the main constituent of all body tissues and the excretion of waste products through faeces and urine. The body water plays an important role in the animal's regulatory mechanism. Most of the water requirement is usually satisfied by the water in the feed during the season when green feed is available and not much water is consumed by sheep. Water intake consideration is highly correlated with both air temperature and absolute humidity. Water requirement increases with growth, gestation and lactation and when sheep are subjected to heat stress or when salt content of the diet is more or they are made to travel long distances. On dry feed, an adult sheep will require about 2 litre water per day during winter and 3.5 to 4 litre during summer. Normally, sheep will drink approximately 2 to 3 litres of water for every kg of dry feed consumed. Sheep in desert areas can stand water deprivation upto 3 days. However, watering, free choice every alternate day has been found to have no deleterious effect. Sheep can also tolerate salt content upto 1 percent in the drinking water.

Energy

The soluble carbohydrates, fibers, fats and oils available in the feed have a common function of supplying energy. The energy feeds are to the animal what diesel is to a tractor, a source of energy and are used to supply heat to keep the body warm, to provide energy for physical activities and other life processes and to help growth, reproduction and milk production. The carbohydrates are also required for the growth and development of rumen microorganisms. Lack of energy is the most common nutrient deficiency in sheep and may result from lack of feed consumption or because of low feed quality, In general, adult ewes consume dry matter 2.5 to 3 percent of their body weight both under stall feeding and grazing conditions. Lambs achieve satisfactory growth rate at a dry matter intake level of 4 to 5 percent of body weight. Different systems are in vogue all over the world to express the energy requirements. In India the total digestible nutrients based system is in practice. In recent times metabolizable energy (ME) based system is also being followed. The TDN requirements of lambs are highter than adult sheep. Similarly the pregnant, nonlactating ewe requires 0/1 kg of TDN per kg live weight for maintenance and wool production. This requirement would be about 50 percent more during first 10 weeks of lactation. The deficeincy of energy in the diet of sheep may result in reproductive failure, poor growth and loss of body weight. The progressive decrease in body weight may ultimately lead to death. The dry matter should contain about 50-55 percent TDN for adults and 60-65 percent for growing lambs.

Protein

Protein is the basic structural material of all the body tissues. This includes not only the muscles, nerves, skin, connective tissues and vital organs but also the blood cells, animal's hair/wool, hooves and horns. Since all the living tissues are in a dynamic state and are undergoing constant wear and tear, protein is essentially required for their regeneration. Thus, proteins are essential for growth and development of the body Table 12.3. The breeding animals need protein for prenatal growth and development of the foetus and to produce milk for postnatal growth and development of young ones. Clean scoured wool is almost pure protein in the form of keratin. If the ration does not contain enough energy, the protein that might be otherwise used for maintenance, growth and production, will be used as energy, source. But protein can not be , replaced by any other nutrient in the ration. Protein deficiency results, in reduced appetite and feed intake and poor feed efficien-cy. This in turn results in poor growth and development of muscles as well as reduced reproductive efficiency and wool production directly as well as through lowered secretion of hormones. The animals on low protein feed have lower disease resistance. The lambs require more protein than mature ewes Table 12.2. It is higher during breeding, later part of pregnancy and lactation than during other periods.

Table 12.2 : Energy and protein requirements at different body weights of lambs

	Body weight (kg)		
	11	15	25
Age group (month)	3.6	6.9	9.12
Daily gain (g)	41	79	97
DM consumed (g)	495	656	990
TDN consumed (g)	346	458	675
DCP consumed (g)	44	59	74
DM/gain	12.1	8.3	10.2
TDN/gain	8.4	5.8	6.96
TDN/DCP	7.9	7.8	9.1
DE consumed (Kcal)	1522	2015	2970
ME consumed (Kcal)	1218	1612	2376
GDCP/TIME	36.12	36.60	31.14

Kg DM/100 kg	4.5	4.4	3.96
Kg TDN/100 kg	3.1	3.05	2.7
g DCP/100 kg	400	393.3	296

Source: Maheshwari, 1977

The present trend in other countries is to express the crude protein requirements for ruminants in terms of digestible crude protein only. Under Indian conditions, it may not, however be applicable because of many unconventional feeds being used which have very low digestibility of protein, although the crude protein content is reasonably high, tree leaves are such an example. The approximate daily requirement of DCP for maintenance is 1/10th of the TDN or 1 g for every kg live weight. This requirement will increase by about 50 percent during pregnancy and by 100 percent during lactation and growth. A 30 kg sheep will thus require 30 g DCP during dry non-pregnant stage, 45 g during last 6 weeks of gestation and 60 g during first 10 weeks of lactation. Protein quantity is more important than quality in the sheep ration because the microbial action in the rumen will convert low quality amino acids although the availability of some sulphur containing amino acids e.g. methionine and cystine could become limiting factor in wool sheep.

Minerals

When a feed stuff is burnt, a white ash is left. This contains minerals. The minerals are used for a number of different purposes. Their greatest use is in the building and maintenance of the skeleton and teeth. They also provide acidic and alkaline reactions to digestive juices and play an important role in digestion. Minerals are also helpful for the maintenance of osmotic pressure in different body fluids. They are also used in wool growth, as the wool proteins contain sulphur. Thus availability of a number of minerals in the diet of sheep is essential Table 12.4. If one or more of these minerals are deficient in the diet, the animals will ultimately exhibit clinical symptoms of deficiency. The role of the minerals in sheep nutrition is complicated because excess of some of them may result in poor feed intake and digestion, hamper utilization of other minerals and can even cause toxicity.

Many deficiencies result in anorexia (reduced appetite), reduced gain or loss in body weight unthriftiness, abnormal hair/

Table 12.3 : Digestible crude protein (DCP) and Total digestible nutrient (TDN) requirement (g/day) for maintenance and growth of lambs

Body wt.	Maintenance		Weight gain							
			50 g/day		100 g/day		150 g/day		200 g/day	
(Kg)	DCP	TDN	DCP	TDN	DCP	TDN	DCP	TDN	DCP	TDN
5	10	120	15	150	20	210	25	250	30	300
10	16	168	20	211	25	254	30	296	35	339
15	21	222	28	279	34	335	40	391	47	447
20	26	271	34	340	42	408	50	477	58	546
25	31	330	41	414	50	497	59	581	69	664
30	36	374	47	469	57	564	68	658	79	753
35	40	425	52	476	64	640	76	747	89	855
40	44	474	58	594	71	714	84	834	98	954

wool coat, dullness of skin, bone deformation and staggered gait. Calcium and phosphorus are necessary for bone formation and its maintenance. Deficiencies of these minerals are indicated by rickets in young ones and osteoporesis in adults. Deficiency of copper and cobalt may result in tetany and in doggy wool. Selenium deficiency may lead to 'White muscle' disease in lambs. Sulphur is present in wool and hair. Inadequate supply of iron, copper and cobalt results in enaemia and deficiency of iodine in goitre Table. 12.5.

Until 1959, 16 mineral elements were considered as essential which comprised macro-elements: calcium, phosphorus, potassium, sodium, chlorine, sulphur, magnesium and the trace elements: copper, cobalt, chromium, iron, iodine, manganese, molybdenum, selenium and zinc. Some evidence is available from feeding trials that bromine, barium and tin can give better growth responses, but conclusive proof of the essential role of these elements is still lacking.

All the essential elements are believed to have one or more catalytic functions in the cell. Sodium, potassium and chlorine have electrochemical function and are concerned with the maintenance of acid base balance and the osmotic control of water distribution within the body. Calcium and phosphorus are essential parts of skeleton. Sulphur is necessary for the synthesis of structural proteins. Magnesium functions catalytically, electrochemically and structurally. Iron is important as a constituent of haemoglobin, cobalt a component of vitamin B12 and iodine forms a part of thyroxin hormone.

Table 12.4 : Approximate concentration of essential macro and micro-elements in the animal body.

Macro element	Percentage	Micro element	mg/kg
Magnesium	0.04	Chromium	0.08
Chlorine	0.11	Cobalt	0.02-0.10
Sulphur	0.15	Manganese	0.2-0.5
Sodium	0.16	Iodine	O.3-0.6
Potassium	0.20	Selenium	1.7
Phosphorus	1.00	Molybdenum	1.4
Calcium	1.50	Copper	1.5
		Zinc	10.15
		Iron	20.80

Major Elements

Calcium

About 99% of the total calcium found in body is present in skeleton and teeth. It is also an essential constituent of living cells and tissue fluids. It is essential for the action of some enzymes and for contractile properties of muscles. It plays role in the coagulation of blood. Its concentration in plasma varies between 8 and 11 mg per 100 ml.

Calcium is found in leguminous crops and in green leaf crops, but cereals and roots are poor sources. Fish meal, bone meal and meat meal are excellent sources of calcium.

Phosphorus

It is present in bones, phosphoproteins, nucleic acids and phospholipids. Eighty percent of the body phosphorus is found in bones. The serum phosphorus concentration varies from 4 to 10 mg per 100 ml. Fish meal, cereal grains and meat are its good sources. Phosphorus is present in the form of phytates in cereal grains which is hydrolysed by bacterial phytases in the rumen of sheep.

Diseases of bones are common in sheep. The bones may become soft and weak. Rickets is a common disease of lambs. These can be treated by giving Vit. D. injections and by feeding wheat bran (Chakrabarti *et al.*, 1975; Clark *et al.*, 1975, Fitch, 1954).

Kidney stones are formed of calcium oxalate, carbonate or phosphate, silica or magnesium, ammonium phosphate. These can be present in kidney or urinary bladder and cause obstruction in urinary passage. This disease is prevalent in areas where there is very hot or cold climate and lot of sweating. This disease occurs due to imbalance of calcium, phosphorus or magnesium contents in ration or due to deficiency of sodium and potassium. There is frequent micturition, pain in kidney area. If diagnosed at an early stage it can be removed surgically.(Field, 1969; Me Intosh, 1978).

Sodium

Major part of body sodium is *present* in body fluids and soft tissues. It is concerned with acid base balance and maintenance of osmotic pressure of body fluids. Much of it is present in extra cellular fluid.

Most of the straws are deficient in sodium. Meat meal and fish meal are richer sources. Its higher concentration in water (1.3 percent) may result in dysentry (Smith and Gawthorn, 1975).

Potassium

It plays importan trole in the regulation of osmotic pressure. It is mostly found inside the cells. It plays an important role in nerve and muscle excitability and plays role in carbohydrate metabolism. If potassium is absorbed in excess in body, it is excreted in urine. Excess absorption of potassium interferes with the metabolism of magnesium, which may result in hypomagnesaemia tetany.

Chlorine

Like sodium and potassium, it also plays role in acid base balance and osmosis. It is present in gastric secretions as hydrochloric acid as well as chloride salts. It is excreted in urine as well as in sweat as sodium chloride. It is present in fish and meat meal. The chlorine content of pasture may vary from 30ppm to 340ppm.

Sulphur

It is present in cystine, cysteine and methionine. It is part of biotine and thiamine vitamins, and insuline hormone. Sheep wool contains sulphur, hence the sulphur requirement is very high. Animals get sufficient amount of sulphur if it is getting enough protein in diet. Its deficiency may result, if protein source is replaced by urea as a partial nitrogen replacement, sulphur may be limiting factor for the synthesis of cysteine, cystine and methionine. The sulphur requirement increases if the diet of sheep is substituted with proteinous nitrogen.

Magnesium

About 70% of total body magnesium is found in the skeleton. The remainder is distributed in fluids and soft tissues. It is catalyst for phosphate transfereases, decarboxylases and acyltransferases. It is closely associated with phosphorus and calcium.

It is present in wheat bran, cotton seed cake and linseed cake etc. The magnesium content of forages varies widely. The daily prophylactic dose for hypomagnesia in sheep is 7g.

Trace Elements

The essential trace elements are copper, cobalt, iron, zinc, iodine, sulphur and manganese. In addition are few which are considered as probably essential elements like, chromium and nickel. Besides these, there are few toxic elements which are also found in water and soil such as molybdenum, arsenic, flourine, lead etc. The minerals that need consideration in feeding sheep are common salt, calcium, phosphorus, copper etc.

Table 12.5 : Requirements for different nutrients for maintenance of an adult sheep weighing 35 kg.

S. No.	Nutrient	Required/day
1.	Total feed	900 g
2.	T.D.N.	450 g
3.	D.C.P.	45 g
4.	Calcium	2.4 g
5.	Phosphorus	2.0 g
6.	Magnesium	1.2 g
7.	Common salt	5.0 g
8.	Potassium	4.5 g
9.	Sulpher	1.5 g
10.	Copper	8.0 mg
11.	Cobalt	0.2 mg
12.	Zinc	75 mg
13.	Iron	60 mg
14.	Manganese	30 mg
15.	Iodine	0.1 mg
16.	Molybdenum	1.5 mg
17.	Selenium	0.1 mg
18.	Carotene	1.5 mg
19.	Vitamin A	750 IU
20.	Vitamin D	200 IU

* Air dry basis

Copper

It is an essential component of the mature red blood cells. Copper is present in several enzymes whose role ranges from utilization of iron to the pigmentation of skin. Some of the enzymes are cerailoplasmin, cytochrome oxidase, lysine oxidase, tyrosinara plastocyranin and haemocynin. This element is essential for normal augmentation of hair, fur and wool. In sheep, disease called swayback is related to its deficiency. The symptoms are that the lambs are born to apparently healthy ewes with some degree of ataxia. Some lambs are unable to stand, some will rise and fall backwards while some lambs show slight weakness of hindquarters. Another deficiency symptom of this element is "steely wool" or "stringy wool". In this condition wool loses its crimps because of copper deficiency. Deficient ewes come in heat late, there may be paralysis of hind quarters. Deficiency occurs, if it is less than 5ppm in soil. The optimum requirement of copper for sheep is 5mg/day. The salt licks containing 0.25 to 0.5% copper sulphate have been found to be effective in grazing sheep.

The copper salts in high doses are toxic as copper can act as cummulative poison and sheep are particularly very succeptible to copper sulphate poisoning.

Chronic copper poisoning in sheep has been reported under natural conditions in some countries. If due to accident, sheep eats copper salt, it may result in toxicity. The liver concentration in such cases may go upto 0.8 percent (Marston, 1952; Allcroft, 1963; Smith and Gawthorn, 1975).

Liver is the store house of body for copper element. Its concentration in sheep liver is 0.01 to 0.04 percent and in blood is 0.007 and 0.012 percent. Copper is widely distributed in foods. Seeds and seed byproducts are usually rich in copper, straws contain little copper. Milk is also a poor source of this element.

Cobalt

A deficiency of cobalt in the soil is reflected in a deficiency in herbage and sheep grazing such pasture develop deficiency 'pine'. The cobalt is a part of vitamin B and is present in enzymes involved in the synthesis of RNA and the metabolism of amino acids. Deficiency of cobalt leads to insufficient production of B_{12}. Therefore, cobalt deficiency symptoms are actually B_{12} deficiency symptoms. Cobalt sulphate and cobalt chlorides are good sources of cobalt. Drenching with an aqueous solution of cobalt chloride, at brief intervals has proved reliable as prophylatic measures, an intake of 8mg/week/sheep will be sufficient for prevention of deficiency symptoms.

The depletion of this element is a gradual process but when it is depleted, there is gradual decrease in appetite with consequent loss of weight, followed by muscular wasting, pica, anaemia and death. This element is activating ion in certain enzyme reactions.

Adult sheep needs about 0.1 mg cobalt daily. This requirement is higher in growing lambs (Lee, 1956, 1963, 1976; Marston, 1952; Smith and Gawlhorn, 1975). Most of the foods contain traces of cobalt. Normal pastures have a cobalt content within a range of 0.1 - 0.25 mg/kg on dry matter basis.

This element is poorly retained by the body tissues and excess is eliminated soon. The sheep are less susceptible to cobalt toxicity and have been shown to tolerate levels upto 3.5 mg/kg body weight.

Iodine

A deficiency of Iodine has been recorded in ewes leading to the birth of lambs with goiter. Indirectly goiter may be caused due to eating goitrogenic substances containing feeds. Iodine deficiency in sheep during pregnancy can also lead to foetal hypothyroidism, therefore, it is necessary to feed iodised salt in certain areas to avoid iodine deficiency. Since this element is stored in thyroid glands, ontime dosing would be sufficient even at monthly intervals. A sheep needs 0.2 mg iodine daily(Mayo and Mulheam, 1969; De-Groot and Stanbury, 1975; Willson, 1975).

The concentration of iodine present in the animals body is very small and in the adult usually less than 0.6 mg/kg. It is a constituent of thyroxin produced by thyroid gland. It is also a constituent of di iodotyrosine which is intermediate product in the formation of thyroxine from the amino acid, tyrosine.

It is found in traces in most of the feeds. Some sea weeds may contain upto 0.6 percent of iodine. Fish meal is also a rich source of iodine. In goitre prone areas iodised salt can be given.

Iron

It is an integral part of haemoglobin, hence its deficiency leads to subnormal haemoglobin levels. More than 90 percent of the body iron is combined with protein. It is an integrated part of many enzymes. Thirty five percent of the iron is present in haemosiderin. It also occurs in the form of transferrin in serum, which is responsible for iron transfer from one part to another.

About 20 percent of body iron is present in liver, spleen and bone marrow.

It is present in various green leafy materials and seed coats. Some materials are good sources of iron, but the body can not use these e.g. blood meal. The absorption is largely controlled by body requirements. Excessive amounts of iron are prevented from entering the body by regulating mechanism mediated by the mucosal cells of intestinal tract. Iron deficiency is caused more due to the feeding of certain feeds which are high in phytic acid.

Molybdenum

It is classified as an essential trace element. It is a component of Xanthine oxidase which plays part in purine metabolism. Its deficiency symptoms have not been reported. A nutritious role of molybdenum has been demonstrated in young lambs. It promotes growth rate. The excess of this element in the body results in the deficency (Marston, 1952; Allcroft, 1963).

Chromium

It plays role in lipid and protein synthesis and serum cholesterol. Its role has been established in sheep.

Zinc

High concentration has been found in wool. It is present in yeast, bran and germs of cereal grains. It is found in every tissue of animal body . This element accumulates in the bones. High concentrations have been found in the skin, hair and wool. It is a component of several enzymes such as carbonic anhydrase (respiratory enzyme), pancreatic Carboxypeptidase and glutamic dehydrogenase. In addition to this, it acts as cofactor for many other enzymes. It is closely inter-related with calcium and vitamin 'A'. Parakartosis in lambs has also been found in zinc deficient diet i.e. less than 30 ppm dietary zinc level. In rams zinc deficiency causes impaired testicular growth. Zinc toxicity causes anorexia. Lambs need fodder, which should contain 7-33 ppm of zinc to prevent disease (Smith and Gawthorn, 1975).

Manganese

It is present in very small amounts in the body. Its highest concentration is found in liver, kidney, bones and pituitary gland and pancreas. It acts as enzyme activator. Forages deficient in manganese cause deficiency symptoms. Most of the pastures contain 20-40 mg/kg in dry matter. Rich sources are rice bran, wheat and offals.

Selenium

This element is interrelated with vitamin E. The high doses lead to toxicity. In grazing animals the distinct selenium responsive syndroms have been found, a muscular dystrophy, called white muscle disease, in newborn or young lambs also known as stiff lamb disease, unthriftiness with poor growth rates and infertility in ewes. Blind stragger is an acute form of selenium poisoning which occurs in sheep after consumption of certain weeds such as *Astragalies bisuecatus,* which contains upto 0.4% selenium. In seleniferous plants, selenium replaces sulphur in amino acids methionine and cystine. Hair, wool and hooves which are rich in sulphur containing amino acids are affected adversely in selenium toxicity. It is known that the toxic effect of selenium is reduced when high protein foods are given to animals. Affected sheep may be given 1-5 mg sodium selenite at a few days interval (Godwin, 1975).

Vitamins Requirements

Sheep require vitamin A, D and E. Since these vitamins are not likely to be lacking, deficiency symptoms are not often seen. How-ever, in famine situation when the pasture does not provide sufficient green carotenoid material for longer period, the symptom of vitamin A deficiency in the form of blindness is usually found. Pasture and ranges usually supply all the vitamin A needed, as sheep have a great capacity for storing it in the liver. As the sheep are nearly always kept out of doors, sunshine provide the vitamin D required, but it may have to be supplied to lambs being kept in sheds during winter. The ordinary rations fed to sheep contain ample vitamin E, but lack of this vitamin in the diet of pregnant ewes may result in lamb disease. To reduce the incidence of this condition, oral dose of 5 mg selenium or one oral dose of 300 mg of alpha tocopherol is recommended. In general sheep do not require B complex and C vitamins. Healthy sheep synthesize the B complex vitamins but the synthesis of cynocobalamin is very limited in sheep on cobalt deficient diet. It has been suggested that debilitated sheep with a high worm burden may benefit from the administration of B vitamins.

Vitamins are organic substances which are required by animal tissues in minute amounts. They are metabolically essential but sheep are able to synthesize certain vitamins in their tissues and they are able to utilize vitamins synthesised by micro-organisms in their gastro-intestinal tract. Vitamins are divided into two types viz. water soluble and fat soluble. The fat soluble vitamin are A, D, E and K.

Many of the symptoms of vitamin deficiencies are similar. These include anorexia, reduced growth, dermatitis, weakness and staggering gait. In sheep vitamin A is more important and its deficiency can cause various kinds of blindness. Vitamin deficiency also leads to abnormal bone development, weak and stiil born lambs and respiratory problems. Vitamin D is essential for the growth and development of bones and metabolism of calcium and phosphorus while vitamin E is concerned with reproduction.

Deficiency Diseases of Sheep

Deficiency diseases often denote conditions associated with faulty diets. Varying degrees of starvation results in multiple deficiencies and are manifested as generalized malnutrition. Specific, well defined deficiency diseases of vitamins are not common in sheep. Mineral disorders are more common in several parts of the world. There is not much documentation from India.

Chief deficiency diseases of livestock are of trace elements. Extensive established knowledge of these disorders has been recently reviewed (Underwood, 1977).

Protein

Protein deficiency is found in areas where green fodder is scanty and poor quality of roughages are fed. Its deficiency results in poor growth, depressed appetite, emaciation, delayed maturity and failure to come in oestrous.

This deficiency can be corrected by giving green leguminous fodder, gram, guar etc.

Vitamin A

Vitamin A deficiency occurs under drought conditions. Its deficiency results in karatinization of respiratory, elementary, reproductive, urinary and occular epithelium. Deficiency results in lowering resistance to diseases. The lambs born are weak, malformed and dead. Severe deficiency results in night blindness, convultions, incoordination in gait, persistent diarrhoea and anoestrous.

Lambs should be given colostrum at least for three days. They should be given cod liver oil and green fodder.

Vitamin B complex

The deficiency occurs in lambs deprived of green fodder, or good quality fodder. It is required by lambs upto 3 months of age. Thiamine, pyridoxine, pentothenic acid and riboflavin deficiencies do not occur in adult sheep but are essential for lambs with under developed rumen.

Deficiency symptoms can be relieved by feeding green leafy feeds and fishmeal in their ration. They can be given orally or through intra muscular route.

Vitamin D

Deficiency occurs in lambs deprived of colostrum. Its deficiency results in rickets in lambs. Lambs are born weak and with congenital malformations. The lambs can be given injections and sun cured hay.

Vitamin E

The deficiency of vitamin E results in white muscle disease and muscular distrophy. 500 mg alpha tocopherol acetate given orally cures the lamb.

Calcium

Calcium deficiency in young lambs produces a disease known as Rickets. The joints become enlarged and bones of legs become bent. There is lameness and stiffness. Deficiency may result in osteomalacia in adults, the calcium removed from bones is not replaced. They become weak and fracture easily. A number of factors are responsible for calcium deficiency. Hypocalcaemia may be caused by parathyroid glands being unable to mobilize calcium to meet the high requirements of sheep. Hypocalcaemia may be one of the causes for non dilation of cervix in parturient ewes (Popoff, 1978). Feeding of concentrate has been reported to induce hypocalcaemia.

Injections of calcium gluconate may restore calcium level. Annual application of lime to pasture land improves dental health (Suckling *et al.*, 1974).

Phosphorus

Like calcium, phosphorus is required for bone formation and its deficiency can cause osteomalacia or rickets. In chronic phosphorus deficiency the sheep may have stiff joints and muscular weakness. It may result in low fertility. Subnormal growth in young lambs and low body weights in adults are symptoms of phosphorus deficiency.

Intravenous infusion of phosphorus has been shown to be effective in improving the calcium and phosphorus balance of sheep (Towns *et al.*, 1978). The ration containing fishmeal,cereal grains may also relieve deficiency symptoms.

Sodium

Its deficiency retards growth and reduces the utilization of digested protein and energy.

Common Salt

A single experimental report from Rajasthan (Singh and Taneja, 1979) showed the adverse effects of high salt levels in water on lambing percentage, birth weight of lambs and parturition of ewes of Marwari breed. Surveys of the drinking water sources in the grazing areas of sheep and their effect on performance will be helpful.

Plants are low in sodium and chlorine. Deficiency results in-norexia and subsequent loss in weight. The addition of salt to the diet may provide an immediate cure.

Sulphur

Wool is rich in cystine and contains about 4 percent of sulphur. Deficiency may result if sheep is given non protenous nitrogen as a source of protein. Under these conditions addition of sulphur to urea containing rations may be beneficial.

Copper

Its deficiency may result in anaemia, stunted growth, scours, bone disorders, depigmentation of wool, lesions in the brain stem and spinal cord and gastrointestinal distrubances. Muscular incoordination occurs in young lambs. Low copper content (2 to 4 mg/kg)in the dry matter may cause copper deficiency. Some genetic factors influence the concentration of copper in the blood, liver and brain of sheep and that the incidence of sway back can be affected by genotype. Number of crimps in wool is affected by copper deficiency.

Lewis *et al.*, (1974) reported that the lesions of sway back developed after birth in lambs born to copper-deficient ewes. Parental administration of copper during gestation prevented the disease. Isolated myelin showed low copper content in deficient animals as compared to the controls (Patterson *et al.*, 1974). Liver copper levels were better indicators of copper status in ewes and lambs, in comparison to brain slices. A linear relationship was established between serum levels of copper and erythrocyte levels of superoxide dismutase in copper deficient rams (Andrewartha and Caple, 1980).

Several reports from different countries are available about copper toxicity in sheep on pasture and on copper toxicity produced experimentally (Bath, 1979; Dzhurov, 1979; Edelsten, 1980;Tontis *et al.*, 1980).

Zinc had been shown to reduce the plasma copper concentration, but it had no effect on hepatic copper concentrations (Saylor and Leach, 1980).

Cobalt

Deficiency symptoms are likely to occur where, levels of cobalt in dry matter of pastures are below o.1 mg/kg. The deficiency of cobalt is gradual and may result in muscular wasting, pica, severe anaemia and death.

Cobalt supplementation has proved to be effective in raising the blood level of vitamin B_{12} and improve the blood picture of sheep (Tressoe and Lamond, 1979; Gyang and Sarror, 1979).

Fluorine

Fluorine is distributed in traces throughout the body, but is concentrated in the bones and teeth. The importance of flourine as a trace element in preventing dental caries has been well established. Physical differences were noted in the inner and outer parts of the tibial cortex of sheep, drinking water containing 100 ppm of fluoride. Fluoride might act as a possible mutagen in the ova of animals taking excess of it in feed or water. High fluoride content of water (7 mg/1) may not depress growth but it cetainly causes fluorosis in animals (Mandrik *et al.,* 1974). A report from Nairobi (Said *et al.,* 1977) suggested that the accepted safe level of fluorine in the form of NaFl (70-100 ppm/l) was too high in the case of sheep. Tiwari *et al.,* (1978) had discribed the pathology of chronic and acute fluoride poisoning in sheep. High accumulation of the 'element in the mandible of sheep (1240ppm) did not cause any depression in growth (Milhand and Cazieuz, 1980). Bone and joint abnormalities also occur. Fluorine is a cummulative poison and ingestion of small amounts over long periods of time may produce toxic symptoms. Incisors and molar teeth were affected by high fluorine feed and water, plasma fluorine content was elevated, but the reproductive performance of adults and the growth of young ones were unaffected (Milhand *et al.,* 1980).

Iodine

Iodine deficiency in sheep manifests itself as a swelling of neck. Reproductive failure is one of the most outstanding consequences of reduced thyroid function; sheep gives birth to hairless, weak or dead lambs. Diseases caused by iodine deficiency in certain parts of the world, their pathogenesis, symptoms and treatments are well documented. A report from Australia (Andrewartha *et al.,* 1980) suggests that the thyroxine concentration in the serum of new born lambs provides a more sensitive indication of hypothyrodism associated with iodine deficiency than the levels in the ewes.

Iron

If the need for iron increases, as it would, after prolonged haemorrhage or during pregnancy, then haemoglobin synthesis may be affected and anaemia may result. Anaemia due to iron deficiency most commonly occurs in rapidly growing sucklings, since the iron content of milk is very low. The association of iron with heamoglobin is well established. Parentel administration of iron dextran certainly improves the blood picture of lambs but its effect on growth performance remains uncertain (Tait and Dubeski, 1979; Wanner and Bose, 1978).

Magnesium

Hypomagnesaemic tetany has been recognised since early thirties. It is also known as magnesium tetany, lactation tetany and grass tetany. Tetany is usually preceded by a fall in blood serum magnesium to about 0.5mg/100ml. The true availability of magnesium is enhanced by increasing the levels of sodium and potassium; the effect of the former is more than the latter (Powley *et al.,* 1977). Cold stress has been found to stimulate urinary excretion of magnesium and induce a drop in the levels of serum magnesium and phosphorus (Shiga *et al.,* 1979). Magnesium concentration of plasma has no relation with its absorption through the ruminal wall. The deficiency symptoms are, impaired growth, reproductive dysfunctioning, skeletal deformities and ataxia in new born lambs. Injections of magnesium sulphate or magnesium lactate, sub-cutaneously can generally be expected to cure the animal if given early.

Selenium

Like molybdenum, selenium has always been regarded as an important element in animal nutrition because of its toxicity. Several out-breaks of selenium deficiency have been reported with paralysis, dullness, stiffness of joints, lameness and hoof deformities, reproductive failure and white muscle disease in lambs and ewes; they were cured by selenium supplemention (Sahebi, 1974; Whanger *et al.,* 1977b). The relationship between selenium status of the animal and its glutathione peroxidase activity in different tissues has been further substantiated (Dotta *et al.,* 1978, 1979; Paynter, 1979; Whanger *et al.,* 1977a). Sulphur intake has not been found to affect the selenium status of sheep (White and Somers, 1977). Selenium pellets containing 95 percent iron, implanted in rumen can maintain the selenium status of sheep,but it is not a life long remedy. After three months of the treatment, there was a gradual decline in the availability of the element, (Paynter, 1979). Soft coal fly ash was shown to be a good source of selenium for sheep also (Hogue *et al.,* 1980). The corticosteriod concentration increased but the progesterone level was unaffected in selenium supplemented ewes during pregnancy (Walker *et al.,* 1979). Equivalent doses of sodium selenite and sodium selenate were equally effective. Parentel injections gave protection for a longer time as compared to oral dosing. The resistance of sheep erythrocytes to copper was not affected by the selenium status of the animal. The toxic effect of seleneum is reduced when high protein foods are given to animals. Trace amounts of arsenic compounds also have a protective action.

Zinc

A significant report of interest to Rajasthan is of Dwaraknath and Ghosal (1981) who recorded subnormal levels of zinc in the serum of Nali sheep. Excess of zinc in the sheep rations had been responsible to induce severe copper deficiency, abortion, still-birth, reduced weight gains and feed intake. Copper supplementation did prevent copper deficiency symptoms, but failed to prevent the adverse effect of excess zinc. Lactating ewes have been shown to be prone to zinc deficiency, causing weight loss and deterioration in health, but the role of zinc during pregnancy was not understood (Apgar and Travis,1979).

Nutrients for Wool Production

The wool production varies from breed to breed. It may be as low as 300 g in some Indian breeds and may be as high as 5 kg in some exotic breeds. If a sheep weighing 40 kg produces 3 kg of wool, 2.25 kg is actual wool fiber and 0.75 kg is, suint and dirt. Suint is produced by Budoriferous glands, it is a mixture of potassium soaps, inorganic salts and potassium salts of lower fatty acids. Wool consists of esters of cholesterol etc. Wool fiber is made up of protein and wool keratine.

Keratine contains sulphur containing amino acid, cystine which in turn is synthesised from essential amino acid methionine.

Keratine contains 10-12 percent of these amino acids. Maximum wool growth takes place at a level of nutrition only slightly above the maintenance level. At sub-maintenance levels, when the sheep is losing weight, its wool continues to grow. The main factors determining the rate of growth of wool appears to be the quantity of protein digested and absorbed in the small intenstine.

The highest rates of wool growth are achieved with foods, which promote high intakes of energy and protein. Wool quality is influenced by the nutrition of the sheep. High levels of nutrition increase fiber diameter and lower levels result in the production of fine wools. Finer wools are produced in areas which are less favourable nutritionally. Copper deficiency of fodder results in loss of crimps in wool.

Nutrients for Reproduction

Nutritional considerations for augmentation of reproductive performance (Table 12.7) focus attention on meeting the nutrient requirements of lambs for achieving early puberty for proper spermatogenesis, semen quality and libido of rams and also of pregnant and lactating ewes for prenatal and postnatal development of the lamb (Table 12.8). The effect of nutrition on the reproductive process are mediated through hormones specific for sexual development and for different phases of reproduction. High plane feeding may advance puberty considerably in small ruminants but over fatiness has an adverse effect on the reproductive performance of both males and females. Early nutrition of the lambs needs to be regulated depending upon whether the young one is to be reared for meat production or fibre production. Under nutrition or malnutrition of the male lamb delays sexual maturity and affect androgenic activity, spermatogenesis as well as reduced viability and motility of the spermatozoa. Additional supplementation of energy to ewes maintained on low plane of nutrition increases tupping rate, while dietary nutrients as protein, phosphorus and vitamin-A enhance ovulation and conception rates. Pregnant and growing ewes should be provided nutritional allowances both for growth and gestation. Meeting nutrient demand during last 6 weeks of gestation is of critical importance for the growth and development of the foetus as well as the lambs. Under nutrition may result in depression in the activity of the hypothlamic pituitary axis accompained by decrease in the release of various hormones responsible for repoductive performance (Table 12.8).

Table 12.6 : Output, nutrient digestibility and intake of nutrients in Marwari ewes grazing on natural rangeland in arid region.

	Season I (Jan-Mar.)	Season II (April-June)	Season III (July-Sept.)	Season IV (Oct.-Dec)
Average body weight of ewes (kg)	28.2	26.5	28.6	31.8
Faecal output (g/day)	508.15	468.97	617.74	715.87
Digestibility coefficient (%)				
Dry matter	44.26	40.55	37.80	28.38
Crude protein	14.47	16.36	16.88	17.28
Crude fiber	61.47	58.63	64.14	37.93

Nitrogen free extract	53.34	48.15	54.36	41.89
Intake of nutrients (g/day)				
Dry matter	906.12	792.44	987.35	1013.67
Crude protein	53.01	49.22	93.51	73.19
Crude fiber	297.31	261.35	260.75	304.01
Ether extract	6.53	2.22	9.58	15.21
Nitrogen free extract	472.00	422.39	454.52	491.85
Ash	77.30	57.62	140.01	129.45
TDN	442.08	364.50	445.44	333.98
DCP	7.57	8.05	15.78	12.65
DM intake/100 kg wt	3.21	2.99	3.45	3.19

Under feeding in late pregnancy may lead to abortion, mortality and loss of milk in ewes. Ewes require DM containing 12 per cent CP and 52 percent TDN @ 4 kg/100 kg body weight during advanced stage of pregnancy. Ewes require DM @ 100 g/kg $W^{0.75}$ containing 12 percent CP and 55-58 percent TDN during lactating phase. The lactating ewes require 255 Kcal ME and 7.85 g DCP/kg $W^{0.75}$. It is suggested to supplement high quality dry roughages or concentrate mixture to the grazing sheep to meet their nutritional requirements during advanced stage of pregnancy and early lactation.

Feed Additives

Feed additives are compounds which stimulate growth, improve efficiency of feed utilization and may be beneficial to the health of the animal. Antibiotics, sulpha drugs, hormones, buffers, activated carbons, antioxidants, bentonites and bloat controlling compounds are in use as feed additives.

Antibiotics

Chlortetracycline, oxytetracycline and bacitracine affect various enzyme systems such as oxidative phosphorylation reactions. Lambs fed a milk replacer, which had a supplement of antibiotic- vitamins and iron added to a basal replacer had a slight response. Antibiotics given to lambs in creep ration may improve weight gain and feed conversion. Cholortetracycline, oxytetracycline and oleandomycine may be beneficial in terms of greater weight gain and improved feed conversion. Feeding of antibiotics reduces incidences of diarrhoea, off feed and enterotoxaemia.

Feeding of antibiotics reduces the incidences of urinary calculi formation.

Activated charcoal

Five percent activated carbon had no effect on weight gain or feed efficiency of lambs fed a ration based on apple pomac contaminated with D.D.T. Feeding of activated charcoal has been found to be effective in reducing tissue contamination of ronnel when the pesticide and carbon were fed together.

Bentonite

Sodium bentonite is used as a pellet binder. 2% level of bentonite produces an apparent improvement in nitrogen retention, weight gain and feed consumption in lambs fed high roughage rations but if the concentration of bentonite is increased to 4 or 8% it reduced phosphorus retention.

Bloat controlling compounds

Surface acting agents are effective in controlling pasture bloat if a regular intake is ensured. Poloxalene prevents pasture bloat when given as a top dressing on supplementary concentrate. Poloxalene in grain or molasses salt blocks reduces the severity of bloat in lambs. Pluronic L62, a polyoxypropylene polymet controls bloat when given in water.

Buffers

Carbonates, bicarbonates, hydroxides, oxides and salts like sodium sulphite and ammonium chloride are being used. Feeding of these compounds change the pH of the ruminal contents. In case of feeding of high concentrate rations, the need for feeding of these compounds occurs, because of less chewing and regurgitation, which in turn results in less salivation and a reduced amount of bicarbonate coming in the rumen.

Including 3% buffers (2:1 maganesium and potassium bicarbonates) resulted in weight gain in lambs even though roughage was reduced suddenly from 60 to 10%.

Feeding of sodium and potassium bicarbonate throughout finishing period in lambs improved gains whereas, feeding of hydroxides of aluminium, calcium and magnesium have no effect. Sodium bicarbonate feeding may or may not improve lamb performance.

Table 12.7 : Requirements of digestible nutrients and minerals of lambs raised for mutton

Body weight (kg)	Daily weight gain (kg)	Daily intake per lamb	% of Diet								
			T.D.N.	M.E. Meal/kg	Protein	D.C.P.	Calcium	Phosphorus	Salt	Vit. A. µ.g.	Vit. D. i.u.
1.	2.	3.	4.	5.	6.	7.	8.	9.	10.	11.	12.
27.2	0.16	1.22	55	2.42	12.0	6.6	0.23	0.21	0.6	450	121
31.8	0.18	1.41	58	2.56	11.0	6.1	0.21	0.18	0.6	470	126
36.3	0.20	1.54	62	2.72	10.7	5.9	0.19	0.18	0.6	498	132
40.8	0.20	1.68	62	2.70	9.5	5.3	0.18	0.16	0.6	507	134
45.4	0.18	1.77	62	2.73	9.4	5.2	0.18	0.16	0.6	529	146

Source: NRC feeding standards 1966.

Table 12.8 : Nutritional requirements of sheep on dry matter basis

Body weight (kg)	Daily weight gain (kg)	Daily intake per animal (kg)	% of Diet								
			T.D.N. %	M.E. %	Protein %	D.C.P. %	Calcium %	Phosphorus %	Salt%	Vit. A. µ.g.	Vit. D. i.u.
1.	2.	3.	4.	5.	6.	7.	8.	9.	10.	11.	12.
DRY AND FIRST 15 WEEKS OF GESTATION											
45.4	0.03	1.18	50	2.2	8.0	4.4	0.27	0.21	0.8	238	212
54.4	0.03	1.36	50	2.2	8.0	4.4	0.24	0.19	0.7	242	220
63.5	0.03	1.54	50	2.2	8.0	4.4	0.22	0.17	0.7	256	227
72.5	0.03	1.72	50	2.2	8.0	4.4	0.20	0.16	0.7	258	231
LAST 6 WEEKS OF GESTATION											
45.4	0.17	1.72	52	2.29	8.4	4.6	0.24	0.18	0.6	403	146
54.4	0.17	1.90	52	2.29	8.2	4.5	0.23	0.17	0.6	428	156
63.5	0.17	2.09	52	2.29	8.0	4.4	0.22	0.16	0.6	454	168
72.6	0.17	2.18	52	2.29	7.8	4.3	0.22	0.16	0.6	503	183
8 TO 10 WEEKS AFTER PARTURITION											
45.4	0.04	2.09	59	2.60	8.7	4.8	0.30	0.22	0.5	333	119
54.4	0.04	2.27	58	2.56	8.4	4.6	0.28	0.21	0.5	359	132
63.5	0.04	2.49	56	2.47	8.0	4.4	0.27	0.20	0.5	379	141
72.6	0.04	2.58	55	2.42	8.0	4.4	0.27	0.20	0.5	423	154

Table 12.8 Contd. ....

1.	2.	3.	4.	5.	6.	7.	8.	9.	10.	11.	12.
					FEMALE LAMBS						
27.2	0.14	1.22	55	2.42	11.0	6.0	0.21	0.19	0.6	187	110
36.3	0.09	1.45	50	2.20	8.7	4.8	0.20	0.18	0.6	214	137
45.4	0.06	1.54	50	2.20	7.6	4.2	0.20	0.18	0.6	243	163
54.4	0.03	1.54	50	2.20	7.0	3.9	0.20	0.18	0.7	298	194
					MALE LAMBS						
36.3	0.18	1.45	62	2.73	10.0	5.5	0.20	0.18	0.6	214	137
45.4	0.14	1.68	57	2.51	8.6	4.7	0.18	0.16	0.6	225	150
54.4	0.09	1.90	50	2.29	7.6	4.2	0.17	0.15	0.6	240	156
63.5	0.05	2.09	50	2.29	6.9	3.8	0.16	0.14	0.5	258	168

Hormones

During the past few decades, growth hormones, natural and synthetic adrenal cortical hormones, natural and synthetic estrogens, androgens, natural and synthetic progestogens, androgen estrogen combinations, thyroid and antithyroid compounds have been used for fattening the lambs.

Thyroid active hormones

Studies with lambs have not indicated favourable responses when they were fed methimazole, a growth active hormone. There is very little current interest in feed lot use of growth active compounds.

Growth stimulating hormones

Diethylstilbestrol and hexosteral (estrogen); melengestrol acetate (progesterone); have been fed orally or have been used as implant for feed lot lambs. The response had been poor with growth hormone and cortisone. Estrogens and progestogens cause increased retention of nitrogen, which results in better feed efficiency and growth. Recent researches have indicated that subcutaneous implants (15-30 mg) will produce the same responses as feeding 10-20mg/day. Common levels for lambs are implants of 2-3 mg. These hormones are effective in young lambs.

Diethystilbestorl and other hormones give better response when the quality of the ration is relatively high, produce better response than those that are inadequate.

Use of melengestrol acetate, a synthetic progestogen did not indicate much promise for feed lot experiments in lambs.

Methane inhibitors

Researches with some methane inhibitors such as dichlorovinyl dimethyl phosphate, sodium sulphite, hemiacetal derivatives of chloral and starch and a vide variety of halogenated, analogues of methane such as chloroform, carbon tetrachloride, methylene chloride and bromochloromethane have been carried out.

13

Pastures and Forages

The average cultivated area devoted for fodder production is only 4.4 percent pastures and cultivable wastelands are approximately 13 and 15 million hectares, respectively. The total area under forests is 25 million hectares and that open to grazing is 21 million hectares. These resources are able to meet the fodder requirements of grazing animals only during the monsoon season. Sheep is the only animal which can utilize such pastures effeciently. Cross breeding of native sheep is being adopted for improving productivity in most of the states. The nutrient requirements of cross breds are more because of higher production.

Only 2-3 sheep per hectare can survive on our pastures. The livestock population is increasing at alarming rate and the area of pastures is also decreasing. Under these circumstances the only alternative left is increasing production from our pasture lands. This is possible by better management of rangelands and by growing good quality fodders. *Cenchrus* and *Lasiurus* species perform well under arid conditions. Crude Protine and Total Digestible Nutrient contents of these grasses are 5.26 and 43.48 and 3.02 and 41.38 percent, respectively. Hundred young sheep need about 300 quintal fodder in a year which can be obtained from 10 hectare of land by better management.

General Tree Plantation

There occurs shortage of fodder for sheep from March to June in arid and semi arid areas. The lopping of fodder trees is a common practice in most parts of this country especially in the hills, for feeding sheep, as well as goats and cattle. The growing of fodder trees should constitute an important programme in developing fodder resources for sheep. The availability of fodder trees is particularly useful during scarcity periods. This shortage can be met by planting fodder trees and the sheep will get nutritious fodder during lean period. Sheep relish leaves of subabool *(Leucaena leucocephala),* Ardu (*Ailanthus excelsa),* Khejri (*Prosopis cineraria)* Babool (*Acacia arabica),* Kachnar (*Bauhinia variegata)* etc. These trees can be lopped in November/December and April/May. The dried preserved leaves of these trees can also be fed during lean periods. One hundred trees should be planted in a hectare of land. Plant to plant distance may be 10 metre. A 0.5 × 0.5 × 0.5 m pit is dug for planting a tree sapling. Ten kilogram organic manure and well rotten tree leaves are put in each pit. BHC or DDT is also mixed with soil, before planting a sapling. It is watered immediately, thereafter the plant should be watered every fortnight.

These trees are not lopped upto 3 years. The branches more than 1 inch in diameter should not be cut, otherwise, the tree may die. The mature trees can yield upto 8-10 quintals of green fodder per hectare. In addition to it, we get timber for household consumption also. The chemical composition of some of the fodder tree leaves are given in Table 13.1. The details of some fodder trees are given below:-

Ailanthus Excelsa Roxb :

Vernacular names : Ardu, Maharukh, Mahanim. Pinari, Peddamanu.

Distribution: It grows in Gujarat, Rajasthan, Haryana, Punjab, Uttar Pradesh, Bihar, Orissa and the Deccan plateau, (Troup, 1921).

Habitat : It grows well in semi arid and semi-moist regions (Seth *et al.,* 1962).It grows in dry areas where the annual rainfall is 400 mm (Bhimaya *et al.,* 1963). It grows on a wide variety of soils, but thrives best on porous sandy loams (Chaturvedi, 1956a). Leaf shedding takes place in February and new leaves appear in March-April.

Planting; The seed is sown in well prepared nursery beds. The sowing is done in lines 20 cm apart. About 15 g seed is sufficient

Table 13.1 : Chemical composition of some fodder trees and shrubs

Botanical names	C.P. %	C.F. %	N.F.E. %	E.E. %	Ash %	Reference
Acacia aneura	12.8	32.9	47.2	1.1	6.0	Everist *et al.*, 1958.
Acacia brachystachya	10.4	41.9	38.2	5.7	4.1	Everist, 1969.
Acacia brevispica	18.8	18.0	51.0	5.7	6.5	Dougal and Bogdan, 1958.
Acacia cambagei	13.3	15.9	55.2	3.1	11.0	Everist, 1969.
Acacia casia	11.3	30.2	48.6	3.2	6.0	Everist, 1969.
Acacia catechu	13.0-18.8	21.9-22.6	46.7-51.0	3.05-4.6	9.7-9.8	Sen and Ray, 1971; Pal *et al.*, 1979.
Acacia catenulata	13.9	21.9	58.4	2.9	2.9	Everist, 1969
Acacia deanei	14.3	18.9	55.6	48	3.4	Everist, 1969
Acacia excelsa	11.8	21.4	56.7	3.6	6.5	Everist, 1969
Acacia farnesiana	18.4	18.5	56.5	2.0	4.4	Everist, 1969
Acacia gerrardii	17.7	24.2	50.5	1.6	6.0	Dougal and Bodgan, 1958.
Acacia girciffae	12.6	34.2	47.9	1.8	3.6	West, 1950
Acacia narpophylla	26.7	19.3	45.7	1.1	7.2	Everist, 1969.
Acacia mellifera	42.9	16.2	32.2	2.6	6.2	Dougal and Bodgan, 1958.
Acacia nilotica	13.9	9.2	69.8	—	7.1	Ganguli *et al.*, 1964.
Acacia nubica	32.5	15.8	41.2	1.7	8.9	Dougal and Bodgan, 1958.
Acacia pendula	13.2	29.6	46.2	3.4	7.8	Everist, 1969.
Acacia sparsiflora	17.8	25.4	47.7	3.3	5.8	Everist, 1969.
Adina cordifolia	8.7-12.5	12.1-13.8	—	3.1-5.6	7.7-8.2	Majumdar *et al.*, 1967 a
Aegle marmelos	15.1-15.3	16.5-18.1	48.4-52.8	1.4-1.5	14.0-16.8	Sen and Ray, 1971.

Table 13.1 Contd. .....

Botanical names	C.P. %	C.F. %	N.F.E. %	E.E. %	Ash %	Reference
Ailanthus excelsa	19.6	13.5	47.7	3.7	15.5	Singh and Patnayak, 1977a.
Albizia amara	26.8	26.8	38.4	1.9	6.1	Dougal and Bogdan, 1958.
Albizia basaltica	12.8	21.5	56.5	4.5	4.5	Everist, 1969.
Albizia chinensis	15.1	31.6	43.6	4.4	—	Sharma *et al.*, 1969.
Albizia lebbek	16.8-26.5	26.5-37.5	36.09-42.8	2.8-4.7	7.1-11.5	Sen and Ray, 1971; Sharma *et al.*, 1966a; Pal *et al.*, 1979.
Anogeissus latifolia	7.5-11.5	16.4-24.1	55.4	2.7-4.4	8.7-10.9	Majumdar *et al.*, 1967 a, b; Sen Ray, 1971.
Anogeissus pendula	7.6	19.0	65.3	—	8.1	Ganguli et al., 1964.
Artocarpus heterophyllus	11.2-14.2	18.7-22.8	46.3-5.7	2.3-3.5	8.3-14.2	Sen & Ray, 1971.
Azardirachta indica	12.4-18.3	11.4-23.1	42.3-66.6	2.3-6.2	7.7-18.8	Patel and Patel, 1957, 1962; Ganguli *et al.*, 1964; Sen and Ray, 1971.
Baphia nitida	21.2	26.8	—	—	5.4	Rose Innes and Mabey, 1964.
Bauhinia carronii	13.7	13.5	64.8	1.6	6.3	Everist, 1969.
Bauhinia hookeri	11.8	13.5	64.3	2.3	8.1	Everist, 1969.
Bauhinia variegata	10.7-15.9	25.3-33.0	40.9-51.8	1.3	1.37-12.3	Sharma *et al.*, 1966a, 1968, Majumdar *et al.*, 1967a.
Butea monosperma	11.5	—	—	—	9.3	Gupta *et al.*, 1975.
Cajanus cajan	16.9	29.2	36.8	9.9	7.1	Otero, 1952.
Careya arborea	10.4	25.9	42.6	7.7	7.5	Sen & Ray, 1971.

Table 13.1 Contd. .....

Botanical names	C.P. %	C.F. %	N.F.E. %	E.E. %	Ash %	Reference
Celtis australis	14.5-15.3	19.5-21.5	42.7-49.0	2.5-5.6	11.6-17.8	Sharma *et al.*, 1966b; Pal *et al.*, 1979.
Cordia dichotoma	12.4-15.1	16.5-26.8	41.9-52.8	1.5-2.9	12.6-17.1	Sharma *et al.*, 1966a; Sen and Ray, 1971; Pal *et al.*, 1979.
Dalbergia sissoo	16.7	22.2	49.4	3.2	8.6	Momin and Ray, 1943.
Desmanthus virgatus	22.4	—	—	—	—	Takahashi and Repperton, 1949.
Desmodium discolor	18.5	25.8	39.1	1.8	1.5	Otero, 1952.
Desmodium molle	16.9	23.6	46.1	4.4	9.1	Otero, 1952
Desmodium pabulare	20.2	22.6	41.6	5.8	9.8	Otero, 1952.
Dichrostachys cinerea	15.3	28.1	50.5	1.2	4.9	Wilson and Bredon, 1963.
Diospyros melanoxylon	7.1	25.3	56.0	2.2	9.4	Sen and Ray, 1971.
Ehretia laevis	13.5	18.0	51.2	6.0	11.4	Sen and Ray, 1971.
Eriosema englerasium	18.5	—	—	—	—	Lawton, 1968.
Ficus bengalensis	7.7-11.2	13.9-30.0	37.6-51.6	2.5-5.1	9.3-16.9	Patel and Patel, 1957; Mia *et al.*, 1960; Pal, *et al.*, 1979.
Ficus glomerata	11.2-15.2	12.3-16.5	56.5-59.0	2.2-2.9	11.9-18.4	Majumdar and Momin, 1960; Majumder, *et al.*, 1967a.
Ficus lacor	8.5-15.7	12.8-25.8	44.8-61.1	1.8-3.4	—	Momin and Ray 1943; Majumdar *et al.*, 1967b.
Ficus religiosa	15.6	26.9	39.0	3.8	14.5	Singh and Patnayak, 1977 a.
Ficus roxburghii	12.3-13.4	7.7-17.8	50.0-64.9	4.5-4.7	9.4-15.4	Joshi and Ludri, 1966, Sharma *et al.*, 1966a.

Table 13.1 Contd. .....

Botanical names	C.P. %	C.F. %	N.F.E. %	E.E. %	Ash %	Reference
Griffonia simplicifolia	15.7	—	—	29.3	8.5	Rose Innes and Mabey, 1964.
Grewia optiva	15.6	20.9	47.0	6.7	9.8	Johri and Talpatra, 1971.
Hardwickia binata	9.0	—	49.7	'4.0	—	Prasad *et al.*, 1974
Holoptalea integrifolia	13.7	—	—	1.9	14.5	Sen and Ray, 1971.
Kydia calycina	11.1-13.6	14.5-23.7	46.1	3.3-6.1	15.3-19.8	Majumdar *et al*, 1967a; Sen and Ray, 1971.
Lagerstoraemia parviflora	7.8	17.3	58.2	5.7	11.1	Sen and Ray, 1971.
Lannea coromandelica	11.4-16.2	16.1-19.8	53.1-59.1	2.5-4.2	8.5-9.2	Sen and Ray, 1971.
Leucaena leycocephala	24.2	13.3	—	4.4	10.8	International Consultation on Ipil Res. 1976.
Madhuca longifolia	9.1-9.8	18.7-20.3	59.4-60.7	3.8-4.1	6.7-7.8	Sen and Ray, 1971.
Mallotus philippinensis	13.4	29.7	44.6	3.7	8.7	Sen and Ray, 1971.
Melia azedarach	—	13.3	—	—	—	Gupta, *et al.*, 1975.
Milletlia thonningii	19.0	29.5	—	—	8.8	Rose Innes and Mabey, 1964.
Mitragyna parvifolia	7.7	19.6	60.7	—	8.7	Sen and Ray, 1971.
Moringa oleifera	15.3-20.7	7.1-17.9	48.7	4.4-11.1	11.8-14.2	Majumdar *et al.*, 1967a, Sen and Ray, 1971.
Morus alba	15.0-27.6	9.1-15.3	40.1-49.1	2.3-8.0	14.3-22.9	Jayal and Kehar, 1962; Majumdar *et al.*, 1967a; Narayan and Setty, 1977, Pal, *et al.*, 1979.
Ougeinia oojeinensis	11.6-15.1	21.9-26.5	47.6	1.4-4.3	10.0-14.9	Majumdar *et al.*, 1967a; Sen and Ray, 1971.

Table 13.1 Contd. .....

Botanical names	C.P. %	C.F. %	N.F.E. %	E.E. %	Ash %	Reference
Parkinsonia ciculeata	13.1	24.9	47.9	1.9	11.9	Everist, 1969.
Piliostigma thonningii	9.9	22.1	59.1	1.9	7.1	vanRensburg, 1968.
Pongamia pinnata	17.6	–	–	–	–	Gupta *et al.*, 1975.
Prosopis cineraria	13.9	20.3	59.2	–	6.5	Ganguli *et al.*, 1964.
Quercus dilatata	9.6	29.1	51.8	4.5	5.1	Sen and Ray, 1971.
Quercus glauca	9.6	29.0	49.6	4.1	7.6	Sen and Ray, 1976.
Quercus leucotrichophora	10.2-11.4	31.3-32.1	46.7-48.4	3.5-4.8	5.1-6.4	Sharma *et al.*, 1966a, Sen and Ray, 1971.
Robinia pseudoacacia	25.5	17.2	46.5	3.3	7.5	Singh *et al.*, 1959.
Schleichera oleosa	10.4	32.3	49.2	1.9	6.1	Sen and Ray, 1971.
Shorea robusta	10.1	27.4	55.4	3.2	3.9	Sen and Ray, 1971.
Syzygium cumini	7.9	20.7	61.7	2.6	7.0	Momin and Ray, 1943.
Tamarindus indica	13.4-14.4	15.9-20.2	46.0-60.5	1.9-7.0	7.9-10.4	Patel and Patel, 1957; Sen and Ray, 1971.
Terminalia alata	8.3-12.7	13.1-21.8	54.1	1.4-4.9	9.1-12.1	Majumdar *et al.*, 1967a; Sen and Ray, 1971.
Terminalia belerica	8.6-14.0	16.7-18.6	54.5-60.1	3.1-4.7	7.6-11.7	Sharma *et al.*, 1966a, Sen and Ray, 1971, Pal *et al.*, 1979.
Ziziphus mauratiana	15.4	15.8	–	–	6.7	Kohli *et al.*, 1969.

for sowing each square metre of nursery area. Germination commences in about 10-15 days and is completed in about 45 days. Planting is done during monsoon rains. Plants are uprooted in the morning with balls of earth and are planted the same day, on the bunds of the fields and pasture plots in pits of the size of 0.5 m^3 dug one month in advance. The watering is done once a month during summer in first year and once in 2 months in second year.

Nutrients : Its leaves are highly palatable and nutritious. An average tree yields about 5-7 quintals of green leaves twice a year (Bhandari and Gupta, 1972). The leaves of this plant are rich in crude protein, ether extract and calcium, but poor in phosphorus when dry. The digestibility coefficients are fairly high for the nutrients. The crude protein digestibility from green leaves is fairly high and nearly equals berseem, while nitrogen free extract digestibility compares well with mulberry leaves and berseem. Green leaves can serve as a maintenance ration for livestock. Dry or green leaves can be mixed with dry fodder as protein supplement. Its dry leaves contain 19.9 percent crude protine (CP), 12.8 percent crude fiber (CF), 5.5 percent ether extract (EE) and 51.8 percent nigrtogen free extract (NFE). A young sheep can consume green and dry leaves @ 3.5 to 4.0 and 2.3 percent respectively, of its body weight. Young sheep can drive 130 g digestible protein and 510 g total digestible nutrients from its leaves.

Albizzia lebbeck (Linn.) Benth

Vernacular names : Siris, Nirith, Chichola, Sirin, Vagai, Dirsinam.

Distribution : It grows wild in Andaman Islands, Chhota Nagpur, Kerala, Rajasthan, Tamil Nadu and West Bengal (Troup, 1921) and in sub Himalyas upto an elevation of about 1500 m. (Venkataramany, 1968).

Habitat: In its natural habitat, the maximum shade temperature reaches about 48°C and minium temperature varies from about 2 to 8°C. The annual rainfall varies from about 600 to 2500 mm. It is suitable species for planting in dry areas of Rajasthan with annual rainfall of less than 300 mm (Bhimaya *et al.*, 1963).

Planting: It grows well on black cotton soil (Troup, 1921) and deep loam soil. It can tolerate salts in soil (Hussain, 1977). Seedlings

Table 13.2 : Package of practices for increased fodder production

Name of crop	Varieties	Time of sowing	Seed rate (kg/ha)	Manuring (kg/ha)	No. of Irrigation	No. of cuts	Fodder yield (q/ha)
Cowpea	H.F.G. 41–1 No. 10 F.O.S.–1	March to July	40–50	N 25 P_2O_5 62.5 K_2O 100	3–4	2	300–350
Guar	Guar–2 Guar–227 H.F.G. 119	April to mid July	40–50	N 20 P_2O_5 50	3–4	1	250–300
Bajra	2nd genera	March to Aug.	8–10	N75	2–3	2–3	400
Berseem	mescabi	October	20–25	N 25 P_2O_5 70 K_2O_5 6O	5–6	5–6	700–800
Jowar	JS–20 JS–263 JS 29/1 HC 136	March to July	50–60	N 50 P_2O_5 15	2	13–4 (Multicut)	400–500
Lucerne	Lucerne–9	October to November	12–15	N 20 K_2O 30 P_2O_5 60	5–6	8–10	700–750

Table 13.2 Contd.........

Name of crop	Varieties	Time of sowing	Seed rate (kg/ha)	Manuring (kg/ha)	No. of Irrigation	No. of cuts	Fodder yield (q/ha)
Maize		April to August		N 50 K_2O 20 P_2O_5 30	5–6	1	350–400
Mustard		September	6–8	N 30	2	1	300–350
Napier Hybrid		March to August	33000 root slips.	N 35 K_2O 15 P_2O_5 25	Fortnightly	6–8	1100–1300
Oats	HFO–114 Weston–110	October to November	75–100	N 60 K_2O 15 P_2O_5 30	3–4	2–3	300–400

attain a height of about l-2 m during first year and 3-4 m in second year in irrigated nurseries. The tree is moderate in height. It can be raised either by direct sowing or planting of nursery raised seedlings. The latter gives better results (Muthana *et al.*, 1976). Stump planting, gives better success and is the easiest method for raising this species (Kadambi and Dabral, 1955; Mathadua, 1956; Venkataramany, 1968). The seed sowing is done in lines which are 3 m apart. About 20 kg seed is sufficient to sow one hectare. Seedlings can also be raised by sowing the seeds either in nursery beds or in containers in March-April. Germination commences in about a week and takes about a month to complete (Seth *et al.*, 1962). Planting is done in July in 30 cubic centimeter pits. Entire plants are uprooted from the nursery with balls of earth.

Nutrients : Its leaves are reported to be good fodder. The leaves contain 0.23 percent tannins (Lohan *et al.*, 1979). The average digestibility coefficient for CP, EE, CF and NFE are reported to be 67.9, 56.1, 41 and 48, respectively and total digestible nutrients per quintal of leaves on dry weight basis are 11.6 kg crude protein, 37.2 kg of carbohydrate and 0.36 kg of ether extract (Sen and Ray, 1971).

Azadirachta indica A. juss

Vernacular names : Neem, Nim, Limba, Bavina, Veppu, Vepa, Veppam.

Distribution : It grows in Andhra Pradesh, Bihar, Gujarat Haryana, Karnataka, Maharashtra, Madhya Pradesh, Orissa, Punjab, Rajasthan, and Uttar Pradesh. It grows wild in Irawadi valley (Brandis, 1906).

Habitat : The temperature ranges between 2 and 40°C in the areas where it grows. It is suscepsible to frost. It is commonly found where the annual rainfall varies from 250 to 1150 mm. It grows well on a wide variety of soils ranging from sandy in Rajasthan to clayey in Maharashtra. It can persist under adverse soil conditions also, where very few tree species can grow (Chaturvedi, 1956 b).

Planting : The seeds germinate quickly and attain a height of 20 to 30 cm by the end of the first season. While a limited supply of soil moisture restricts plant growth, excessive soil moisture causes the taproot to rot. A full grown tree is of moderate size and with a short straight trunk and long spreading branches forming fairly dense rounded crown. It is a moderately salt resistant tree which can grow on alkaline *usar* soils. It can be raised successfully through direct sowing.

Nutrients : It is a good fodder tree, lopped to feed small ruminants (Chaturvedi, 1948). The best lopping time varies with the climatic conditions of the locality. In general, feeding of the leaves during winter is preferred under North Indian conditions. Digestibility coefficient for organic matter, CP, EE, CF, NFE and total carbohydrates are 53.2, 49.0, 65.0, 51.2 and 53.9, respectively.

Bauhinia variegata Linn

Vernacular names : Kachnar, Raktakanchan, Kanchivola, Bodantam.

Distribution : It grows in Assam, Madhya Pradesh and Western parts of Indian Peninsula (Brandis, 1906). It is cultivated widely throughout the greater part of the country and in Himalayas upto height of 1500 m.

Hibitat : It grows in areas where the shade temperature varies between 20°C and 44°C and rainfall between 500 and 2500 mm. It grows in soils ranging from gravelly soil in mountains to sandy loam and loamy soil in the valleys.

Plantation : Entire plants are planted out with balls of earth and care is taken that the roots are not injured during digging out of the plants from nursery. The spacing is about 2.5 × 2.5m.

Nutrients : The leaves are regarded as good fodder and the cultivation is advocated for leaf fodder production (Sagreiya, 1940). Average fodder yield is reported to be 15-20 kg/tree (Negi *et al.*, 1979). The chemical composition of leaves varies with the locality and season (Majumder *et al.*, 1967a,b; Sinha and Jolly, 1971; Negi *et al.*, 1979). The leaves are fairly palatable. The digestibility coefficients and DCP, TDN and SE values differ with animal fed. The DCP, TDN and SE values of leaves fed to rams are 4.98, 47.92 to 55.54 and 32.48, respectively.

Cordia dichotoma Forst

Vernacular names : Buhal, Bohari, Lasura, Borla, Challe, Bhoker, Vidi.

Distribution : It grows in sub Himalayan tract and outer ranges, upto an elevation of 1500 m and in Rajasthan, Madhya Pradesh and Khasi hills (Brandis, 1906).

Habitat : It grows in areas where the minimum and maximum temperature is 1° to 15°C and 35° to 49° C and the normal rainfall 250 to 300 mm. In areas with annual rainfall less than 500 mm, it thrives along streams or depressions where the moisture may be available. Deep moist sandy loam soils support its good growth while gravelly shallow and dry soils are not suitable.

Planting : Seeds are sown in June-July at a depth of 2 cm in lines spaced about 20 cm apart. Irrigation is done frequently. The germination starts in about 3-4 weeks and may take about 3 or 4 months to complete (Troup, 1921). The seedlings are kept in nursery for about 1 year before being planted in the following year with the commencement of monsoon rains. Planting is done in 0.3 m^3 pits dug in advance.

Nutrients : The leaves yield good fodder and are lopped for this purpose. The digestibility coefficients are 71, 30,58 and 76 for CP, EE, CF, NFE, respectively. The tannins account for 0.84 percent of the DM in leaves (Lohan, *et al.*, 1979).

Dalbergia sissoo Roxb

Vernacular names : Sisam, Shisham, Agra, Tali, Sissoo, Sissu-karra.

Distribution : It grows in Himalayan tract upto an elevation of about 1500 m and in Indo-Gangetic plain and Rajasthan.

Habitat : In its natural habitat the annual rainfall varies from about 750 to 4500 mm. The absolute maximum shade temperature varies from 39° to 49°C and absolute minimum shade temperature from about 4°C to 6°C. It has been found to be a suitable species for planting in sand dune areas of Rajasthan with only 400 mm annual rainfall (Bhimaya *et al.*, 1963).

It prefers well drained sandy loam soil with good moisture supply. It prefers freshly deposited and well drained sandy soil with pebbles and boulders in river bed (Chaturvedi, 1956 b). It can tolerate salinity to some extent.

Planting : Natural reproduction takes place through root suckers and seedlings. It can be raised by either 1) direct sowing, 2) planting out entire plant raised in the nursery, 3) stump planting or 4) planting out root suckers. Stump planting gives better results than planting out entire plants (Kadambi and Dabral, 1955; Mathauda, 1956). The planting of one year old nursery raised seedlings or of pre sprouted stumps gives good results (Singh, 1963). Direct sowing is done in June-July with the commencement of monsoon rains. Sowing is done in lines about 2-3 m apart. About 35 kg of broken pods are needed to sow one hectare with a spacing of about 2 m (Homfray, 1937). Irrespective of the type of planting stock, planting is done in July-August in pits of 0.5 × 0.5 m dug in advance. Planting in trenches is advocated in dry areas (Singh, 1963). Full grown trees may attain a height of about 20 to 25 m. Some well grown trees may attain a height of about 30 m.

Nutrients: The leaves are more palatable when succulent and the palatability decreases as the leaves mature. Average digestibility coefficient for CP, EE, CF, NFE is 56, 27, 41 and 68, respectively (Sen and Ray, 1971). The leaves are fed in" April-May. Once the sheep become accustomed to this fodder, their health is very well maintained (Lander and Dharmani, 1925).

Leucaena leucocephala

Vernacular names : Subabool, Koobabool.

Distribution : It is distributed in Andhra Pradesh, Karnataka, Himachal Pradesh, Tamil Nadu and Uttar Pradesh (Lohani, 1979).

Habitat : It can withstand large variations in temperature and rainfall, wind and drought. It grows in tropics and subtropics. The best growth is obtained in areas with 600 to 1700 mm annual rainfall. It grows in a wide range of soil types. It grows best in deep, fertile soils which can supply adequate moisture and abundant nutrients. It can also grow in shallow soils. It is a leguminous fodder plant which fixes nitrogen in the soil.

Planting : The seeds can be either broadcast or in drilled or patches. Seeds can also be raised in polythene bags. Planting out is done in July-August in 0.5 m pits dug during summer. Spacing is determined by the purpose of raising the plantation. Spacing between the rows may vary from as narrow as 0.3 m for forage to as wide as 3-4 m in case of intercropping.

Nutrients : The foliage is relished by livestock. High mimosine content limits its use as sole feed for sheep. It can be fed upto 15 percent of the fodder to sheep. The chemical composition of the leaves varies in different months and is influenced by edaphic and climatic changes (Singh and Mudgal, 1967). On the basis of chemical composition and digestibility of different nutrients *Leucaena* leaves are slightly inferior to lucerne leaves (Kharat *et al.*, 1980).

Prosopis cineraria (Linn) Druce

Vernacular names : Khejari, Khejra, Jand, Shami, Jambi, Shema.

Distribution : It is found in Gujarat, Haryana, Punjab, Rajasthan and Western Uttar Pradesh (Kaul and Ganguli, 1962).

Habitat : This plant can grow very well in sandy soils, where the annual rainfall is less than 500 mm, the maximum shade temperature varies from about 48° to 52° C and the minimum temperature during winter touches about-4° C. It can grow upto a height of 10-13 m. In its natural range, it grows on coarse sandy soils (Sharma, 1966; Bawa *et al.*, 1988).

Planting : The plants are raised in the nursery in polythene bags of 10 × 30 cm size, containing following mixture :–

Pulverised soil	2.5 kg
Compost manure	2.5 kg
Single super phosphate	0.5 kg
D.D.T.	0.05 kg
Urea	0.125 kg.

The seed is sown about 1 cm deep. The growth of the seedling is slow and even under irrigated conditions they may seldom attain a height of 30 cm in one year (Wadhani, 1953). It is planted during rainy season when the plant is 25-30 cm tall. About 40-50 trees per hectare are planted. It drains moisture not only from the upper 2m soil layer, but also from lower layer below kankar zone (Krishnan *et al.*, 1968). It is watered twice in a month. Light prunning is done when the plant is 2 to 2½ year old.

Nutrients : It is an evergreen or nearly so. Its pods ripen from May to June. A mature plant can yield 40 to 50 kg of green leaves, which on dry matter basis contains CP 31.1 percent, CF 12.8 percent, EE 2.2 percent and NFE 53.0 percent. An adult ewe can consume green leaves @ 4-5 percent of its body weight. The dry matter consumption in case of adult Chokla rams was reported to be about 2.18 percent of its body weight (Bhandari *et al.*, 1979).

Ziziphus mauritiana

Vernacular names : Beri, Ber, Bordi, Elangi, Barboli, Reegu.

Distribution : It grows throughout India except temperate Himalayas and very humid areas of North East India.

Habitat : It is a hardy tree and can grow under varying climatic conditions and even in conditions where most other trees will fail to grow (Singh *et al.*, 1967). The average maximum and minimum temperature range in which it grows is 8 to 39°C and -7 to 13°C and annual rainfall 125 to 2225 mm (Troup, 1921). It grows in laterite, black cotton and moderately saline soils (Hussain, 1977). Its best growth is obtained on deep sandy loam soil which is neutral or slightly alkaline (Singh *et al.*, 1967).

Planting : The seed can be sown either in furrows or in dug up lines, spaced 3 to 5 m apart in July after the commencement of monsoon rains.

Nutrients : The leaves are a good fodder for cattle, sheep and goats. The average digestibility coefficient for different chemical constituents are CP, 36; EE, 62, CF, 27; and EE, 34. The total digestible nutrients per quintal of dry fodder is 30.65 kg and the nutritive ratio is 9.9. The leaves thus have fairly high nutritive value. The composition of Z. *nummularia* leaves has been estimated by Malik and Nath (1970).

Plant Poisoning

Sometimes, heavy mortality can occur amongst sheep owing to the grazing of poisonous plants which they are forced to consume during scarcity periods or when new animals not accustomed to local environment, are introduced. Katiyar (1962-1968) reported 233 (6.3 percent) deaths due to poisonous plants and listed 32 such plants and grasses as were reported poisonous for sheep. Some of these are new leaves and buds of Oak, new shoots of *Pieris ovalifolia,* leaves of *Storbilanthes wallichii* Nees, new leaves of *Rhododendron campanulatum* and flowers of *Kalnchoe spathulata.* The tender leaves and honey of *Rhododendron arboreum* are toxic and the symptoms observed are frothy saliva and grinding of teeth (Kehar and Rao, 1944).

Lal and Kalra (1960), Sastry and Mahadevan (1963) and Katiyar (1964) reported that *Lantana camara* causes poisoning in sheep resulting in the symptoms of photosensitization and severe icterus. Katiyar (1961a) also reported that mature arunja pods contained hydrocyanic acid. Case of hydrocyanic acid poisoning in sheep owing to the ingestion of leaves of *Sorghum halepense* (L) Pers were also reported by Bawa (1943) and Katiyar (1961b).

Pasture establishment

Some pasture species can be established without seedbed preparation and tropical legumes have been successfully established in this way. Generally, however, some seed bed preparation is needed prior to seeding and the better seedbed preparation, the better the result.

Preparation of land

Pasture can be established on prepared seedbed. Better the seed bed preparation, the better the pasture establishment. Finer seed bed is required for small seeds. The larger seeded legume species such as *Lablab purpureus* and *Siratro* can be satisfactorily established with a minimum of seedbed preparation.

Pasture establishment is usually more successful on new land because of the higher initial fertility after recent clearing and less competition from weed growth. The first cultivation should be given by a heavy duty stumpjump disc plough which can remove the remaining small stumps and roots and prevent a good deal of bush regrowth. Then disc harrows are used to chop small pieces of root and stems. If a finer and more level seed bed is desired, a subsequent harrowing can be given. For small areas, row seeding may be accomplished with a hand planting machine. Canvas bags holding 2-4 Kg seed hanging from shoulder and seed covering the chest are attached from which the seed can be taken out and broad cast.

Selection of seeds

Cenchrus, Lasiurus and Elephant grasses are good for sowing in arid and semi arid areas. The nutrient contents of these can be increased by sowing leguminous fodders such as *Dolichos, Siratro* etc. Mixed pastures have been found to be good for sheep.

Seed rate :

About 6-7.5 kg of *Cenchrus* seed and 30 kg of Butter fly pea is sufficient for one hectare.

Sowing: Depth of planting depends on seed size. Very small seeds should never be covered but broadcast on the surface; large seeds can germinate from deep planting. The usual planting depth is 1 to 2.5 cm for most species. Most seeds are planted dry just ahead of the expected rainy season. Seeding after rain is more successful with large seeded legumes. *Cenchrus* is sown in July-August. There should be a distance of 0.5 m between rows and between plants incase of *Cenchrus* and butter fly pea. The sowing is done in strips of 2-3 rows.

Irrigation :

It is impossible in pastures. Only water can be conserved; small bunds can be made against slopes. If there is depression in the field, it can be deepend further to store water for future irrigation.

Harvesting and grazing

The sheep are not allowed to graze upto one year of the establishment of the pasture. The grass should be harvested after the seed shedding in first year. This helps in the establishment of the pasture. The grass should be cut 10 cm above the ground. This helps in profuse tillering. The sheep should be allowed to graze in the second year in the month of August when the grass is 25 to 30 cm in height. Rotational grazing is very useful. This saves sheep from diseases which spread through dung. If the production falis, grazing should be stopped for a year or two.

Lambs should be allowed to graze first followed by lactating ewes and dry ewes should be allowed to graze in the end.

Crop Rotation

As the pasture lands are shrinking day by day and an average farmer does not have more than 5 acres of land the only method left is growing nutritious fodder for livestock. By adopting any of the following crop rotations sheep farmers can get green nutritious fodder throughout the year.

Napier-Lucerne

Napier × Bajra cross No. 21 and Lucerne can be sown in strips. This will give fodder throughout the year.

Sorghum-Berseem-Maize

Improved varieties of sorghum such as J. L-263 and J.S-29/1 can be sown in the last week of June. It can be harvested after 80 to 90 days. Then Berseem can be sown in mid October. After taking 5 -6 cuts, maize can be sown in April, which can be harvested by June.

Sorghum-Oats-Bajra

After harvesting sorghum in mid October, H.F.O-114 variety of oats can be sown. Two cuttings of oats can be taken. Then fast growing varieties of bajra (S 530 or S 1 /3) can be sown.

Sudan grass- Berseem- Maize

Sudan grass is a high yielding crop which can give 5-6 cuttings. Then berseem can be sown in October. Maize can be sown in April.

In addition to these there are more fodder crop rotations by following these sheep farmers can get green fodder for their animals throughout the year. Package of practices for increasing fodder production are given in Table 13.3.

Table 13.3 : Some suggested crop rotations

Month/s	Crops which can be sown	Crops which can be harvested
December & January	Oats	Berseem, Lucerne, Sarson
February & April	Sudan grass, Cowpea, Maize	Berseem, Lucerne, Oats
May & June	Maize, Sorgham, Bajra	Lucerne, Maize, Sudan grass
July & August	Sorgham, Maize, Cowpea, Sudan grass, Bajra	Maize, Sorgham, Gowpea, Sudan grass, Bajra
September	Lucerne, Sarson	Maize, Sorghum, Bajra Sudan grass
October & Nov.	Lucerne, Berseem, Oats, Sason.	Sorgham, Sudan grass

Avena sativa

Vernacular name : Oats.

Habitat : It is grown in Andhra Pradesh, Bihar, Gujarat, Haryana, Himachal Pradesh, Jammu and Kashmir, Maharashtra, Punjab, Rajasthan, Tamil Nadu and Uttar Pradesh. It can be grown in irrigated and unirrigated places. It can be grown in places where the general atmospheric temperature varies from 21 to 24° C.

Sowing : It is sown in October in Northern India. Irrigation is done 3 weeks after sowing and subsequent irrigations are done at an interval of 2 to 3 weeks. Irrigation is done after every cutting.

Fodder : First cutting is done 50 to 60 days after sowing. It yields 400 to 450 quintal green fodder in two cuttings.

Nutrients : It has 10 to 20 percent protein. It has 32 to 38, 36 to 44, 1.60 to 2.30 and 9.30 to 13.90 percent CF, NFE, EE and total ash, respectively on dry matter basis.

Brachiaria ramosa

Vernacular name : Kurra

It is a very good fodder crop. Its grains are also consumed by human beings. It is cultivated in Ajmer, Pali, Sirohi and Udaipur districts of Rajasthan. It can be used in crop rotation as follows :-

Wheat - Kurra - Methi

Barley - Kurra - Methi

Cotton - Kurra - Wheat

Cotton - Kurra - Methi.

It is normally sown after the harvesting of wheat crop. It can be sown at the onset of rainy season also.

It can be sown in sandy or loam soil. Accumulation of water is harmful for the crop. One or two ploughings are sufficient for this crop. 20 to 25 cart loads of sheep manure is needed per hectare.

Method of sowing : After broadcasting, the field is ploughed and small plots are made for irrigation. Eight to ten kilogram seed is sufficient for a hectare if it is sown for dry fodder and grains. It is sown for green fodder, 13 to 28 kg seed is required per hectare.

Irrigation : If the crop is sown in April/May, first irrigation should be done immediately after sowing. Irrigation is done every fortnight.

Fodder: It becomes available for fodder after 1 to 1 1/2 months of sowing. Three cuts can be taken in three months. For crop sown in May harvesting is completed by July, if it is sown in July, harvesting is completed by September. One hectare plot yields 300 quintals of green fodder.

Brachiaria mutica

Vernacular name : Para grass

Habitat : It is also called Buffel. This was imported from Ceylon in 1894. Humid climate of Bombay, Kerala and Karnataka suits it well. It grows well on the banks of rivers, in humid climate where the atmospheric temperature is 32°C and above. It can grow on saline land also.

Sowing : Cuttings having 2 to 3 inter nodes are sown at a distance of 15 cm in spring in irrigated areas and in the beginning of rainy season in unirrigated areas. 30 to 40 cart loads of organic manure is added to ordinary soil before sowing and 70 kg ammonium sulphate after second or third cutting.

Fodder : It gives fodder throughout the year upto 5 to 7 years. Fodder production per hectare is 375 quintals. After every cutting 25 to 30 kg nitrogen containing fertiliser per hectare is applied.

Nutrients : Its nutrients differ according to locality in which it is grown. Its is 2.8 to 16.1 per cent, CF 28 to 34 percent, NFE 41 to 57 percent and EE 0.9 to 3.9 percent. Singh *et al.. (1963) determined the chemical composition of B.mutica* leaves.

MARWARI FLOCK-GRAZING IN NATURAL RANGE LAND

Chloris gayana Kunth

Vernacular name : Rhodes grass

Habitat : This was first brought from South Africa in 1915. It has become very popular. It can grow well under drought conditions and can survive in cold also. It can tolerate saline soil to some extent.

Sowing : It can be sown in spring if irrigation facilities are available, otherwise it can be sown in the beginning of rainy season by using stem cuttings having roots. About twenty five thousand stem cuttings are needed per hectare. It should be irrigated every 2-3 weeks. It can also be grown by sowing seeds.

Fodder : First cutting can be taken after 2½ months after sowing, thereafter it can be cut at monthly intervals upto 7-8 times. Fodder production per hectare is 275 quintal.

Nutrients : The grass is leafy and palatable. It can be fed as chaffed fodder or can be grazed. It contains 5 percent protein.

Clitoria ternatea **L.**

Vernacular name : Butterfly pea, Kordofan pea.

Distribution : It is a native of tropical America, widely grown as leguminous fodder (Bermudez *et al.*, 1968).

Habitat : It is summer growing perennial, low frost bite tolerance, can climb tall grasses. Grows in areas where the annual rainfall varies between 400 and 1500 mm. It is fairly drought tolerant (van Rensburg, 1967), but also performs well under irrigation (Barran, 1953; Parbery, 1967). Adapted to a wide range of soil conditions from sandy to deep alluvial loams and heavy black cracking clays (Lee, 1954). It is tolerant to high pH soils.

Yield : First cutting is done 2 1/2 months after sowing, van Rensburg (1967) obtained a yield of 33.3 quintal/ha of dry matter. Parbery (1967) obtained yield of 133.50 quintal/ha/year on Cununnura clay under irrigation. It is very palatable. Its drought resistance is also an asset.

Lablab purpureus **(L) Sweet Syn.**

Vernacular name : Carpet bean, Sem.

Distribution : Widespread throughout the tropics.

Habitat : It grows as summer growing annual, biennial or short term perennial. It requires warm temperatures for good growth. Minimum temperature for growth is about 3°C (Murtagh and Dougherty, 1968). It is more tolerant to cold than velvet bean. It is quite drought tolerant when established (Luck, 1965). It is sown in early summer.

Yield : Dry fodder yield varies from 25-40t/ha (Crowder, 1960; van Rensburg, 1967).

Nutrients : French (1937) reported 11.74 percent crude protein, 37.67 percent crude fibre and 39.47 percent carbohydrates. Luck (1965) reported 25 to 26 percent crude protein in leaf. Digestibility figures for sheep are more than 50 percent. The palataibility of the hay (French, 1957) and silage (Ryley, 1966) for sheep has been recroded. Sheep ate the silage readily at approximately 1 kg/head/day.

Table 13.4 : Digestible nutrients of grasses

Botanical Name	Common Name	DCP	TDN	SE	References
Tribullus terristriss	Gokhru	3.2	12.9	8.9	Shukla and Ranjan (1969)
Brachiaria mutica	Para grass.	1.2	18.0	13.5	Joshi and Ludri (1967)
Panicum Kabulabula		1.3	26.2	17.8	Joshi and Ludri, 1967
Brigival		2.1	12.5	11.5	Joshi, 1966.
Garthamus tinctorius	Kerr	2.0	12.7	10.4	Ranjan *et al.*, 1959
Ghiabate	Ipomoea	1.5	10.3	9.0	Talpatra and Bansal, 1959.

Macroptilium atrepurpurium (DC) urb.

Vernacular name : Siratro.

Habitat : It is a summer growing perennial with greatest growth in mid summer to autumn. The temperature for optimum growth varies from 26.0 to 30°C. It requires atleast 615 mm and preferably more than 850 mm of rainfall.

Yield : van Rensburg (1967) obtained an average dry matter yield of 79.6 quintal /ha.

Nutrients : Milford (1967) gave the analysis of mature Siratro at the seed shedding stage. He recorded figures of 35 percent dry mater, 16.8 percent crude protein, 33.4 percent crude fibre, 1.2 percent ether extract, 38.8 percent nitrogen free extract and 9.8 percent ash.

Table 13.5 : Chemical composition of grasses on dry matter basis

S. No.	Botanical Name	Common Name	DM	CP	CF	EE	NFE	Total ash	Cal-cium	Phos-phorus	References
1.	*Amaranthus spinosa*	Goja	—	16.13	20.29	2.51	42.12	18.89	2.28	2.28	Singh and Joshi, 1957.
2.	*Andropogon annulatus*		young	9.70	33.06	1.81	42.51	13.58	—	—	Mathur, 1960
3.	*Andropogon contortus*	Speargrass	50.0	1.87	11.12	0.26		2.00	0.22	0.12	Chucko, 1954
4.	*Andropogon Muricatus*		Ripe	4.84	36.60	1.18	44.18	12.60	—	—	Mathur, 1960
5.	*Briachiaria ramosa*	Kura	—	9.43	28.53	0.78	43.92	—	—	—	—
6.	*Cenchrus ciliaris*	Dhaman	Ripe	11.42	27.20	1.87	41.14	18.10	—	—	Mathur, 1960
7.	*Cynodon dactylon*	Doob	Ripe	4.84	36.60	1.18	44.18	12.6	—	—	Mathur, 1960
8.	*Cyperus iria*	—	Ripe	6.04	32.77	6.89	51.69	8.61	—	—	Mathur, 1960
9.	*Dicanthium annulatum*	Karad	—	2.40	35.77	2.01	49.84	—	—	—	—
10.	*Carthamus tinctoris*	Kerr	—	12.40	30.89	1.94	44.37	—	—	—	Ranjan *et al.*; 1959.
11.	*Heteropogon contotus*	—	young	5.00	34.24	1.36	47.70	11.64	—	—	Mathur, 1960
12.	*Ischeum rengachunianum*	—	60.0	1.90	13.27	0.87	—	2.22	0.20	0.05	Chucko, 1954
13.	*Lasiurus sindicus*	Seven	—	5.98	30.08	—	44.73	—	—	—	—
14.	*Panicum sa uilatis*			7.47	31.64	2.01	59.19			—	Mathur, 1960

Table 13.5 Contd. .....

S. No.	Botanical Name	Common Name	DM	CP	CF	EE	NFE	Total ash	Cal-cium	Phos-phorus	References
15.	*Panicum antidotale*	Blue panic	Ripe	7.21	40.07	1.91	43.11	7.07		—	Mathur, 1960
16.	*Panicum colonum*		Ripe	7.26	40.47	1.19	43.11	7.79	—	—	—
17.	*Panicum maximum*	—	Ripe	7.69	37.33	1.67	39.44	13.89	—	—	Mathur, 1960, Saxena *et al.*, 1972
18.	*Pennisetum flacidum*		21.15	11.36	30.40	1.97	45.07	11.21	0.34	0.30	Venkatachalam, 1960
19.	*Tribulus terestris*	Gokhru	—	12.10	27.80	2.60	40.80	—	—	—	—

The digestibility figures were 50.4 percent for dry matter, 53.4 percent for organic matter, 67.6 percent for protein, 50.9 percent for fibre and 50.6 percent for nitrogen free extract. It is very palatable.

Mucuna prurients **(L) D.C. var** *utilis* **(Wight) Burck**

Vernacular names : Velvet bean, Bengal bean.

Distribution : Probably native to southern Asia and Malaysia, now widely distributed in tropics.

Velvet beans require a hot, moist climate for maximum growth. They grow from sea level to 2100 m. The rainfall range of the species is from 650 to 2500 mm. Tolerates a wide range of soils from sands to clays and will grow on soils of appreciable acidity.

Yield : King et al. (1965) stated that a crop of velvet beans yielded 17.4 t of green fodder per hectare.

Nutrients : Axtmayer *et al.* (1938) had the plants analysed when in flower and also the pods. The plant contained 15.65 percent crude protein and 34.47 percent crude fibre. French (1935, 1937) found that the velvet bean hay contained 17.84 and 12.95 percent crude protein and 10.88 and 7.12 percent digestible protein, respectively.

Pennisetum perpureum **Schum**

Vernacular names : Elephant grass, Napier, Hathi grass.

It grows well in loam or sandy loam soil with good drainage. Saline soil is not good for it. It is a good fodder crop which can give green fodder throughout the year, for several years. It attains a height of 2.5 to 3.0 metre. Its leaves resemble those of maize.

Sowing : It can be sown from February to July. Its small stem pieces are sown like sugarcane. These pieces are sown at a distance of 0.5 m, row to row distance is 1.0 m. It can also be sown, with lucerne or velvet beans. About 20-25 thousand pieces are needed per hectare. Humid and hot humid climate is good for this crop. It continues to give fodder upto 5-6 years. About 6 cuttings per year produce 250-300 quintal of green fodder. If the soil is fertile and climate favourable, it can yield 600 quintal of fodder in 6-8 cuttings.

Nutrients : It has 22 percent dry matter, 6.3 percent protein and 12.5 percent digestible nutrients.

Pusa giant Napier

It has been produced by crossing Napier with Bajra. It is better than Napier. The height of the plant is 3.0 m.

Sowing : Fifty carts of organic manure, 2.5 quintal superphosphate, 1.5 quintal calcium ammonium nitrate and ammonium sulphate is put per hectare. Then the field is ploughed 2-3 times. Pusa giant Napier are sown at a distance of 75 cm in February. Then they are covered with soil and watered immediately. Two to three irrigations are done at weekly intervals and there after at fortnightly intervals. About 15-20 thousand stem cuttings are required per hectare.

Fodder : The first cutting can be taken by the middle of May. Thereafter cuttings can be taken every 40-50 days. Fertilizer should be added to soil and irrigation should be done 10-15 days after second cutting for better growth. A plant can give fodder for 2-3 years.

Nutrients : The fodder is nutritious and palatable.

Pennisetum pedicellatum

Vernacular name : Dinanath grass

Habitat : It is an annual which grows in arid and semiarid areas where the rainfall varies between 500 and 1000 mm. It is grown in Andhra Pradesh, Bihar, Bengal, Orissa and Tripura. It is also grown in parts of Assam, Gujarat, Haryana, Himachal Pradesh, Karnataka, Madhya Pradesh, Tamil Nadu and Uttar Pradesh. Loam and sandy loam soil is good for its growth. It can be grown in poor soils also but growing in fertile soils is beneficial.

Sowing : It is best sown in June-July. It can be sown in March-April in irrigated places. It is sown in rows. Plant to plant distance is 25 cm. Seed can be broadcasted. About 10-12 kg seed is required for a hectare. It needs one irrigion in November if sown in rainy season. Summer sown crop needs irrigation at an interval of 10-15 days.

Fodder : It is first harvested 70-80 days after sowing. It can be cut at flowering stage for second cutting. It can yield 600-700 quintal of green fodder per hectare.

Nutrients : It is palatable fodder, grazed by sheep, goat and cattle. Nutrients vary according to age of the grass. At 30, 45, 60 and 80 days it has CP 19.0, 10.0, 5.7 and 3.4, percent respectively. CP may be 25-33 percent. Similarly DP, EE, CP, NFE are 50, 63, 65 and 62 percent respectively.

Vigna unguiculate **(L) Walp**

Vernacular names : Cow pea, Lobia,

Distribution : The country of origin is uncertain. Vavilov (1951) thought it might be India with secondary centers in China and Ethiopia. It is widespread throughout tropics and subtropical areas. It is grown in summer. It is herbaceous annual with twining stems.

Yield : Heavy grazing should be avoided. At no time should be grazed or cut before flowering. Milford and Minson (1968) found that cowpea yielded 23.1 quintal DM/ha in summer and 40.7 quintal DM/ha in autumn and early winter.

Nutrients : The crude protein contents of cowpea were 13.01 percent and digestible crude protein 7.92 percent. For first day or two the sheep may shun the crop, but when they become accustomed to it they will eat it readily.

In addition, the chemical composition of *Anagallis arvensis* (Upadhyaya *et al.,* 1971); Bajra Orientale (Bhatia and Munshi, 1972); *Brachiaria mutica* (Singh *et al.,* 1963); *Cassia tirra* (Gupta *et al.,* 1970); *Chicorium intybus* (Shukla *et al.,* 1970 b); *Clitoria ternatea* (Katiyar *et al.,* 1970); *Cynodon dactylon* (Pachauri *et al.,* 1970); *Dactyloctenium aegyptium* linn (Katiyar & Ranjhan, 1969a); *Dendrocalamus strictus* (Jayal, 1966); *Digitaria decumbens* (Kulshreshta *et al.,* 1971); *Grewia oppositifolia* (Sharma *et al.,* 1966); *Indigofera cordifolia* and I. *en-neaphyla;* (Nath & Malik, 1968); and *Morus indica* (Narayana & Setty, 1977); *Pennisetum clandestinum* (Katiyar & Ranjhan, 1969b); *Pennisetum purpureum* (Singh, *et al.,* 1965); *Panicum maximum* (Saxena *et al.,* 1972); *Phaseolus atropurpureus* (Saxena *et al.,* 1971); *Rhynchosia minima* (Shukla *et al.,* 1970a); *Glycine max* (Johri, *et al.,* 1971); *Vigna sinensis* (Singh & Patnayak 1977b) and *Ziziphus nummularia* (Malik and Nath, 1970) has been determined.

Table 13.6 : Digestibility of dry matter and crude protein contents of some grasses and legumes.

Species	Dig DM (%)	C. P. (%)	References
GRASSES :			
Brachiaria mutica	55–61	6.5–132	Combellas and Gonzalez, 1973.
Cenchrus ciliaris	60–77	8.6–17.3	Combellas and Gonzalez, 1973.
C. ciliaris	44	9.7	Singh, 1975.
Chloris gayana	62–67	–	Mc Leord and Minson, 1974.
C. gayana	48–63	4.9–13.1	Minson, 1972.
C. gayana	–	8.2–17.0	Ruxton, 1975.
Cynodon dactyton	47–53	9.1–11.7	Combellas *et al.*, 1972
C. dactylon	–	8.4–9.6	Ruxton, 1975.
Digitaria decumbens	39–63	–	Funes, 1975.
D. decumbens	66	7.2	Kulshreshtha *et al., 1972.*
D. decumbens	49–67	6.0–14.9	Minson, 1972.
Hyparrhenia spp.	54	6.5	Gihad, 1976.
Panicum maximum	47–64	6.4–16.0	Minson, 1972.
Paspalum dilatatum	43–59	5.9–13.3	Minson, 1972.
Pennisetum clandestinum	47–62	5.9–17.3	Minson, 1972.
P. purpureum	61	4.5	Lima *et al.*, 1972 a

Table 13.6 Contd..........

Species	Dig DM (%)	C. P. (%)	References
Setaria anceps	60	10.5	Sharma *et al.*, 1972.
S. splendido	63–66	–	Mc Leord and Minson, 1974.
S. splendida	51–67	5.7–13.6	Minson, 1972.
LEGUMES :			
Desmodium uncinatum	47–54	10.9–18.2	Milford, 1967.
Lablab purpureus	56	14.4	Hendriksen *et al.*, 1981.
Lotononis bamesii	60	20.0	Milford, 1967.
Macropitilum atropurpureum	42–71	13.8–18.0	Lima *et al.*, 1972 b.
M. atropurpureum	50	16.8	Milford, 1967.
M. atropurpureum	55	16.7–23.0	*Saxena et al.*, 1971.
M. lathyroides	42–62	7.6–19.2	Milford, 1967.
Stylosanthes guianesis	57–64	12.4–13.7	Ademosun, 1970.
S. guianesis	48	11.8	Milford, 1967.
Trifolium alexandrium	68	20.6	Sharma and Murdia, 1974.
Vigna vexillate	58–69	16.2–20.3	Milford, 1967.

Table 13.7 : Different varieties of fodder crops recommended for different parts of the country

Name of crop	Varieties	Recommended in	Green fodder yield (q/h)	Remarks
1	2	3	4	5
Berseem	Mescabi	All parts where berseem is sown.	800–900	Can tolerate high temperature.
	Pusa giant	–do–	900–1000	Gives more production in lower atmospheric temperature, leaves bigger in size, succullent.
	B.L.1	Punjab	800–900	Capable of giving fodder upto late stage.
	S. 91–1	All regions where berseem is grown	1000	Grows well even in places where the temperature is low.
	J.B.1	–do–	1000	Regeneration good, can be grown in places where the temperatureis either low or very high.
	C. 3–94	M.P., U.P. Haryana, Punjab	1000–1050	Fast growing, gives fodder for longer period.
	U.P.O.–102	Throughout India	900–1000	Can tolerate high or low temperature, gives fodder for longer periods.
	U.P.O.–103	–do–	900–1000	–do–
	B.L. 10	Punjab	1000	Higher fodder and seed yield.
Cowpea	Russian giant	Plains of Northern India and Assam.	350–400	Late maturing, good– height and branching.
	E.G 4216	plains of Northern India	300	Medium duration, growth as for Russian giant.

Table 13.7 Contd. .....

1	2	3	4	5
	F.O.S.–1	Haryana, Panjab, Delhi.	250	Medium duration, stem erect.
	H.F.C. –1	Throughout India	300	Matures earlier, stem erect, bushy.
	No. 10	Haryana, Punjab, Delhi	200	Stem erect, medium duration.
	S.450	Througout India	250	Fast growing, good for summer, early maturing.
	S. 457	Throughout India	250	Fast growing, early maturing.
	C. 26	Throughout India	350	Multicuts, resistant to diseases.
	J.C.1	M.P., U.P.	300	Dark green, Geaper type.
	U.P.C	Southern India	350	Medium duration, dark green, resistant to diseases.
	U.P.C. 5286	Throughout India except S. India	400	Medium duration, darkgreen, resistant to diseases.
Dina Nath	P.P. 3	Haryana, Punjab, M.P., U.P.	400	Fast growing, succullent, leafy, early maturing.
	P.P. 10	Warm climate provinces	450	Late maturing.
	P.P. 15	Warm climate provinces	450	Medium duration, leafy, fast growing
	Pusa 38	Warm climate provinces	500	Fast growing.
	Pusa 3	Warm climate provinces	500	Fast growing.
	P.P. 47	Warm climate provinces	500	Fast growing.
	S. 3808	Warm climate provinces	500	Fast growing.

Table 13.7 Contd. .....

1	2	3	4	5
	JP 12	Warm climate provinces	510	Fast growing.
Guar	No. 2	All guar growing areas	175	Plants dwarf, more grains and more branching.
	F S 277	All guar growing areas	225	Plants, tall, more fodder, grain production less.
	Durgapura white	All guar growing areas	175	plants dwarf, branching, good for grains.
	T–3	All guar growing areas	175	Early maturing, not good for grains.
	HF.G. 119	All guar growing areas	250	Plants tall, leafy, good for fodder and grains
	G.L. 6	All guar growing areas	250	Medium height, good for fodder and grains.
Lucerne	T. 8	Northern, Northwest and central India	600–650	Good for places where it is sown every year, leafy, fast growing.
	T. 9	Throughout India	800–1000	Fast growing, disease resistant.
	Anand 1	Gujarat, Maharashtra	800–1000	Fast growing & disease resistant.
	Anand 2	Gujarat, Maharashtra, Haryana	700–800	Fast growing, disease resistant and regeneration.
	Coimbatore	Tamil Nadu, A.P.	800–1000	Disease resistant, fast growing after cutting good for seed production.
Senji	F.O.S.–1	Dry regions of Northern India	400	Good leguminous fodder.
	No. 76	Haryana, Punjab, Delhi, western U.P.	450–500	Late maturing, leafy, dark green colour.

Table 13.7 Contd. .....

1	2	3	4	5
Metha	T.–8	Punjab, Haryana, Western U.P.	350–400	Leaves broad, good for late sowing.
Makchari	–	Throughout India	500–600	Multicut.
Bajra	H.B.S	Throughout millet growing areas.	500–600	Plants tall, leafy, profuse tillering.
	H.B.S	Throughout millet growing areas.	500–550	Leafy, tall, dark green.
	T. 55	Semi arid and arid parts	500–500	Leafy, as good as native varieties.
	S.530	Throughout millet growing areas.	450	Leafy, darkgreen, multicut, profuse tillering
	K 675	Throughout millet growing areas.	600	Tall, dark green.
	K 677	Throughout millet growing areas.	600	Tall, dark green.
Napier	Pusa giant	Hot humid areas	1000–1200	Fast growing, leafy.
	Gujrat	Hot humid areas	800–900	Leafy, Tall.
	N.B. 21	Hot humid areas	1200–1500	Leafy, thin stem, Fast growing.
	Hybrid 2	Hot humid areas	700–800	Leafy, thin stem.
	Hybrid 3 and 6	Hot humid areas	700–800	Leafy.
	N.B.5	Hot humid areas	800	Tall, profuse tillering
	N.B. 9	Hot humid areas	800	Leafy soft stem, profuse tillering.
Sorghum	J.S. 20	Punjab, Haryana, U.P., Rajasthan.	300	Early maturing, thin stems.
	Vidisha 60–1	Bihar, Haryana, M.P., U.P., Punjab.	400–450	Late maturing, thick stems, Tall plants
	M.P. Chari	Bihar, Haryana, M.P., U.P., Rajasthan, Gujrat.	400	Early maturing, succullent, better for warm climate areas.

Table 13.7 Contd. .....

1	2	3	4	5
	Hundi	Maharashtra	330	Late maturing, good for grains and fodder.
	Pusa chari–1	UP, Haryana, Gujrat, A.P., Bihar, Maharashtra.	400–425	Medium duration, Sweet succulent stems.
	Pusa chari–6	Almost throughout India	450–500	Stems thin and succullent, early maturing.
	Haryana chari	Gujrat, Haryana, Punjab.	350	Early maturing.
	S.C. 44	Punjab, UP, Haryana, Bihar	350	Early maturing, stems thin.
Sudan grass	Sweet Sudan (SSG 69–3)	Throughout India	600–650	Thin succulent, multicut, profuse tillering, leafy good for summer early maturing.
	J.S. 4776	Southern India, Maharashtra	400	Early maturing.
	J.S. 6	M.P., UP., Bihar.	500	Late maturing.
	S. 136	Haryana Punjab, UP.	450–500	Medium duration crop.
	J. 69	U.P, M.P., Bihar, Rajasthan	450–500	Early maturing, multicut, good for summer cultivation, beter than M.P. chary in disease resistance.
	Ganga– 2	Plains	450	Medium duration crop.
	Ganga– 5	Plains of Northern India	450	Medium duration crop.
	Ganga– 7	Plains of Northern India	450	Medium duration.
	Vijay	Throughout India	425	Medium duration, Disease resistance.

14

Economics

Indian breeds of sheep are poor producers of wool and mutton than their counterparts in some other parts of the world, yet they contribute substantially to our national income. The entire quantity of wool is used in manufacturing warm clothing, blankets and carpets. Most of the wool produced in our country is of the carpet wool variety. Apart from wool, sheep provide mutton, manure, milk and hides (Sahani *et al.*, 1981; Kaushish and Jain, 1987). They are also used as pack animals in some parts of the country over the high mountain ranges, where other means of transportation are hard to use. The return from sheep depends on the type of sheep i.e. the breed and other economic characters inherent in the breed such as fleece weight, birth weight, growth rate and lambing percentage.

Utility of sheep

Sheep is an important source of livelihood to a large population in arid and semi-arid areas and provides it with food and fiber. In addition to wool, meat and milk, sheep provide employment to a large section of the population in these areas in the form of self-employment of persons attending their own flocks or those hired for attending flocks during migration. Economic characters of some natural and cross bred sheep are given in Table 14.1. Besides grazing and management of sheep many people are engaged in the work of sheep shearing and in wool and skin based industries. According to 1972 census, there were 40.39 millon sheep producing 34.5 million kg of wool, 101.1 million kg of mutton, 14.6 million pieces of skin, in addition to manure, casings, offals etc and contributes Rs. 140 crores per annum to the national income.

Economic analysis of a flock of 100 adult females based on the data recorded and survey conducted during 1985-86 in semi arid areas of Rajasthan is presented below :-

Sources of income

1. The flock consisted of 100 adult females and 2 to 3 rams at the beginning of the year. Considering 7 percent adult mortality, the adult strength at the end of the year will be 93. Of these 83 ewes were available for breeding as 10 ewes may not be available due to low body weight, debility, old age etc. Estimated number of live lambs born is 75 considering 10 percent abortions still births etc. The number of lambs reaching yearling age of the young stock will be 62 assuming 20 percent young mortality. Of these, 31 will be male lambs and 31 will be female lambs. The male lambs are disposed off at marketing age which varies from 5 to 12 months. Lambs were marketed in two lots i.e. about two fifth males during April-May (at the age of 6 months) @ Rs. 110/- each and rest males during October-November (at the age of 12 months) @ Rs. 130/- each giving an income of Rs. 3790.00 from sale of 31 male lambs.

2. The next important source of income is from sale of washed wool. Sheep are shorn twice or thrice in the area. The average annual washed wool production is 0.90 kg per adult and 0.70 kg per lamb. The total wool production from adults (90 × 0.9 = 81.00 kg), female lambs (31 × 0.70 = 21.70 kg) and male lambs (31 × 0.50 = 15.5 kg) will be 118.2 kg. Market rates for washed wool ranged from Rs. 35.00 to 43.00 per kg. Considering average rate of Rs. 39/- per kg. the income from sale of wool will be Rs. 4609.80 per year.

3. Manure is also a source of income. A flock of 100 adult ewes will provide 156 to 170 quintals manure per year. Considering average collection of 165 quintals of manure and sale at the rate of Rs. 70/- per cart load of 5 quintals, the income from sale of manure is Rs. 2310/- per year.

4. In the area sheep is also milked and milk is sold. Considering average production of 250 gm during the first fifteen days and 130

gm during next 30 days, 65 adult females in milking would give 496 kg of milk. The income from sale of milk at the rate of Rs. 2.50/kg will be Rs. 1240/- per year.

5. The income from the sale of 7 hides from adult sheep and 13 hides from young ones at the rate of Rs. 12.00 and Rs. 6.00, respectively comes to be Rs. 162.00 per annum.

Total income from all the five sources listed above works out to Rs. 12111.80 per year.

Sources of expenditure

Following expenditures normally occur on the given flock size:–

i. The cost of vaccinating 170 sheep against sheep pox and enterotoxaemia @ 10 paise each will be Rs. 34.00 per year.

ii. The cost of health cover including drenching was Rs. 2.00 per sheep in the project area. The expenditure on health coverage will be Rs. 340/- per year.

iii. The shearing charges per sheep are Rs. 1.00 and the expenditure on shearing will be Rs. 320.00 per year.

iv. Cost of purchase of patri (Acacia pods) for feeding sheep during lean period will be Rs. 600/- per year, cash payment of wages Rs. 600/-, two dresses (*dhoties* and shirts) Rs. 100/-, one pair shoes Rs. 30.00 and expenditure on feeding Rs. 1190/-.

Income

Thus the total expenditure on maintenance of a flock of 100 adult female sheep will be Rs. 3234/- and the net income will be approximately (Rs. 12111.80 - 3234.00 = 8877.80) say Rs. 9000/- per year.

This analysis however has not taken into account expenditure on investment, repayment of loan interest, depriciation and income through increase in flock addition of 30 yearling females.

Socio economic survey

Detailed economics of sheep farming worked on the basis of results obtained from socio-economic survey conducted on sheep farming in Malpura sub-division is presented in Table 14.2 The net annual income from owner grazed flock of 100 breedable ewes, adult rams and followers was Rs. 3472. Supplementary feeding is rarely practised for which a provision of Rs. 810 has been made. The

Table 14.1 : Average for various economic characters in some natives and crossbred sheep.

							Percent Mortality					
	6 month wt. (kg)			Adult animal			Pre–weaning		Post–weaning		Adult	
	Native	Cross breed	% improv -ement	Native	Cross breed	% improv -ement	Native	Cross breed	Native	Cross breed	Native	Cross breed
Rambouillet												
Nilgiri	15.9	19.8	24.7	0.84	2.34	179.6	21.9	10.6	5.3	8.0	8.5	3.0
Gaddi	10.4	11.7	12.6	0.90	1.47	68.7	16.2	21.5	14.4	11.4	10.3	10.2
Chokla	12.7	15.2	19.0	*1.34	1.22	8.7	*38.3	36.7	–	–	8.5	5.4
Nali	13.5	14.6	8.5	*1.45	1.25	12.3	*51.0	33.3	–	–	9.1	7.3
Chokla	–	–	–	*1.32	1.49	14.3	28.5	10.1	48.5	10.3	10.4	7.0
Malpura	–	–	–	0.56	1.14	76.7	14.0	4.6	31.8	4.8	1.3	3.8
Merino												
Nilgiri	15.9	18.4	16.0	0.84	2.41	187.5	21.9	5.5	5.3	2.9	8.5	3.3
Gaddi	10.4	12.4	19.3	0.90	1.53	70.6	16.2	21.2	14.4	16.9	10.3	9.3
Chokla	12.7	14.8	16.2	*1.36	1.20	10.5	*38.4	45.0	–	–	8.3	5.3
Nali	13.5	14.6	18.2	*1.45	1.80	11.9	*51.0	35.0	–	–	9.1	6.0
Dorset X												
Mandya	15.9	18.1	13.8	0.28	0.39	39.8	9.9	9.7	11.2	12.4	16.4	9.8
Nellore	16.6	20.4	23.0	–	0.33	–	7.8	6.0	11.3	11.5	13.6	9.9

Table 14.1 Contd........

							Percent Mortality					
	6 month wt. (kg)			Adult animal			Pre–weaning		Post–weaning		Adult	
	Native	Cross breed	% improv -ement	Native	Cross breed	% improv -ement	Native	Cross breed	Native	Cross breed	Native	Cross breed
Muzafar-nagri	16.2	19.2	18.3	*0.55	0.63	15.6	*31.0	16.7	-	-	13.9	15.4
Malpura	12.6	16.4	31.0	*0.54	0.58	6.3	*30.2	37.5	-	-	8.9	16.5
Sonadi	12.2	16.1	21.6	*0.43	0.51	18.5	*38.1	39.8	-	-	8.1	15.6
Suffolk X												
Mandya	15.9	18.3	15.1	0.28	0.34	20.1	9.2	15.4	11.2	15.2	16.4	7.4
Nellore	16.0	18.7	12.6	-	0.32	-	7.8	3.9	11.3	44.0	13.0	9.5
Muzafar-nagri	16.2	19.8	22.2	*0.54	0.68	23.7	*31.1	14.3	-	-	13.9	8.4
Malpura	12.6	16.9	34.9	*0.54	0.54	0.9	*30.2	44.6	-	-	8.9	33.3
Sonadi	13.2	17.2	30.4	*0.43	0.48	11.4	*30.1	47.1	-	-	8.1	34.2

* 6 Monthly   * Young mortality up to one year.

Table 14.1 Contd. .....

	6 month wt. (kg)			Adult animal			Percent Mortality: Pre–weaning		Post–weaning		Adult	
	Native	Cross breed	% improv -ement	Native	Cross breed	% improv -ement	Native	Cross breed	Native	Cross breed	Native	Cross breed
Muzafar-nagri	16.2	19.2	18.3	*0.55	0.63	15.6	*31.0	16.7	-	-	13.9	15.4
Malpura	12.6	16.4	31.0	*0.54	0.58	6.3	*30.2	37.5	-	-	8.9	16.5
Sonadi	12.2	16.1	21.6	*0.43	0.51	18.5	*38.1	39.8	-	-	8.1	15.6
Suffolk X												
Mandya	15.9	18.3	15.1	0.28	0.34	20.1	9.2	15.4	11.2	15.2	16.4	7.4
Nellore	16.0	18.7	12.6	-	0.32	-	7.8	3.9	11.3	44.0	13.0	9.5
Muzafar-nagri	16.2	19.8	22.2	*0.54	0.68	23.7	*31.1	14.3	-	-	13.9	8.4
Malpura	12.6	16.9	34.9	*0.54	0.54	0.9	*30.2	44.6	-	-	8.9	33.3
Sonadi	13.2	17.2	30.4	*0.43	0.48	11.4	*30.1	47.1	-	-	8.1	34.2

* 6 Monthly   * Young mortality up to one year.

Table 14.2 : Estimated cost and return for a flock of 100 adult units.

A.	**Non-recurring expenditure**			
1.	Cost of 100 ewes @ Rs. 110.00 each		Rs.	11,000.00
2.	Cost of 4 rams @ Rs. 200.00 each		Rs.	800.00
3.	Construction of pens with thached roofing of 250 sq. ft.		Rs.	300.00
4.	Equipments		Rs.	200.00
		Total (A)	Rs.	12,300.00
B.	**Recurring expenditure**			
1.	Cost of feedings concentrate @ 200g/ animal/day for two months at the time of advance pregnancy and early lactation i.e. 9 quintals @ Rs. 90 per quintal		Rs.	810.00
2.	Medicines		Rs.	100.00
3.	Shearing charges @ Rs. 0.30/animal twice in year		Rs.	75.00
4.	Interest on capital investment @ 11% per year		Rs.	1,333.00
5.	Depriciation		Rs.	100.00
		Total (B)	Rs.	2,438.00
C.	**Income**			
1.	Sale of wool @ Rs. 18.00 per kg. 0.95kg/adult/year and 0.4 kg/ lamb		Rs.	2,170.80
2.	Sale of ram lambs 32 @ Rs. 65.00 each		Rs.	2,080.00
3.	Sale of skin of dead animals @ Rs. 2.00 per skin		Rs.	32.00
4.	Sale of manure 20 carts @ Rs. 25.00 per cact		Rs.	500.00
5.	Capital increment in farm strength **		Rs.	1,127.50
		Total (C)	Rs.	5,910.30

Net income = Rs. 5910.30-Rs. 2438.00 = Rs. 3432

** Stock value

a)	Balance of old stock @ 5 per cent mortality and 5 percent depriciation in the price of animals	Rs.	10,687.00
b)	Balance of new stock 32 ewes lambs @ Rs. 70.00 each	Rs.	2,240.00
	Total	**Rs.**	**12,927.00**

Capital increament in farm strength = Rs. (12,927-11, 800.00 = 1,127.50)

Note : 1. Mortality in adult stock and lambs has been considered @ 5 and 15 per cent respectively.

2. Grazing will be available free of cost.

3. Annual lambing percentage has been considered as 75 percent.

income from sale of milk and milk products was also estimated at Rs. 1536 per year considering surplus saleable milk production of 400 ml/ewe/day for 60 days i.e. 1536 litres milk sold @ Rs. 1.00 per kg. Annual income from sale of wool, male lambs, culled sheep, hide, manure, milk and products was estimated to be Rs. 5,819. Sale of ram lambs, wool and capital gains through increase in flock size are the major sources of income each contributing 35.2, 36.7 and 19.1 percent, respectively to the total income.

A survey conducted by Central Sheep and Wool Research Institute, Avikanagar (Raj.) on economics of sheep rearing in Chokla and Nali region in Rajasthan indicated the average income from a flock of 100 breedable ewes, adult rams and followers to be Rs. 3,861 for Chokla and Rs. 5645 for Nali per annum. The number of lambs born per 100 ewes, lamb and adult survival and market weight determines the return from sale of ewes and ram lambs, greasy fleece weight and quality determines returns from sale of wool. The Nali ewes gave birth to more lambs per 100 ewes (67 vs 60), survived more lambs upto one year of age (83.2 vs 68.8) and produced more wool per year (1.5 vs 1.2 kg) resulting in almost 50 percent higher income from a similar flock. The profit from sheep flock is influenced by herbage production, ewe, ram, and management, breeding, feeding and health.

15

Farm Records

Importance of keeping records

Sheep breeding is an integral part of farm enterprise. Using memory to plan breeding policy has its drawbacks, one cannot remember good and bad points of the flock. It becomes difficult to remember about each and every sheep of the flock. Sheep farmer can find out from farm records, where his present weakness lie and which aspect of this enterprise should be improved. The records of near relatives will give some idea of the genetic potential of the animal, since occasionally a good animal may come from a poor family and *vice versa.* The sheep farmer should be able to see from his records whether or not is being successful. Very often a farmer will find it simple and more convenient to use records on specially designed farms, but there is nothing to prevent him from making and using his own.

The recording system should be simple, accurate capable of collecting the required information and finally kept upto date. The data thus collected should not be kept but should be used up. In sheep growth rate, food conversion, wool quality dressing percentage are highly heritable characters. Prolificacy, milking capacity and fleece weight are characters of low heretability. More profits can be obtained if more attention is paid to characters of high heretability. Having decided his breeding programme a farmer must then choose his parent stock.

Method of selection on the basis of appearance gives good results. As the knowledge in this field increases, sound relationship between visible characteristics in the live animal and important economic factors may be found. Choosing sheep without any regard for their confirmation can be as disastrous as choosing them on look alone.

Good record should identify the offspring with the parents. They should also give date of birth, sex and final disposal of individual. This information together with desirable confirmation is the basis upon which selection for flock improvement should be made.

There should be a system for ear marking, tattooing and or tagging. In case of short ear breeds, tattooing on inside of thighs or tails be practiced. A combination of tattooing and tagging is desirable.

Following data observations are of great help at an organised farm :

Livestock inventory : This register gives information about the individual identification, parentage and date and reason for sale or culling.

Wool weight and quality : Wool evaluation should be limited to greasy fleece weight of each dip, staple length at a particular site, average fiber diameter, medullation percentage, visual assessment of fineness and making careful examination of undesirable characters such as black fibers or coarse fibers.

Growth records : Body weights at birth 3, 6, 9 and 12 months of age are important. Birth weight is related with vigour, survivability and growth rate. Weaning weight is used to measure total lamb production of an ewe. Six month weight is correlated with market and yearling weight is highly related to adult weight and ewe productivity. Body weights of each lamb are required to be correlated to age, sex, type of birth, season and age of dam etc.

Prolificacy : Twin producing ewes are more profitable than single producing ewes excepting in areas where grazing conditions are poor.

Health : Information on causes of death and incidences of diseases helps in planning prophylactic health programme.

Importance of Ram : Ram is half of the flock. The improvement in a character in a flock can be made rapidly by introducing a ram with that character. The choice of a suitable ram needs a careful thought. In improving a flock the ram is more likely to have greater effect. The method of evaluating the true breeding worth of a sire are performance testing and progeny testing.

Mutton production : Efficiency of feed conversion is an economic trait with high heretability.

Breeding plan for flock improvement : All the sheep should be identified by tagging or tattooing. Record the performance on individual lamb, ewe and ram. Adjust weight of each lamb for age, sex, season, twin age of dam etc. Cull ewes with poor performance records, defects, wean light lambs and produce light fleece at shearing. When more than one ram is used in the flock compare the records of the off spring and cull those rams whose off springs are below average for desired traits. Ram lambs should be selected on the basis of their own weaning weights, growth rate, quantity and quality of their wool.

Recording should be encouraged not only as an aid to management, but also to provide more accurate base for culling and selection of ewes for further breeding. Recording is essential to constructive ram breeding. The performa for recording are given in Tables below:

Table 15.1 : 1. DAILY DIARY

Place Name of incharge

Date

i. Flock structure :-

Age	Breed and number				
	A	B	C	D	Total
Adult					
above 6 months					
3-6 months					
below 3 months					
Total					

ii. Shepherds :—

Attributes	Name of shepherd
Number of sheep in flock	
Time of sending out	
Breed of sheep	
Time of return	
Pasture plot No,	
Situation of plot	
Number of sheep lost	

iii on Farm :-

Work	Name of shepherd
Watering of sheep cleaning watchman care of lambs care of sick sheep	
Total No of shepherds — on leave — on rest — on rest — Balance —	

iv. Birth of lambs :

S.No.	Tag No. of lamb	Bread	Sex	Birth weight	% colour	dam number	Weight

v. Breeding:

S.No.	dam No.	dam breed	Sire no.	Sire breed	weight of Remarks dam

vi. Weight:

S.No.	Ram /Ewe/ lamb number	Breed	Sex	Weight	Remarks

vii. Drenching and vaccination

Vaccination
Drenching
Others

viii. Treatment

Sheep/ Lamb No.	Disease	Treatment	Rectal temp.

x. Death /culling:

x. Other operations

Docking
Castration
Hoof pairing

Table 15.2 : LAMBING REGISTER

S.No.	Tag No.	Sex	Birth weight	Type of birth		Ewe		Ram		Weaning		Culling		Re-marks
				Single	Twin	Tag No.	Breed	Tag No.	Breed	Age	weight	Date	Method	
1	2	3	4	5	6	7	8	9	10	11	12	13	14	15

Table 15.3 : RAM RECORD REGISTER

S.No.	Tag No.	Breed	Sire Tag No.	Dam Tag No.	Date of birth	Nutrition		Birth		Birth weight	Weaning Wt.	Yearly Wt.	First clip Wt.	Wool quantity	Re-marks
						Natural	Artificial	Single	Twin						
1	2	3	4	5	6	7	8	9	10	11	12	13	14	15	16

Table 15.4 : SHEEP REGISTER

S.No.	Tag No.	Breed	Tag No.		Date of birth	Nutrition		Type of Birth		Weight at			First clip		Re-marks
			Sire	Dam		Natural	Artificial	Single	Twin	Birth	3 month	12 month	Weight	quality	
1	2	3	4	5	6	7	8	9	10	11	12	13	14	15	16

Table 15.5 : LAMBS REGISTER

S.No.	Date of birth	Tag No.	Breed	Sex	Sire No.	Dam No.	Type of Birth		Weight at			Date of castration	Colour	Colour%	Date of culling	Re-marks
							Single	Twin	Birth	3 month	6 month					
1	2	3	4	5	6	7	8	9	10	11	12	13	14	15	16	17

Table 15.6 : BREEDING REGISTER

S. No.	Dam No.	Weight of dam	First service		Second service		Third service		Date of Lamb-ing	Weight of dam at lamb-ing	Lamb			Birth type		Remarks Aborti-ons etc.
			Ram No.	Date of service	Ram No.	Date of service	Ram No.	Date of service			Tag No.	Sex	Wt.	Single	Twin	
1	2	3	4	5	6	7	8	9	10	11	12	13	14	15	16	17

Table 15.7 : GROWTH REGISTER

S. No.	Tag No.	Sex	Date birth	Date of Sex	Breed of dam	Weight of dam at lambing	Weight of 1amb							Rem-arks
							at birth	1 month	2 month	3 month	6 month	9 month	12 month	
1	2	3	4	5	6	7	8	9	10	11	12	13	14	15

Table 15.8 : WOOL SHEARING REGISTER

S.No.	Tag No.	Sire Tag no.	Dam Tag No.	Wool Weight	Measurements		Heart Girth	Ponch girth	Remarks
					Length	Height			
1	2	3	4	5	6	7	8	9	10
1.									
2.									
3.									
4.									
5.									
6.									

Table 15.9 : Feeding and weighing records on the feed lot experiment.

Lamb No.		Sire No.				Dam No.		Date of putting in feed lot			
I		II		III		IV		V		VI	
Initial weight (kg)											
Initial Measurements:											
L.H. H.G. PG		L.H. H.G. PG		L.H. H.G. PG		L.H. H.G. PG		L.H. H.G. PG		L.H. H.G. PG	
Date	Feed	Date	Feed	Date	Feed	Date	Feed	Date	Feed	Date	Feed
1											
2											
3											
4											
5											
6											
7											
8											
9											
10											
Total		Total		Total		Total		Total		Total	
Left overs (1):		Height		Height		Paunch girth				Feed eficiency:	
Net feed consumed											
Final weight (kg)				Heart girth gain during the Fortnight (kg)							
Final measurements:											

Table 15.10 : Live stock strength register

Breed/Breed Cross								
S.No.	**Animal indent**	**Sire indent**	**Sire breed**	**Dam indent**	**Dam breed**	**Date of purchase**	**Date of disposal**	**Re-marks**

Table 15.11 : Mortality/Disposal Register

S.No.	**Animal indent**	**Breed/ breed-cross**	**Sex**	**Age at death/ disposal**	**Date of death/ disposal**	**Reason (s) of death/ disposal**	**Re-marks**

Table 15.12 : Daily feed register

S. No.	**Date**	**Breed /Breed -cross**	**Category of animals**	**No. of animals**	**Rate of concen-trate feeding**	**Indent of concent mixture**	**Total Conc. fed**	**Re-marks**

Table 15.13 : Proforma for slaughter data

Tag No. Breed/Cross: Date of birth:

Sire No. Dam No. Date of slaughter:

A. MEASUREMENTS (cm):

Length: Height: Heart girth: Paunch girth:

B. WEIGHTS (kg)

1. Live weight before slaughter :
2. Empty live weight :
3. Hot carcass weight with kidneys and kidney fat :
4. Dressing out percentage on live weight :
5. Dressing out percentage on empty live weight. :
6. Weight of the fore quarters :
7. Weight of the hind quarters :
8. Composition of the cuts from half carcass :

Name of the cut	Weight of the cut	Weight of the Bone	Percent Bones.
Leg			
Loin			
Rack			
Neck & shoulder			
Breast & fore shank			
Total			

9. Loin eye area (mm^2) :

10. Offals (kg) :

a. Blood :

b. Head :

c. Skin :

d. Fore cannons with hooves :

e. Hind cannons with hooves :

f. Lungs with Trachea & diapharm :

g. Gallbladder :

		FULL	EMPTY	INGESTA
h. Stomachs with Oesophagus	:			
i. Small Intestines	:			
j. Large Intestines	:			

k. Testicles :

l. Spleen :

m. Pancreas :

n. Caul fat :

o. Kidney fat :

p. Kidneys :

q. Heart :

r. Liver :

APPENDIX-I

GLOSSARY OF TECHNICAL TERMS

Abscess : A localized collection of pus surrounded by inflamed tissue.

Abortion : Premature expulsion of a foetus before it has matured enough to survive.

Acute : A rapid and often severe development of a disease. Compare with chronic condition, which is continuous and long term.

Additive : A substance included in a diet or ration, usually in small quantities to fortify them with nutrients, stimulants or medicines.

Ad libitum : Feeding without limitation; also called free choice feeding.

Adsorption : The process of being taken up and held on the surface of a substance.

Aerobe : A microorganism that requires the presence of oxygen to live and reproduce.

After birth : The placenta and fetal membranes expelled from the uterus after parturition.

Air dry : Feed dried by natural air movement is termed as air dry. Most feeds are air dry and contain aproximately 90% dry matter (DM).

Ambient temperature : Temperature surrounding the animal at any given time. The same as environmental temperature.

Amino acids : Organic compounds containing nitrogen that are the building blocks in the formation of proteins.

Anabolism : The buildup of complex substances such as body tissues from simple compounds.

Anaerobe : A microorganism that can exist without free oxygen.

Anaemia : A condition in which the blood is deficient in red blood cells or haemoglobin.

Anestrous : The period of sexual quiescence between periods of sexual activity in cyclically breeding mammals.

Anorexia : A lack or loss of appetite for feed.

Anthelmintic : A drug or chemical agent used to destroy or control internal parasites.

Antibiotic : A substance that destroys or inhibits the growth and activity of microorganisms.

Antibody	: A protein substance (modified type of blood serum globulin) development or synthesized by lymphoid tissue of the body in response to an antigenic stimulus. It provides protection against specific disease by inter-acting and neutralizing a disease causing agent.
Antigen	: An enzyme, toxin, or other "foreign" protein to which an animal's body reacts by producing antibodies specific to the invading antigen.
Antioxidant	: A compound that prevents oxidative rancidity of polyunsaturated fats. Used in feeds and foods to prevent rancidity.
Antiseptic	: A chemical substance that prevents growth and development of micro-organisms.
Apparel wools	: All wool suitable for manufacture into apparel fabrics.
Appetite	: The craving for food; the desire to eat.
Artificial insemination (AI)	: Depositing semen into the female reproductive tract by other than natural means (i.e., by a technician).
Ash	: That mineral matter derived from incinerating organic matter.
Assay	: To determine purity or potency of a substance. To determine the level of a substance in a feed.
Astringent	: A substance that causes shrinkage or contracting of localized blood vessels to reduce blood flow and any discharge.
Atrophy	: Wasting away or decrease in size, usually muscular.
Average daily gain (ADG)	: Average daily increase in body weight.
Bacteria	: Single-cell plant microorganisms widely found in manure. Many are beneficial; others are pathogenic.
Bactericide	: A product that destroys bacteria.
Balanced ration	: A daily allowance of feed that provides proper amounts and proportions of nutrients.
Barren	: Description of ewe which fails to give birth.
Basal metabolic rate (BMR)	: The energy required to maintain the vital functions of an animal (i.e., cellular activity, respiration, and circulation). BMR is measured in a thermoneutral environment, resting, postabsorbtive state, consciousness, quiescence, and sexual repose.
Bellies, or belly wool	: Short and often less desirable wool from the belly of the sheep.

Biological control : Control of undesirable insects, other animals, or plants by the use of natural enemies of the pests.

Biosynthesis : The production of a chemical compound by an organism.

Black wool : Any fleece from sheep that is not completely white. It may contain fibers that are black, grey, or brown in colour.

Bleaching : Removal of colouring material in natural wool or yarn.

Bleat : The sound or cry of a sheep.

Blending : Mixing of different types of wool and thread for reducing cost or for lending some specific properties.

Bloat : An abnormal condition found in ruminants characterized by distension of the rumen on the upper left side by excessive accumulation of gas.

Blocky : A term indicating a conformation that is deep, wide short, and low set.

Blood : A method of grading wool with the standards based on the amount of Merino breeding in the animal. Thus, fine for a pure Merino, half-blood, etc. It also is used to denote the amount of improved breeding in an animal.

Bolus : The undigested feed regurgitated during rumination. Also the name of a large pill.

Brand : Numbers or letters stamped on animals using special paint.

Bred : Synonymous with pregnant. Also used to indicate mating.

Breed : A group of individuals that as a result of selection and breeding have similar characteristics that distinguish them from other groups of the same species.

Breed character : Those characteristics that distinguish the breed from other breeds.

Breed type : Those characteristics that are commonly accepted as ideal for a particular breed.

Britch wool : Wool from the hindquarters of a sheep. It is usually the coarsest wool on the body.

Broad tail : The pelt from the slaughter of newborn Karakul lambs.

Broken mouth : Description of a sheep which has lost some or all of its teeth.

Browse : To consume vegetation from trees or shrubs, food which is browsed.

Burry wool : Wool heavily contaminated with vegetable matter, including burrs, seeds, leaves, and twigs. Such wool is discounted because of the expense of removing the foreign matter.

By-product : An output from a system, but not a primary product (e.g. hooves are a by-product of meat production, and straw is a by-product of cereal production).

Caesarian section : Removal of the foetus by surgical means at or near term. Caesarean sections are necessary if the pelvic opening is too small or when the foetus is hopelessly malpositioned.

Calcification : Hardening of tissues by deposits of calcium salts. Caloric : Relating to heat energy.

Caracul : The pelt from newborn Karakul lambs.

Carbonizing : removal of vegetable matter of wool by using sulphuric acid.

Carcass : The dressed body of a lamb of sheep after the offal, pelt, feet, and head are removed.

Carcass weight : Weight of the carcass on the rail, either hot or chilled.

Carcass yield : The percent of live weight in carcass (carcass weight + live weight × 100). Also called dressing percent.

Card : A machine for processing wool before spinning. Also a hand instrument for disentangling wool when trimming show sheep.

Carding : It is opening, separating and straightening of individual fibers of wool.

Carding wool : Fine wool with 1 1/4 - in. staple and coarse wool of 2 in. may be carded for the production of woolen yarn. Synonymous with clothing wools.

Carotene : Substance occurring naturally in green plants that is the precursor of vitamin A.

Carpet wool : Long staple, strong, coarse wool suitable for carpet manufacture.

Carrying capacity : The number of animals that can be supported on unit area of land.

Case mortality : The number of animals which die divided by the number of animals which are sick.

Castrate : (a) To render sterile by removal or blocking the activity of testicles or the ovaries.

(b) An animal that has been castrated.

Catabolism : The degradation of a complex substance into simple compounds by living cells.

Cauterize : To burn away or seal with a hot instrument or caustic substance to prevent infection and supress blood flow.

Celsius (C) : The metric expression of temperature (formerly, centigrade). Conversion to Fahrenheit may be done by multiplying the Celsius reading by 9/5 + 32.

Cereal : A plant of the grass family which produces seeds used to feed humans and livestock.

Cervix : The structure located between the vagina and uterus that protects the uterus during gestation.

Chronic : A disease condition that continues over a long period of time.

Clip : Total annual production of wool produced by a flock.

Clitoris : The reproductive organ in the female that corresponds to the penis in the male.

Clothing wool : Short staple wool suitable for woollen yarn production. Also called carding wool.

Coarse wool : Wool grading one-quarter or 46's to 48's under the Bradford system.

Coccidiosis : An intestinal disease caused by protozoan coccidium parasites.

Coefficient of digestibility : The amount of nutrient absorbed by an animal.

Colostrum : The first milk secreted by mammals after parturition. It is rich in antibodies and is laxative.

Combing : Process of removing short fibers from longer ones.

Complete ration : A portion of a feed or combination of feeds that satisfies the nutritional needs of an animal for one day.

Concentrate : Food with a high concentration of energy and little fibre (e.g. cereal).

Conception : (a) The union of sperm and ovum to form a zygote. (b) To become pregnant.

Conception rate : The number of females which conceive divided by the number of females mated.

Conformation : The structure or shape of an animal.

Contagious : A disease transmitted by direct or indirect contact.

Corpus luteum : The yellow body on the ovary that replaces the follicle and produces progesterone to maintain pregnancy.

Cortical cells : Inner section of the wool fiber.

Cotyledons : Buttonlike structures on the placenta that attach to the maternal caruncles in the uterus of a ewe.

Creep : An area within a pen, corral, or pasture constructed so that it excludes ewes and allows lambs access to special feed.

Crimp : The natural waves in wool fiber that allow it to have resilience and elasticity.

Crossbred : Genotype derived by crossing one breed with another (often an indigenous breed crossed with an imported breed).

Crossbreeding : The practice of mating parents of two or more breeds.

Crude fat : Ether extract from moisture - free feeds. It contains 2.25 times the energy of carbohydrates. It consists mostly of fats and oils, with small amounts of waxes, resins, and other organic materials.

Crude protein : Nitrogen content of food x 6.25.

Crutching : The trimming of wool from around the vulva and udder of a ewe before lambing.

Cryptorchid A male animal in which one or both testicles have failed to descend into the scrotum. A unilateral cryptorchid (one testicle descended) may be fertile. Bilateral (neither descended) is invariably sterile.

Culling : Removal of a ewe or ram from the flock, usually because of some defect.

Cuticle : Outer scale like surface of the wool fiber as seen under the microscope.

Cutting chute : A narrow chute with gates that allows individuals to be sorted into groups.

Dagging : Removal of soiled wool from a sheep.

Dam : Female parent or mother of an animal.

Dead wool : Wool pulled from dead sheep.

Deficiency disease : Disease resulting from lack of an essential nutrient. Usually it is a mineral or vitamin.

Dehydration : Loss of water from the body of an animal due to sickness or lack of drinking water.

Density : Number of wool fibers per unit area. Fine-wool sheep have greater density of wool than breeds with coarser wool.

Dental pad : That cartilaginous part of the front of the upper jaw that the lower incisors press against when the sheep grazes.

Depilatory : Chemical solution applied to the flesh side of a sheep pelt to loosen the fibers for pulling.

Depraved appetite (pica) : Tendency of animals on a nutrient deficient diet to eat unnatural feed.

Dermatitis : Inflammation of the skin.

Diarrhoea : Abnormal frequency of intestinal discharge. Also called scours.

Diet : (a) Feed ingredients including water consumed by an animal, (b) A mixture of feed ingredients fed ad libitum to animals.

Digestible nutrients : That part of each nutrient digested by an animal. Obtained by subtracting fecal nutrient content from amount fed.

Disinfectant : A strong chemical used to kill microorganisms.

Disposition : The temperament or personality of an animal.

Dock : (a) To sever the tail from a lamb, (b) That part of a sheep where the tail was cut.

Docking : Removal of tail.

Dorsal : Pertaining to or located on the back of an animal as opposed to ventral, which is the lower part or belly.

Drench : (a) To mechanically apply oral medication, (b) The medication applied orally.

Dressing percentage : Ratio of dressed carcass weight to live weight, expressed as percentage.

Dry : Nonlactating.

Drying off (up) : The process of causing cessation of lactation when lambs are weaned.

Dry matter basis : Expression of nutrients on a moisture free basis.

Dung locks : Wool from the area of the britch or twist that is contaminated by feces.

Dystocia : Difficulty in giving birth.

Early maturing : Sexually mature at a young age.

Ear notching : Marking an animal by cutting triangular notches in the ear according to a specified system.

Ectoparasite : A parasite that lives on the exterior of the host, for example, flies and mosquitoes.

Edema : Swelling caused by accumulation of water in tissues of the body.

Elasticity : The ability of a wool fiber to return to its original length after being stretched.

Elastrator	: A bloodless means of docking and castrating by use of rubber bands. The band blocks circulation and the member sloughs off.
Electro ejaculation	: Use of a rectal probe and electrical impulses to stimulate ejaculation.
Electrolytes	: Body acids, bases, or salts necessary for proper functions of nerves, respiration, digestion and muscle contraction.
Emasculatome (burdizzo)	: An instrument used for bloodless castration by severing the spermatic cord without injury to the scrotum.
Emasculator	: An instrument used for docking and castrating lambs. It has cutting blades for severing testes and tails and crushing jaws to prevent bleeding.
Endocrine	: Ductless gland that secretes hormones necessary for the function of an animal. The hormones pass directly into the bloodstream from the glands.
Endogenous	: Secretions originating within the body of an animal.
Enteritis	: Inflammation of the intestines.
Enterotoxemia	: A toxemia or blood poisoning caused by the toxin of Clostridium perfringens; also called overeating disease.
Environment	: The aggregate of all the external conditions that affect an animal.
Epididymis	: The convoluted tube at the posterior end of the testicle that serves as a passageway from the seminiferous tubules to the vas deferens.
Epithelium	: Covering of the interior and exterior of the body.
Essential amino acids	: Amino acids required by an animal that cannot be synthesized in the body.
Estrous cycle	: The physiological events occurring in the reproductive system of an ewe between one estrous and another. Average length is approximately 17 days.
Estrus (noun); estrous (adjective)	: The period when a ewe is physiologically and psychologically receptive to sexual advances of a ram. Also known as the heat or estrous period.
Ether extract (EE)	: Substances in feed soluble in ether. Included are fats, oils, waxes, resins, etc. Energy content is 2.25 times that of nitrogen free extract or crude protein.
Ewe	: Female sheep which has given birth.
Excreta	: Substances excreted by an animal, primarily in feces and urine.

Exogenous	:	Substances introduced into the body of an animal.
Exotic breed	:	Breed from another locality (usually from a temperate country).
Expectorant	:	A medicine that promotes the discharge of mucus from the respiratory tract.
Fattening	:	Feeding animals so they will store energy in the form of fat. Also called finishing.
Fecundity	:	Ability to produce offspring.
Feed	:	Material capable of nurturing livestock.
Feed additive	:	An ingredient or substance added to a basic feed, usually in small quantities, to fortify it with nutrients, medicines, or stimulants.
Feed efficiency	:	The relationship between feed consumed per day and increase in weight or other measure of production. Usually expressed as feed per unit of production. For example, a lamb feeder will be much interested in weight of feed required per unit of gain in live weight.
Feeder lambs	:	Lambs under one year of age that need to grow and fatten before being suitable for slaughter.
Feedstuff	:	A product suitable for nurturing livestock.
Felt	:	A dense, lightly woven, heavily filled wool cloth.
Fertility	:	Number of ewes lambing divided by number of ewes mated.
Fertilization	:	The union of the sperm and ovum to form a zygote or independent individual.
Foetus	:	The developing organism in the uterus of the pregnant female.
Fiber content of feed	:	The content of complex carbohydrates (cellulose) and lignin in feed. These are difficult to digest and tend to lower feeding value.
Fine wool	:	The finest grade of wool corresponds to 64's or finer using the spinning count system.
Finish	:	Refers to the degree of fatness of the animal. Also used synonymously with fattening an animal.
Finishing	:	Preparing lambs for slaughter by feeding a good-quality diet.
Fleece	:	The wool from one sheep.
Flock	:	A group of sheep.
Flushing	:	Giving ewes or gimmers extra food before mating, to enhance lambing rate.

Fly strike : A seasonal -situation in which flies irritate or attack livestock blowing sores, open wounds, incisions, or damp wool.

Forage : Vegetable matter consumed by livestock as pasture, silage, or hay.

Fostering : Process of transferring of a lamb from its mother to another ewe. Also called grafting.

Free choice : Allowing uninhibited access to feed same as *adlib* feeding.

Fumigant : A substance that is vaporized to destroy pathogens, insects, or rodents.

Fungl : Saprophytic, parasitic lower plants such as molds, rusts, mildews, and mushrooms.

FCE : Food conversion efficiency. Weight of product per unit weight of food input.

Generation interval : The time span between the, same stage in the life cycle of two successive generations.

Gestation (pregnancy) : Period between fertilization and parturition. Time the fetus is carried in the uterus. Average gestation for sheep is approximately 147 days.

Gimmer : Female sheep aged 6 months (or from weaning) to first parturition.

Gimmer lamb : Female lamb.

Grade : (a) Measure of how well an animal or product meets the standards of a class, (b) Animals that have characteristics of a breed but cannot be registered because a parent or parents are not purebred. Offspring of a purebred and one of inferior breeding, (c) The classifiction system used to describe greasy wool.

Grading up : Increasing the proportion or exotic genotype in a flock by back-crossing ewes to exotic rams for several generations.

Grain : Seeds from cereal plants.

Graze : To consume standing vegetation.

Greasy wool : Wool shorn from live sheep.

Growth : The increase in size of the muscles, bones, and internal organs.

Growth promoter : Substance used in small quantities to increase growth-rate, usually during finishing.

Growthy : An animal large and well developed for its age. Usually large-framed animals.

Gummer : A sheep with no incisor teeth. A sign of old age.

Hand feeding : A method of feeding lambs in which they are fed grain twice daily plus hay.

Hand mating : A method of mating in which the ram occupies a pen and ewes in estrous are brought to him for one service each.

Hank : A standard length of yarn wound on a real. Length 560 yards (512 m).

Heat (estrous) : The period when a ewe is physiologically and psychologically ready to accept service from a ram.

Heavy wool : Wool with a high shrinkage due to a high grease and dirt content.

Haematoma : A swelling or tumor that is filled with blood.

Haemoglobin : The iron-containing protein in red blood cells responsible for oxygen transport in respiration.

Haemorrhage : Loss of blood from blood vessels. Bleeding.

Hogg, hogget : Sheep aged 6 months to 2 years

Hormone : Substance secreted by endocrine glands that regulate the activity of organs in the body.

Hulls : Fibrous covering of grains and seeds. Used as roughage component of feeds.

Hypocalcemia : Below normal levels of calcium in the blood, which may result in convulsions or parturient paresis as in milk fever.

Hypoglycemia : Below normal levels of blood glucose.

Hypomagnesemia : Below normal levels of magnesium in the blood.

Hypothalamus : That part of the brain that regulates body temperature, appetite, hormone releases, and other functions.

ILCA index 2 : Total weight of lamb weaned per unit body weight of ewe, per year.

Immunity : The ability of an animal to resist or overcome a disease by preventing development of pathogens or by counteracting effects of its products.

Immunization : Creating immunity by such means as vaccination.

Impaction : Stoppage of movement of the digestive tract by partially digested or indigestible material wedged into a segment of the system.

Inbreeding : Mating of related individuals.

Incidence of disease : The number of new cases of the disease appearing each year, expressed as a proportion of the total sheep population.

Incisors	:	The sharp cutting or biting teeth located in the front of the lower jaw.
Indigenous breed	:	Breed native to the locality.
Infection	:	Invasion of the body by pathogenic microorganisms.
Inflammation	:	Reaction of tissues to injury characterized by pain, heat, redness, and swelling.
Ingestion	:	The act of eating or drinking.
Ingredient	:	A constituent of a mixture.
Inguinal (ring)	:	In a male, the space or opening through which the spermatic cord passes from the interior of the body to the scrotum.
Instinct	:	A natural ability to accomplish certain tasks without benefit of prior experience. It is probably hereditary.
Intradermal	:	Within or between layers of the skin.
Intramuscular	:	Being within or going into a muscle.
Intravenous	:	By way of a vein.
In vitro	:	Observable in a test tube or other artificial environment.
In vivo	:	Occurring within a living body.
Involution	:	Return of an organ to normal size and condition from an enlargement. For example, the involution of the uterus after parturition.
IU	:	Abbreviation for international unit. A standard of measure potency of a biologic agent as specified by the International Conference for Unification Formulae. Potency is based on a bioassay that produces a particular effect. Potency is also expressed as USP units.
Joining	:	Putting ewes or gimmers to the ram for mating.
Joint ill	:	Also called navel ill. The result of invasion of pathogenic bacteria through the navel cord at birth. May cause umbilical abscesses, diarrhoea, stiff joints, peritonitis, and convulsions and death.
Ked	:	An external parasite of sheep. It is an insect that lives on blood from the skin of the sheep. Often incorrectly called the sheep tick.
Kemp	:	A straight, chalky white, weak, heavy medullated fiber that will not take a dye. It is highly objectionable in wool.

Keratin : A complex protein substance that is the principal constituent of wool.

Ketosis : Also knwon as lambing paralysis, pregnancy disease, or twin-lamb disease. Usually observed in ewes carrying twins late in gestation. Probably caused by inadequate carbohydrate metabolism, as indicated by ketone bodies in the blood and urine.

Lactation : The period of milk production. The period a mother nurses her young.

Lactose : The sugar in milk.

Lamb : Young sheep, usually under 9 months (or before weaning) meat from young sheep.

Lambing interval : Period between successive parturitions (of one ewe).

Lambing paralysis : See ketosis.

Lambing rate : Number of lambs born divided by number of ewes joined.

Lamb's wool : Wool shorn from a lamb 7 to 8 months old. It is softer and has spinninig qualities superior to wool of the same quality from an older sheep.

Lanolin : Refined wool grease. Used in medicines, cosmetics and paints.

Laxative : A drug or feed used to relieve constipation.

Libido : Sexual drive or desire by the males of an animal species.

Litter : (a) Number of young produced in breeds that normally have multiple births, (b) Bedding in barns.

Litter size : Number of lambs born divided by number of ewes lambing.

Live weight : Weight of an animal on foot.

Local breed : Breed native to the locality (indigenous breed).

Lock : A tuft or small finger size bit of wool that tends to cling together when shorn from the sheep.

Lot : The entire offering of wool for sale.

Low set : An animal with short legs.

Luster : The natural gloss or sheen of wool from long wooled sheep.

Macro elements : The major minerals calcium, phosphorus, sodium, chlorine, potassium, magnesium, and sulfur.

Macroscopic	: Large enough to be visible to the naked eye.
Maintenance requirement	: Nutrients required solely to maintain body weight.
Malnutrition	: A ration deficient in required nutrients.
Manure	: Usually a mixture of feaces, urine, bedding, and waste feed.
Mastication	: Chewing feed.
Mastitis	: Inflammation of the mammary gland.
Meal Medicated feeds	: A mixture of ground feed ingredients. Feeds containing drugs for the treatment or prevention of diseases.
Merino	: A breed of sheep known for producing high quality wool.
Metabolic liveweight	: Liveweight 0.75 (previously defined as liveweight 0.73). Usually in units of $kg^{0.75}$.
Microingredient	: A feed ingredient added to a mixture in quantities of milligrams or micrograms or parts per million.
Microorganisms	: Organisms of microscopic size such as bacteria.
Milk let-down	: Release of milk from the alveola into the cisterns of the udder by the action of oxytocin.
Minerals	: Inorganic elements required by animals for their normal function.
Moisture free (MF)	: Oven-dried feed subjected to 212° F (100° C) temperatures until weight becomes static. This is known as dry matter.
Molars	: The grinding jaw teeth.
Molds	: See fungi.
Morbidity rate	: Proportion of animals showing signs of the disease.
Mortality	: Death or death rate.
Mortality rate	: Proportion of the population which dies from the disease (see also case mortality).
Mucous membranes	: The soft membranes lining the passage and tracts of the body.
Mutton	: Meat from (adult) sheep.
Mycotoxins	: Toxic substances produced by some molds.
ME	: Metabolisable energy. Digestible energy of food minus energy lost in methane and urine. Usually expressed in MJ.

National Research Council (NRC) : A division of the National Academy of Sciences that promotes the effective utilization of scientific and technical resources. For example, a subcommittee of the NRC calculates and complies nutrient requirements for animals.

Natural immunity : Inherited immunity to disease or parasites; not acquired immunity.

Navel ill : Same as joint ill.

Necrosis : Death and decay of body tissue that is still in contact with living tissue

Nitrogen fixation : The ability of legumes and bacteria to convert atmospheric nitrogen into organic nitrogen compounds and thus enrich the soil.

Nitrogen-free extract (NFE) : In proximate analysis, the sum of the contents of moisture, crude protein, crude fat, crude fiber, and ash are subtracted from 100 and the result is called nitrogen-free extract. It contains sugars, starches, pentoses, and non nurogenous organic acids.

Non protein nitrogen (NPN) : Simple organic compounds containing nitrogen, which can be utilized by ruminants to satisfy part of their protein requirements. Examples are urea and biuret.

Nulliparous : Females that have never given birth to young.

Nutrient requirements : Values of nutrients considered necessary for maintenance, optimum production, and prevention of signs of nutritional deficiencies of animal.

Nutrients : Chemical substances present in feeds that are used by animals for maintenance, growth, and production.

Oestrous : Period of sexual receptivity of ewe.

Offal : Organs and tissues removed at slaughter.

Offtake : Number of animals removed from the flock (for sale or consumption), divided by the number of animals in the flock.

Open wool : A wool that lacks density and may part down the back.

Ossification : Hardening of the bone tissues by calcification as the animal ages.

Osteomalacia : Softening of the bones caused by vitamin D deficiency or lack of calcium or phosphorus or an improper ratio-between the two.

Ovary : The reproductive organ of the female.

Palpate : To investigate or examine by feeling or touching with the hands.

Parrot mouth : A defect of the mouth caused by a shortness of the lower jaw that does not permit the incisor teeth to come in contact with the dental pad.

Parts per million (ppm) : Equals milligrams per kilogram or milliliters per liter.

Parturition : The act of giving birth.

Pathogenic : Causing or capable of causing disease.

Pedigree : A written record of an animal's ancestry.

Pelt : The skin of a lamb as removed at slaughter.

Peritonitis : Inflammation of the peritoneum, the thin membrane that covers the organs of the abdomen.

Per os : To administer a substance by way of the mouth.

pH : A measure of the acidity or alkalinity of a solution with values from 0 to 14. Most acid is 0 and most alkaline is 14, with 7 as neutral.

Photoperiod : Day length or period of light.

Photosynthesis : The formation of carbohydrates by the action of sunlight on chlorophyll in a plant.

Physiological : Healthy or normal function of a living organism.

Placenta : The vascular organ that surrounds and protects the fetus and provides the means by which nourishment is transferred from the mother to the young.

Pleura : Delicate membranes that line the thorax of an animal.

Pneumonia : A disease of the lungs.

Polled : A naturally hornless animal.

Polyestrous : Animals showing regularly occurring estrous throughout the years.

Posterior : Refers to the back portion of an individual.

Postnatal : Event occurring to the offspring after birth.

Preconditioned : Management before shipping, weaining, etc., to reduce stress. It includes vaccination, use of antibiotics, and controlled feeding.

Precursor : A compound converted by the body to one needed by the animal.

Prehension : A compound converted by the body to one needed by the animal.

Prehension : Grasping grass or feed by the tongue lips and teeth and conveying it to the mouth.

Prenatal : Event occurring to the foetus before birth.

Prevalence of disease : Proportion of animals which show signs of the disease at any time.

Prepotency : The ability of an individual to transmit its characteristics to its offspring.

Produce : The offspring of a female.

Productivity index : Measure of production of flock.

Progeny : The offspring of animals.

Prognosis : An estimate of the progression and final outcome of a disease.

Prolapse : Abnormal protrusion of an organ or part of organ.

Protein : A molecule consisting of chains of amino acids.

Provitamin : A precursor of a vitamin.

Proximate analysis : Chemical evaluation of feedstuff. These include dry matter (DM), crude protein (CP) total digestible nutrients (TDN), ether extract (EE), ash, crude fiber (CF), and nitrogen-free extract (NFE).

Puberty : The time an animal reaches sexual maturity.

Pulled wool : Wool pulled from the skin of sheep after slaughter.

Purebred : An animal eligible for registration in the herd book of the breed to which it belongs.

Pus : Thick, opaque, yellow-white fluid containing leucocytes, tissue debris, and microorganisms.

Putrefaction : Decay as a result of enzyme or bacterial action or both.

Quantitative traits : Traits that do not vary much in phenotypic expression but include most of economic importance. Inheritance is through many pairs of genes.

Quarantine : Segregation of sick, exposed, or susceptible animals to prevent the spread of infectious disease.

Radioactive : Emitting atomic energy as in the form of alpha, beta, and gamma rays.

Ram : Adult male sheep, usually used for breeding.

Range wool : Wool produced by sheep on the range lands.

Rate of passage : Time required for undigested feed to pass through the digestive tract.

Rate of return on capital invested : Net annual return divided by total capital investment.

Ration : The amount of feed supplied to an animal in one day.

Raw wool	:	The same as greasy wool.
Rectum	:	The terminal part of the colon.
Registered	:	Designating animals recorded in the herdbook of their breed association.
Regurgitation	:	The movement of undigested fibrous feed from the rumen to the mouth for further chewing.
Relative humidity	:	The absolute humidity divided by the absolute humidity at saturation at the same temperature.
Repeatabtility	:	Correlation between repeated measurements of the same trait on animals.
Replacements	:	Animals selected to go into the breeding flock.
Reproductive rate	:	Number of lambs weaned per ewe of reproductive age per year.
Retained placenta	:	A placenta not expelled at birth or soon afterward.
Rigor mortis	:	Temporary rigidity of muscles after death.
Rolag	:	The neat roll of wool that comes from the hand card.
Roughage	:	Food with a low concentration of energy and much fiber (e.g. straw).
Rumen	:	The first and largest compartment in the ruminant stomach. Also called the paunch.
Rumenitis	:	Inflammation of the rumen wall.
Sacking	:	Woven fiber bags used for storing and transporting fleecese. usually 2.1 m long and 1.0 m in diameter.
Saliva	:	The clear liquid solution secreted by glands in the mouth that contains enzymes used in digestion of carbohydrates.
Salmonella	:	Aerobic, rod-shaped pathogenic bacteria that are the cause of severe diarrhoea.
Scouring	:	Removal of impurities of wool by washing.
Scrotum	:	The bag or sack in a male that contains the testicles.
Scurs	:	Small, rounded, loose, horny growths attached to the skin at the hornpits of a sheep.
Secondary infection	:	An infection following an infection but with dif-ferent organisms.
Selection	:	The decision whether or not to keep an animal in the flock.
Semen	:	The ejaculate of a ram that contains fluids from the accessory glands and spermatozoa.
Septicaemia	:	Blood poisoning caused by bacteria in the circulatory system.

Serrations : The scaly surface of the outer layer of cells of a wool fiber that causes fibers to cling together in felting. Fine wool generally has more serrations than coarse wool.

Serum : The colourless fluid component of blood.

Service : Mating of male and female.

Settled : Becoming pregnant.

Shearing : Clipping the fleece either by hand or power shears.

Shock : Physical or mental disturbance as a result of major physical injury or stress.

Shoddy : Wool fibers recovered from used fabric. Fibers must be at least 1.3 cm in length.

Shrinkage : (a) Loss of weight in transit or in handling, (b) Loss of carcass weight in the cooler.

Sire The male parent.

Skirting : Removing stained or inferior parts of a fleece on the shearing floor.

Sorting : Dividing wool into grades according to fineness, length, colour and vegetable content.

Spermatic cord The cord that passes from the scrotum to the body and includes the vas deferens, muscle, artery, vein, nerves, and lymphatics leading to and from the testicles.

Spermatogenesis : The formation and development of sperm in the testes.

Spinning : Process of converting carded or combed wool into thread.

Spinning count : A measure of the fineness of a fleece based on the length of yarn that can be spun from 454 g of scoured wool. It is expressed in hanks, which are 512 m in length. Thus 64's would produce 64 hanks of yarn per 454 g of scoured wool.

Spore : A bacterial cell with an extra thick, impervious wall that makes it resistant to sterilization.

Stag : A male castrated after secondary sexual characteristics are well developed.

Stained wool : Wool discoloured by urine, manure, or other materials that will not scour out. This detracts from the value of the fleece.

Standing heat : The highest level of estrous when the female's receptivity to the male is the greatest.

Staple : The lenght of the lock of shorn wool.

Sterile : Incapable of reproduction.

Sterilization : Process of destroying microorganisms by boiling or use of chemicals. Used mainly on surgical tools. Also to render an animal incapable of reproduction.

Stillborn : Born dead.

Stocking rate : Number of sheep per unit area of land.

Straw : The stems and leaves of grains, soybeans, etc., after threshing.

Stress : Any environmental factor causing bodily or emotional tensions that interfere with an animal's internal function.

Store lamb : Lamb which has been growing slowly, and which is ready to enter a finishing unit.

Subcutaneous : Anything that occurs beneath the skin.

Suckle : To draw milk from a teat by sucking. To nurse.

Suint Salts of perspiration in the raw wool fleece.

Superanumerary : Teats in excess of the two normally found on a sheep. They are usually nonfunctional.

Supplement : An ingredient added to basal feeds to improve the nutritional value.

Suppuration : Formation of pus.

Symmetry : Blending together of all parts of a well developed animal.

Synthesis : Production of a substance by combining simpler compounds.

Synthetics : Products similar to natural ones that are produced artifcially.

Tagging : Removing the dung locks from a sheep. Usually done before breeding and shearing.

Tags : Dung locks occurring in wool around and below the anus of a sheep.

Tallow : The fat of a sheep.

Tattoo : A permanent marking of an animal by insertion of indelible ink under the skin, usually in the ear on flank.

Teaser ram : A ram used to locate ewes in estrous, the ram may be vasectomized or equipped with an apron to prevent impregnating the ewes.

Tender : A weak spot or spots along the length of wool fibers caused by the growth of a small, weak fiber during sickness or poor nutrition.

Testosterone : The male sex hormone produced by the testes.

Tetanus : An infectious bacterial disease that results in spasms and rigidity of muscles, especially the jaw. Clostridium tetani enters the body by way of wounds.

Tetany : A condition in an animal characterized by tonic muscle spasms due to a physiologic mineral imbalance. Grass tetany is caused by a magnesium deficiency.

Thrifty : Healthy and vigorous in appearance.

Tocopherol : Fat-soluble chemical compounds with vitamin E activity.

Tonic : An appetite-stimulating agent.

Total digestible nutrients (TDN) : A method of determining the energy value of feeds based on the digestibility of nutrients in the feed. TDN can also be converted to digestible energy (DE). One pound of TDN = 2000 calories DE.

Toxemia : A general, systematic toxification or poisoning caused by absorbing toxins produced by bacteria in a localized infection.

Toxic : Poisonous.

Toxoid : A toxin from a pathogenic bacteria that has been created to destroy the toxicity, but with the capability of causing the formation of antibodies when injected into the body of a healthy animal.

Trace mineral : A mineral required in micro amounts (mg/kg or less) by animals.

Tupping : Mating.

Tupping season : The breeding season.

Twist : The number of turns per inch in a thread of yarn.

Udder : A pendulous organ consisting of two mammary glands enclosed in a common envelope and each provided with a teat also called a bag.

Underfeeding : Providing inadequate energy to maintain weight or production.

Unthriftiness : Apparent lack of vigor and vitality as shown by poor condition of hair, thin body condition, and lack of growth and development.

Unwashed wool : Raw wool, unscoured wool.

Urinary bladder. : The organ in the body that collects and stores urine until it is voided.

Urinary calculi : Stones formed at any place in the urinary tract. Also known as kidney stones or gravel.

USDA : Abbreviation for United States Department of Agriculture.

Uterus : The part of the female reproductive tract where the fertilized ovum attaches and develops into the fetus.

Vaccination : An injection of a biological agent to produce live immunity to a disease.

Vaccine : A suspension of attenuated or killed microorganism that is administered to an individual for the prevention or treatment of an infectious disease.

Vagina : The part of the female reproductive tract between the cervix and vulva where semen is deposited during copulation.

Vasectomy : Sterilization of a male by removal of a section of the spermatic cord (vas deferens). The libido is not affected.

Vectors : An organism that transmits a pathogen. Usually an insect.

Vegetable matter : Any contamination of wool from plants or plant products. Examples are seeds, burrs, chaff, and straw.

Ventral : The lower part or surface of a body near or level with the belly or abdominal area.

Viroid : A particle of nucleic acid that is much smaller than a virus but lacks the tough protein coat of the virus. It is a disease causing particle in animals.

Virulent : Pathogenic or capable of causing disease.

Virus : Small infectious particles that can pass through a filter, they cause disease. They consist of a tough protein coat within which is enclosed either RNA or DNA, but not both.

Viscera : Internal organs of the body, including the heart, liver, and intestines.

Vitamin : Organic substances essential in minute quantities to the nutrition of animals. They function as parts of enzyme systems to regulate metabolic processes, but do not provide energy or serve as building units in the body.

Void : To excrete faeces and urine.

Vomiting : To disgorge stomach contents through the mouth.

Warp : Yarns extending lengthwise in a loom.

Weaning : Removing young animals from their dam to terminate the suckling period.

Weaning percentage : Number of lambs weaned devided by number of ewes joined, expressed as a percentage.

Wether : Castrated male sheep, older than 6 months (or weaned).

Woollens : Cloth made from woollen yarns. The fabric is not tightly woven and tends to have a soft finish. Examples are flannels and tweeds.

Woollen yarn : Yarn spun from wools that are fine but short. The wool is carded to form a weblike configuration. When the wool is spun, the yarn is not tightly twisted and the broken loose ends stick out from the thread.

Wormer : A commercial product containing an anthelmintic to control internal parasites.

Worsted : A tightly woven fabric with a hard finish. Examples are worsted and serge.

Worsted yarn : Yarn made from long fiber wool that is combed. The fibers lie parallel before being spun into tightly twisted, smooth yarn.

Yearling : An animal at least 12 month of age.

Yield : (a) Percent of scoured wool from greasy wool, yield % = wt. of scoured wool/wt. of greasy wool × 100 (b) Percent of live weight that becomes carcass at slaughter. Carcass yield % = wt. of carcass/wt. of live lamb × 100 Also called dressing percentage.

Yolk : Natural grease, oil, and suint in wool.

Zygote : A cell resulting from the union of male and female gametes. After further development and implantation it becomes the fetus.

APPENDIX II

Table : Body weight at different ages of lambs of different breeds

Breed	Sex	Weight at different ages				
		Birth	3 Month	6 Month	9 Month	12 Month
1	2	3	4	5	6	7
Nali	Male	2.85 ± 0.13(11)	12.66 ± 1.10 (10)	17.00 ± 1.29 (6)	22.13 ± 0.97(4)	26.50 ± 0.76 (3)
	Female	2.77 ± 0.17 (10)	12.94 ± 1.02 (10)	18.69 ± 0.60 (8)	19.88 ± 0.39(8)	23.06 ± 0.64 (8)
	Overall	2.81 ± 0.12 (21)	12.80 ± 0.75 (20)	17.97 ± 0.65 (14)	20.63 ± 0.41 (12)	23.99 ± 0.52 (11)
Chokla	Male	2.39 ± 0.26 (6)	12.63 ± 1.52 (5)	19.13 ± 1.13 (4)	20.50 ± 1.32 (3)	26.33 ± 1.33 (3)
	Female	2.35 ± 0.18 (10)	13.98 ± 0.65 (9)	17.63 ± 0.42 (8)	20.25 ± 0.67 (8)	23.31 ± 0.51 (8)
	Overall	2.37 ± 0.15 (16)	13.50 ± 0.68 (14)	18.93 ± 0.50 (12)	20.85 ± 0.60 (11)	24.13 ± 0.50 (11)
Nali Cross	Male	3.16 ± 0.05 (135)	15.28 ± 0.28 (131)	22.10 ± 0.29(121)	23.39 ± 0.28 (114)	26.56 ± 0.49 (45)
	Female	3.13 ± 0.05 (108)	14.57 ± 0.57 (104)	20.51 ± 0.26(100)	20.68 ± 0.22 (100)	23.17 ± 0.25 (99)
	Overall	3.15 ± 0.04 (243)	14.97 ± 0.97 (235)	21.66 ± 0.20(221)	22.13 ± 0.19 (214)	24.23 ± 0.23 (144)
Patan wari (P)	Male	3.37 ± 0.27 (3)	8.57 ± 0.72 (3)	12.60 ± 0.88 (3)	16.67 ± 1.14 (3)	19.67 ± 2.73 (3)
	Female	3.16 ± 0.59 (4)	8.85 ± 0.41 (4)	13.25 ± 0.22 (4)	16.73 ± 0.36 (3)	19.80 ± 0.00 (2)
	Overall	3.28 ± 0.29 (7)	8.73 ± 0.38 (4)	14.97 ± 0.39(7)	16.70 ± 0.72 (6)	19.77 ± 1.73 (5)
Rambouillet (R) XP	Male	3.18 ± 0.20(13)	5.71 ± 0.48 (9)	12.73 ± 0.83 (6)	16.27 ± 1.60 (6)	22.80 ± 0.14 (3)
	Female	2.94 ± 0.09 (21)	7.96 ± 0.39 (16)	12.16 ± 0.38 (14)	15.83 ± 0.74 (14)	20.26 ± 0.74 (14)
	Overall	3.03 ± 0.09 (34)	7.79 ± 0.30 (25)	12.33 ± 0.36 (20)	15.96 ± 0.70(20)	20.71 ± 0.63 (17)

Table Contd. .....

1	2	3	4	5	6	7
Merino (ME) XP	Male	3.16 ± 0.12 (37)	8.95 ± 0.35 (23)	13.72 ± 0.05(22)	17.30 ± 0.72 (22)	23.12 ± 0.95 (8)
	Female	3.20 ± 0.12 (27)	8.83 ± 0.41 (24)	13.06 ± 0.61 (23)	16.75 ± 0.80 (22)	21.54 ± 0.87 (18)
	Overall	3.18 ± 0.07 (58)	8.89 ± 0.27 (47)	13.38 ± 0.40 (45)	17.05 ± 0.54 (44)	22.03 ± 0.67 (26)
¾ R × ¼ P	Male	2.86 ± 0.29 (4)	12.23 ± 2.09 (6)	17.50 ± 3.30 (6)	21.90 ± 4.21 (6)	29.40 ± 8.21 (3)
	Female	2.59 ± 0.18 (8)	9.93 ± 0.53 (8)	13.19 ± 1.95 (5)	16.24 ± 1.35 (5)	19.88 ± 1.41 (5)
	Overall	2.72 ± 0.14 (17)	10.92 ± 0.90 (14)	15.87 ± 1.89 (11)	19.33 ± 2.40 (11)	23.45 ± 3.04 (8)
¾ ME × ¼ P	Male	3.46 ± 0.24 (4)	12.83 ± 1.02 (4)	17.00 ± 1.16 (4)	17.00 ± 1.16 (4)	20.40 ± 1.23(3)
	Female	2.88 ± 0.30 (6)	10.26 ± 0.16 (5)	13.54 ± 1.31 (5)	16.48 ± 1.95 (5)	19.50 ± 1.74 (5)
	Overall	3.11 ± 0.04 (10)	11.40 ±0.75 (9)	15.06 ± 0.00 (9)	18.22 ± 1.22 (9)	21.10 ± 1.36(8)
Nilgiri(Ni)	Male	3.40 ± 0.80 (27)	10.85 ± 0.59 (27)	14.50 ± 0.71 (18)	20.96 ± 2.03 (5)	28.00 ± 3.00 (2)
	Female	2.85 ± 0.15 (57)	10.29 ± 0.39 (29)	13.59 ± 0.46 (17)	17.73 ± 0.61 (12)	20.09 ± 0.60(11)
	Overall	3.03 ± 0.35 (84)	10.56 ± 0.45 (56)	14.06 ± 0.43 (35)	18.68 ± 0.76 (17)	21.31 ± 0.63 (13)
¾ ME × ¼ Ni	Male	3.45 ± 0.05 (2)	11.39 ± 0.44 (44)	16.15 ± 0.47 (26)	19.43 ± 0.83 (11)	24.00 ± 1.87 (4)
	Female	3.20 ± 0.20 (2)	11.84 ± 0.55 (30)	15.75 ± 0.43 (20)	18.47 ± 0.64 (18)	20.94 ± 0.49(18)
	Overall	3.33 ± 0.10 (4)	11.56 ± 0.35 (74)	15.98 ± 0.33(46)	18.83 ± 0.51 (29)	21.50 ± 0.52 (22)
5/8 ME × 3/8 Ni	Male	3.40 ± 0.12 (3)	11.22 ± 0.57 (37)	14.42 ± 0.51 (26)	18.48 ± 0.72 (11)	26.00 ± 2.00 (2)
	Female	2.65 ± 0.15 (2)	10.53 ± 0.32 (39)	14.83 ± 0.48(24)	17.36 ± 0.65 (22)	20.53 ± 0.69 (22)
	Overall	0.90 ± 0.12 (5)	10.87 ± 0.32 (76)	14.44 ± 0.35 (50)	17.74 ± 0.49 (33)	20.79 ± 0.67 (24)

Table Contd. .....

1	2	3	4	5	6	7
3/4 R × 1/4 Ni	Male		12.31 ± 0.55 (32)	17.39 ± 0.68 (13)	21.24 ± 0.88 (5)	26.00 ± 1.00 (3)
	Female		11.13 ± 0.42 (23)	14.79 ± 0.62 (14)	18.07 ± 0.58 (13)	20.92 ± 0.95(12)
	Overall		11.62 ± 0.36 (55)	16.04 ± 0.61 (27)	18.95 ± 0.47 (18)	21.93 ± 0.80 (15)
5/8 R × 3/8 Ni	Male		13.43 ± 0.92 (19)	19.56 ± 0.02 (9)	22.64 ± 1.32 (5)	26.25 ± 1.38 (4)
	Female		11.36 ± 0.74(20)	14.64 ± 0.76 (14)	18.12 ± 0.23(13)	19.48 ± 1.38 (4)
	Overall		12.37 ± 0.59 (39)	16.57 ± 0.61 (23)	19.37 ± 0.97 (18)	21.27 ± 0.70 (15)
Malpura (M)	Male	3.30 ± 0.18 (7)	17.21 ± 0.86 (7)	27.25 ± 0.25 (2)		
	Female	2.98 ± 0.15 (6)	13.78 ± 0.88(6)	21.15 ± 0.55 (15)	22.63 ± 1.39 (6)	24.20 ± 1.77(4)
	Overall	3.15 ± 0.12 (13)	15.63 ± 0.60 (13)	21.87 ± 0.50 (17)		
Sonadi	Male	3.07 ± 0.16 (16)	15.58 ± 0.75 (15)	24.88 ± 0.98(14)		
	Female	2.84 ± 0.12 (16)	13.68 ± 0.92 (14)	21.05 ± 0.67 (13)		
	Overall	2.96 ± 0.10 (32)	14.66 ± 0.59 (29)	23.03 ± 0.60 (27)		
Mutton Mynthetic	Male	3.50 ± 0.56 (86)	18.82 ± 0.39 (53)	28.84 ± 0.48 (46)	31.55 ± 0.89 (29)	
	Female	3.31 ± 0.53 (96)	16.23 ± 0.30 (87)	23.72 ± 0.37 (87)	25.03 ± 0.38 (78)	
	Overall	3.43 ± 0.38 (192)	17.21 ± 0.23 (140)	25.48 ± 0.29 (133)	27.16 ± 0.35 (107)	
Mandya	Male	1.75 ± 0.20 (8)	12.29 ± 0.63 (7)	19.75 ± 0.54 (4)	23.00 ± 0.00 (1)	
	Female	1.50 ± 0.30 (2)	6.50 ± 2.50 (2)	9.50 ± 1.75 (2)	11.00 ± 1.00 (2)	
	Overall	1.70 ± 0.17 (10)	11.00 ± 0.68 (9)	16.33 ± 0.63 (6)	15.00 ± 0.82 (3)	

Table Contd. .....

1	2	3	4	5	6	7
Mandya Mynthetic	Male	2.52 ± 0.09 (62)	15.15 ± 0.40 (58)	20.65 ± 0.88 (20)	30.00 ± 1.10 (16)	
	Female	2.35 ± 0.08 (66)	13.82 ± 0.34 (64)	19.40 ± 0.40 (60)	21.00 ± 0.42 (56)	
	Overall	2.43 ± 0.06 (128)	14.45 ± 0.28 (122)	19.71 ± 0.37 (80)	23.00 ± 0.41 (72)	
Nellore	Male	2.92 ± 0.24 (6)	13.50 ± 0.50 (6)	20.50 ± 1.16 (3)	25.00 ± 1.00 (2)	
	Female	2.19 ± 0.17 (5)	13.40 ± 0.50 (5)	13.40 ± 0.48 (5)	20.13 ± 0.28 (4)	
	Overall	2.59 ± 0.15 (11)	14.45 ± 0.59 (11)	16.06 ± 0.51 (8)	21.75 ± 0.35 (6)	
Nellore Mynthetic	Male	3.18 ± 0.12 (50)	14.75 ± 0.60 (44)	24.40 ± 3.53 (5)	29.25 ± 3.58 (4)	
	Female	3.22 ± 0.08 (51)	14.35 ± 0.45 (50)	20.63 ± 0.49 (49)	22.49 ± 0.44 (47)	
Deccani (De)	Male	3.04 ± 0.08 (43)	11.77 ± 0.56 (35)	18.54 ± 0.88 (32)		
	Female	3.01 ± 0.08 (40)	12.37 ± 0.33 (26)	15.92 ± 1.25 (26)		
	Overall	3.02 ± 0.06 (83)	12.03 ± 0.35 (61)	17.37 ± 0.74 (58)		
Me × De	Male	3.35 ± 0.10 (15)	12.89 ± 0.60 (13)	20.54 ± 1.12 (13)		
	Female	3.12 ± 0.13 (16)	10.97 ± 0.64 (14)	14.75 ± 0.84 (14)		
	Overall	3.23 ± 0.08 (31)	11.90 ± 0.44 (27)	17.54 ± 0.69 (27)		
Dorset (Do) × De	Male	3.68 ± 0.05 (109)	12.83 ± 0.41 (84)	20.06 ± 0.47 (76)		
	Female	3.57 ± 0.45 (136)	13.14 ± 0.27 (02)	17.28 ± 0.32 (100)		
	Overall	3.62 ± 0.35 (245)	13.00 ± 0.24 (186)	18.48 ± 0.37 (176)		

Table Contd. .....

1	2	3	4	5	6	7
Muzzafar Nagri (Mu)	Male	3.51 ± 0.09 (51)	16.76 ± 0.50(50)	22.90 ± 0.74 (32)	28.56 ± 0.79 (33)	40.85 ± 0.82 (11)
	Female	3.42 ± 0.10 (50)	15.82 ± 0.50(46)	19.66 ± 0.45 (42)	24.56 ± 0.45 (53)	29.60 ± 0.55(48)
	Overall	3.46 ± 0.06 (101)	16.31 ± 0.35 (96)	21.06 ± 0.45 (72)	26.16 ± 0.41 (86)	31.70 ± 0.47 (59)
¼ Do × ¼ Suff × ½ Mu	Male	3.63 ± 0.14 (19)	18.00 ± 0.57 (15)	32.20 ± 0.09 (18)	30.91 ± 1.12 B (7)	41.16 ± 1.57(5)
	Female	3.31 ± 0.10(30)	16.24 ± 0.62 (28)	21.43 ± 1.21 (18)	25.00 ± 0.75 (18)	30.15 ± 1.27(17)
	Overall	3.43 ± 0.08 (49)	16.86 ± 0.46 (43)	26.81 ± 0.81 (36)	26.65 ± 0.63(25)	32.65 ± 1.05 (22)

APPENDIX–III

Table : Preweaning growth rates of lambs

Country	Breed	LWG (g/d)	Ape at weaning	References
Brazil	Blackhead Persian	117	112	Figueiredo *et al.*, 1982
	Corriedale × Illde–France	240	30	Surreaux *et al.*, 1980
	Morada Nova	119	112	Figueiredo *et al.*, 1982
	Santa Ines	152	112	Figueiredo *et al.*, 1982
Cyprus	Awassi	240	35	Lawlor *et al.*, 1974
	Chios	230	35	Lawlor *et al.*, 1974
	Chios	183	42	Mavrogenis, 1982
	Cyprus fat Tailed	230	30	Lawlor *et al.*, 1974
India	Deccani	116	28	Gupta *et al.*, 1974
	Madras Red	100	119	Ganesakale, 1975
	Malpura	89	28	Bohra *et al.*, 1979
	Mandya	81	119	Ganesakale, 1974
	Merino × Deccani	105	28	Gupta *et al.*, 1974
	Muzaffarnagari	122	80	Ali *et al.*, 1980
	Nali	90	84	Gaur *et al.*, 1977
	Rambouillet	162	21	Sahni and Tiwari, 1975 b
	Rambouillet × Chokla	156	21	Sahni and Tiwari, 1975 b
	Rambouillet × Deccani	118	28	Gupta *et al.*, 1974

Table Contd. .....

Country	Breed	LWG (g/d)	Age al weaning	References
	Rambouillet × Jaisalmeri	160	21	Sahni and Tiwari, 1975 b
	Rambouillet × Malpura	153	21	Sahni and Tiwari, 1975 b
Indonesia	Javaneses Thintailed	51	91	Hetzel *et al.*, 1982
	Javaneses Thin	173	91	Hetzel *et al.*, 1982
	Tailed × Suffolk			
Iran	Karakul	21	–	Farid *et al.*, 1976
Iraq	Awassi	165	120	Eliya and Juma, 1970
Kenya	Masai	173	153	Wilson *et al.*, 1981
Mexico	Romney Marsh	193	30	Ugalde Orta, 1978
Nigeria	Uda	118	91	Buvanendran *et al.*, 1981
	West African Dwarf	48	30	Adeleye and Oguntona, 1975
	Yankasa	109	91	Buvanendran *et al.*, 1981
Srilanka	Jaffna	45	91	Ravindran *et al.*, 1983
Sudan	Baggara	140	91	Wilson, 1976
	Desert Sudanese	157	42	Pollott and Ahmed, 1976 b
Trinidad	Barbados Blackbelly	102	21	Rastogi *et al.*, 1979
	Blackhead Persian	188	84	Butterworth *et al.*, 1968
	Blackhead Persian	125	21	Rastogi *et al.*, 1979
	West African	116	21	Rastogi *et al.*, 1979
Venezuela	West African	122	–	Combellas, 1978
	West African × Dorset Horn	171	–	Combellas, 1978

N.B. : L. W. G. = Live Weight Gain

APPENDIX – IV : BOTANICAL & COMMON NAMES OF SOME FODDER TREES

S.No	Botanical name	Assamese	Bangali	Gujarati	Hindi Rajasthani	Kannada
1.	*Acacia catechu*	Khoira	——	——	Khair	Kangli
2.	*Acacia nilotica*	——	——	——	Babul	Jali
3.	*Adina cordifolia*	Roghu, torak–epapa, bonglong, phurkhona;	bangka, petipuria, kelikadam	Haladwan, haldarvo	Haldu, hardu, karam, kaim, kalmi	hedde, pettega, yetteda, anavu, yettagal, arasintega, kdambe
4.	*Aegle marmelos*	–	Bela	–	Bel	Bila patri
5.	*Ailanthus excelsa*				Mahrukh, arua, ardu	Peetheral
6.	*Albizia chinensis*	Chakua	Chakua	–	Siran	Kalbage
7.	*Albizia lebbek*	Nirith	Sirisha	pilo sarshio	Siris, Sirs, kan, bage, sirsul	–
8.	*Anogeissus latifolia*	——	——	——	Bakli, dhaura, dhau, dhawa, —— Dindal, dinduga	
9.	*Anogeissus pendula*	–	–	–	Dhokra, kardhai	–
10.	*Artocarpus heterophyllus*	–	–	–	Kathar, Kathar	Halsu
11.	*Azadirachta indica*	–	Nim	Limba	Nim	Bavina
12.	*Bauhinia variegata*	Bogakatra, Kotora	Rakta–kanchan	–	Kachnar	Kanchava–lado, kanchi vala
13.	*Butea monosperma*	–	–	–	Dhak, palas	Muttuga
14.	*Careya arborea*	Dambel, Kumbhi	Kumbi	Kumbi	Kumbi	Doddala, kaval
15.	*Celtis australis*	–	–	–	Khirak	–

Table Contd. .....

Malayalam	Marathi	Nepalese	Oriya	Punjabi	Tamil	Telegu
——	Kadori	——	——	Khair	Karungali	Sundra
Karuvelum	——	——	——	Kikar	Karuvelum	Nilla tumma
Maja–kadambu, yebudijamaram	Heddi, heddu, honangi, kalamb		Holonda, kurum	haldu	Majikadambai, bandaru	pasupu–kadambi, bandaru, dadduga, kanapu, rudraganapa
–	–	–	–	–	Vilva	Maredu
Peenari	Makarukh		Maha nim		Pimaram, pinari, periamaram	Peddamanu
–	Laell	–	–	Ohi	Pilivagai	Konda chigara
Vaga, vaka	chichola		Tinia, shirson	Sirin	Vagei, vagai	Dirsinam
Vella–naga	——	——	Dhau	——	Villay–naga	Chiriman
–	–	–	–	–	–	–
Pilau	Phannas			Kathal	Pilla	Panasa
Veppu	Limba	–	–	Nimb	Veppam	Vepa
Kotidaram	–	–	–	Kachnar	Segapumanchorii	Devakanchaanmu, Bodantam
–	Paras, palas	–	–	–	Parasu	Moduga
Alam, pera, peal	Kuba, kumbia	–	–	–	Ayama kal–kambi	Gadava, arava
–	–	–	–	Kharik	–	–

Table Contd. .....

1	2	3	4	5	6	7
16.	*Cordia dichotoma*	–	Buhal, bohari	–	Lasura, bhokar, borla	Challe
17.	*Dalbergia sissoo*	Sissu	Sisu	Tanach, sisam	Shisham, sissu, sisam	Agara, sishamage
18.	*Diospyros melanoxylon*	–	Kendu	Tembru	Tendu	Balai
19.	*Ehretia laevis*	–	–	–	Chamror, koda, tambolia	Adak
20.	*Ficus bengalensis*	–	Bar bor	Vad, Vadlo	Bar, Bargad	Alada, mara, vata
21.	*Ficus glomerata*	–	–	–	Gular, umar	–
22.	*Ficus lacor*	Prob, Kangji	Pakar, pakur, –	Pakar	Bassari	
23.	*Ficus religiosa*	–	Aswattha	Pipro pipul	Pipal	Araliashwattha
24.	*Ficus roxburghii*	–	–	–	Timla, tirmal	–
25.	*Grewia optiva*	–	–	–	Bihul, bhimal, dhaman	–
26.	*Hardwickia binata*	–	–	–	Anjan	Kamra, karachi
27.	*Holoptelea integrifolia*	–	–	–	Papri, Papar, kanju, karanji, chilla, banchilla, kunj, bisenda	Tapasi
28.	*Kydia calycina*	Pichhola, bankopah, kukuha	Pala, bankopas	Varang, petari	Pula puli, pathera, pola, safed dhaman	Bellaka, bendi, billubhendy
29.	*Lagerstroemia parviflora*	–	Sidha	–	Dhaura, lendia	Channangi
30.	*Lannea coromandelica*	–	Jial	–	Jhingan, jigna, kaimal, Kinjan	Gugul

Table Contd. .....

8	9	10	11	12	13	14
–	Bhokar, shelu	–	–	–	Vidi	Iriki
–	–	–	Sisu, sissua, padimi	Tali, tahli, shin, shisham	Sissoo, yette, nukku kattai, gette	Sissu, karra, sissu–karra
Kari	Temru	–	Kendu	–	Karai, karundumbi	Tumki
Chavandi	Datrang	–	–	–	Addula	Pogadi
Ala, vatam, Wad, Vada	–	–	–	Ala, alam	Marri, peddamari	–
–	Umbar umra	–	–	–	–	Atti
Cherla	Kel	–	–	Palakh, pilkhan	Malai icchi	Jatijuvi
Arasu, ashvatham	Ashvatha	–	–	–	Arasi, ashwaiham	Ashvatthamu, ravi, ragi
–	–	–	–	–	–	–
–	–	–	–	–	–	–
–	Anjan	–	–	–	Acha	Vepi, naryepi, yana
–	Papara, wavuli	–	–	Rajain	Aya, avali	Tapasi
Velukhu, venda	Warung, potari, ranbhendi	–	Bankopasia, bharimo kopasia	Pola, pula	Vendai, bendi, pula, vattakannu	Potri, hondapotari, pandiki, peddapotri
–	Bonda, bhondara	–	–	–	Ventek	Chennangi
–	Moi, shembat	–	–	–	Uthi, odi	Gumpan

Table Contd. .....

1	2	3	4	5	6	7
31.	*Leucaena leucocephala*	–	–	–	Koo, babul	–
32.	*Madhuca longifolia*	–	Maul, mohwa	Mohuda, mahudo, mavdo	Mahua, mohwa, maul	Ippahal tumbri, hippe, honge
33.	*Mallot us philippinensis*	–	–	–	Rohni, roini, rainy	Kapila
34.	*Melia azedarach*	–	–	–	Bakain	Kadbevu, bevu
35.	*Mitragyna parvifolia*	–	Guli kadam	–	Phaldu, karm	Kadawar, yetega
36.	*Moringa oleifera*	Saijna, sohjna	Sujina	–	Sanjana, sohanjna	Nuggi
37.	*Morus alba*	–	Tut	Shetur, setur	Shahtut, tut, tutri, chinni	Kambilimara, kamblihaunu
38.	*Ougeinia oojeinensis*	–	Bandhar	Tanach	Sandan, chanjan, tinsa, tinas	–
39.	*Pongamia pinnata*	–	Karanja	–	Karanj, papar, kanji	Batti, hongay
40.	*Prosopis cineraria*	–	Shami	Semru, sami, khijado, sumr	kehjri, khejra	Jambi

Table Contd. .....

8	9	10	11	12	13	14
–	–	–	–	–	–	–
Illupa	Moho, mohwa, mohwra	–	Madgi, mahula, moha	Mohwa	Illupai, elupa, kat–illipi	Ippa, ippi, yeppa
–	–	–	–	Kamila	Kapila	–
Kattuveppa	Pejri, padrai	–	–	Bakain	Malai vembu	Thurakavepa, yerri vepa
–	Kalamb	–	–	–	Chinna kadambu	Rattaganapa
Moringa	Shevgi	–	Sajina	Suhanjana	Murunga	Saihan
–	Tut, ambat	–	Tuto, tuticoli	Tut, shahtut	Mussuketi, kamblichedi	Reshma chattu
–	Tiwas	–	Bandhano	–	Kallu, mokke	–
Pongu	Karanj	–	Korojo, korjoi	Sukhchain	Ponga, pungam	Kagu, kanuga
Parampu	Shema, shemri, semder	–	Shami	Jand	Perumbay, jambu	Jammi chettu

Table Contd. .....

1	2	3	4	5	6	7
41.	*Quercus dilatata*	–	–	–	Moru, tilonj	–
42.	*Quercus glauca*	–	–	–	Bani, banni, phanat, phaliant	–
43.	*Quercus leucotrichophora*	–	–	–	Ban, banj	–
44.	*Roginia pseudoacacia*	–	–	–	–	–
45.	*Schleichera oleosa*	–	–	–	Kusam	Sagada
46.	*Shorea robusta*	Hal, sal, boreal	Shal	–	Sal, sakhwa, sakhu	–
47.	*Syzygium cumini*	–	–	–	Jamun	Nerlu
48.	*Tamarindus indica*	Teteli	Tintiri	–	Imli	Hunase, hunasemara
49.	*Terminalia alata*	Amari	–	Sadada	Sain, asna	Matti, karimaddi
50.	*Terminalia belerica*	–	–	–	Bahera	Tare, goting, bherda
51.	*Ziziphus mauratiana*	–	Berboroi	Bor, bordi	Ber	Yalachi, elanji
52.	*Acacia lcucophloea*	–	–	–	Safed kikar, reru, rinj, raunj, rhea	–
53.	*Albizia odoratissima*	–	–	–	Kali siris, bansa	Bilkumlu, bilwara
54.	*Atbizia procera*	Medeloa	Koroi	–	Safed siris, gurar, karak	Chikul,bellati
55.	*Anthocephalus chinensis*	Roghu	–	–	Kadam	Kadwal

Table Contd. .....

8	9	10	11	12	13	14
–	–	–	–	–	–	–
–	–	–	–	–	–	–
–	–	–	–	–	–	–
–	–	–	–	–	–	–
–	–	–	–	–	Pava, puvan	–
–	–	–	Salwa, sakwa, soringhi	–	Gugal, thamba, sarjakamu	–
–	Jambul	–	–	–	Naval	Neredu
Puli	Chinch	–	–	Imli	Puliyamaram	–
Thenpavu	Ain, madat	–	–	–	Karimarudu	Maddi, nallamaddi
–	–	–	–	–	Tani	Tandi
Elentha	Bor, bera	–	Barkoli, bodokoli	–	Elandai, yellande	Reegu, gangareegu, Karakandhavu
–	Hewar	–	–	–	Velvaylam	Telia tuma
–	–	–	–	Karmaru	Karu vagai	–
Vellavuka	Kinhai	–	–	–	Velvagal	Chigara, telia chinta
–	–	–	–	–	Vellei kadambu	Kadambu

Table Contd. .....

1	2	3	4	5	6	7
56.	*Artocarpus chapiasha*	Cham, sam	Chaplash, chaplis	–	–	–
57.	*Bauhinia malabarica*	Kattra	Karmai	–	Amli, amlosa	Shadlu
58.	*Bauhinia pur pure*	Kurial	Devakanchan, raktakanchan	Chambali	Khairwal, karar, koliar, kaniar	Sarul, baswanapada, kanchival
59.	*Bauhinia racemosa*	–	Banrai	Ashitri, asintro	Amli, jhijeri, jhanjhora, makuna, jinja	Banne
60.	*Bauhinia semla*	–	–	–	Kandla, semla, kural	–
61.	*Bombax ceiba*	Himala, himale	Simul, shemhel	Simalo, shimul	Semul, semal	Buruga, burla
62.	*Boswellia serrata*	–	–	–	Salai, saler	Maddi
63.	*Bridelia retusa*	–	–	–	Kaj, khaja, karka, kassi	Muljane
64.	*Buchanania latifolia*	–	–	–	Chiroli, piyal, paira, achar, char	Nurkal, murkala
65.	*Dillenia pen tagyna*	Aksi	Tatri, karkotta	Sonchempo	Aggi, kallai	Kaltega, kanagola kangal
66.	*Emblica officinalis*	Ambuki	Amla		Aonla, amla	nelli
67.	*Ficus cunia*	–	–		Khini, khunia, khain, jharphali	–
68.	*Ficus hispida*	–	–		Gabla, kath gular, kagsha	–

Table Contd. .....

8	9	10	11	12	13	14
–	–	–	–	–	–	–
Aram puli	–	–	Koteli	–	Atthi	–
Survannam andaram	Atmatti	–	Sono	Kolar	Mandarai	Kanchan
–	Apta	–	–	Taur, kosundra	Archi, vattathi	Ari
–	–	–	–	–	–	Nirpa
Seemapula, pula	Savar, syar	–	Boro, bush, salmali	Simbal	Ilavam, elevamarum	Buruga, salmali
–	–	–	–	–	Kundrikam	Anduga
–	Asana, kutki	–	–	–	Mullu–maruthu	Koramadi
Munga, perlu	Char, chirauli, chironji	–	–	Chiroli	Morala, sarai	Morli, sara
Punna, kodapunna	Karmal, kanagalu, kari	–	Rai	–	Kanaga1a, kangal, nai, teku	Pedda, kalanga, ravudana, karuva
nelli	–	–	–	Aola, amli	Nelli	Usirikai
–	–	–	–	–	–	–
Eruma nakku	Kalaumber	–	–	–	Chona atthi	Bamori

Table Contd. .....

1	2	3	4	5	6	7
69.	*Ficus palmata*	–	–	–	Pheru, phedu, khemri, anjiri	–
70.	*Ficus retusa*	–	–	–	Barri	Pilala
71.	*Ficus rumphii*	–	Galasvattha	–	Pilkhan, khabar	–
72.	*Flacourtia indica*	–	Benchi, baichi, katai	Galgugal, katamla	Katai, kakein, kukai, bilangra, kaikun, kanker	Hummunki
73.	*Gardenia latifolia*	–	–	–	Papra, pindalu	–
74.	*Gardenia turgida*	–	–	–	Thanela, karamba, ghurgia	Bengeri
75.	*Garuga pinnata*	–	–	–	Kharpat, kaikar	Godda
76.	*Gmelina arborea*	Gomari	Gumbar, gomari	Shewan	Gumhar, sewan	Shivani
77.	*Grewia elastica*	Bijal–goch, man–bijal	Dhaman	–	Dhaman, phalwa	–
78.	*Grewia tiliaefolia*	–	–	Dhamana	Dhaman	Dadsal, butale
79.	*Hymenodictyon excelsum*	–	–	Dondro	Bauranga, bhaulan, kukurkat, barthua	–
80.	*Mangifera indica*	Ghariam	–	Ambo	Aam	Mavu, majinamera

Table Contd. .....

8	9	10	11	12	13	14
–	–	–	–	–	–	–
Kal atthi	Nandruk	–	–	–	Ichi	Yera juvi
–	Pair, sahta	–	–	–	–	–
–	Paker, kaker, bhekal	–	Balibh–ainchol	Kukai, kakoa	Sottaikala, cholhakilai	Gandregu, giamehetu
–	Ghogar	–	–	–	–	Pedda bikki gegar
–	Phetra, pendra	–	–	–	–	Yerra bikki
–	Kudak	–	–	–	–	Garuga, gargu
Kumbil	Shivan	–	Gambari	–	Gumadi	Gumartek, gummadi
–	–	–	Mirgichara	Dhaman, phalsa, pheru	–	–
–	Dhamani	–	–	–	–	–
–	Bhorsal	–	–	–	–	Pottaka
Mavu	Amba	–	–	Amb	Maa, mamaram	Mamidi

Table Contd. .....

1	2	3	4	5	6	7
81.	*Manilkara hexandra*	–	–	–	Khirni	–
82.	*Morus serrata*	–	–	–	Kimu, chimu, kahun	–
83.	*Olea ferruginea*	–	–	–	Kahu	–
84.	*Populus Ciliata*	–	–	–	Pahari pipal, chalaun, chalni, safeda, ban pipal, bagnu	–
85.	*Pterocarpus marsupium*	–	–	Bio	Bija sal, Honne	–
86.	*Quercus semecarpifolia*	–	–	–	Kharsu, maru, kreo, khor	–
87.	*Salix tetrasperma*	Bhe	–	–	Bed, bains, bilsa, birsa	Niranji
88.	*Soymida febrifuga*	–	–	–	Rohan, rohini	Suami
89.	*Spondias pinnata*	–	–	–	Amra, ambara	Amate
90.	*stereospermum suaveolens*	–	Parul, parub	–	Padal, pandri	Pudika, geri, hooday, billa
91.	*Terminalia arjuna*	–	–	Dhaula sadar	Arjun, koha	Holematti
92.	Terminalia chebula	–	–	–	Harad, harra	Anale
93.	*Terminalia paniculata*	–	–	–	–	Bill–mathi, hulve, honal
94.	*Toona ciliata*	Poma	–	–	Tun, tuhi, maha nim	Tundu
95.	*Trema orientalis*	Jhupon, jopang	Jiban chickan	Kargol	–	Gurkulu
96.	*Wrightia tomentosa*	–	–	–	Dudhi, keor, darbela	–
97.	*Zizyphus xylopyrus*	–	–	–	Kat–ber, chitena, ghat–bor, goti	–

Table Contd. .....

8	9	10	11	12	13	14
–	Ranjana, raini	–	–	–	Palla	Pala
–	–	–	–	–	–	–
–	–	–	–	Kau	–	–
–	–	–	–	–	–	–
–	*Bibla, asan*	–	–	–	Vengai	Yegi, peddagi
–	–	–	–	–	–	–
Vanti	*Bocha, bitasa*	–	–	–	–	–
–	*Palara*	–	Sohan	–	Shem	Sumi
–	*Ambada, ambe*	–	Ambota	–	Ambalam	–
Kariyam, vedankorana	*Pond, kala–gari, patola*	–	Boro–pat–uli, padal	padal	padiri	Kala goru
–	*Savimadat*	–	–	–	Kula marutha	Thella maddi
–	*Hirda*	–	–	–	Kadakai	Karaka
–	*Kinjal, kindal*	–	–	–	Admarud, pulavati, vanmarduu, Pumarudu	Nimiri, pula–maddinalla– pulaga
Rachagiri vembu	*Todu*	–	–	–	Thevatharam	–
Ami, ratthi	*Kagol, gol ranambada*	–	Jivani	–	Oman, oma	Gada-nelli
–	*Kola indarijau*	–	–	–	Pala	Telia palla
–	*Goti*	–	–	–	Kottai	–

References

References

Aboul-Naga, A.M. 1977. Location effect on the lamb performance of three indigenous breeds of sheep under subtropical conditions of Egypt. *Indian J. Anim. Sci.*, 47: 29-33.

Aboul-Naga, A.M.; El-Shobokshy, A.S.; Marie, I.F. and Moustafa, M.A. 1981. Milk production from subtropical nondairy sheep. I. Ewe performance. *J. agri. Sci.* (Camb), 97: 297-301

Acharya, R.M. 1978. *Breeding Strategy for Sheep in India.* A discussion paper published by Central Sheep and Wool Research Institute Avikanagar, Rajasthan.

Acharya, R.M. 1982. *Sheep and Goat Breeds of India,* FAO, Rome.

Acharya, R.M. and Bawa, M.S. 1971. Milk production of ewes and its relationship with preweaning growth of lambs. *Indian J. Anim. Sci.*, 41 : 572-76.

Acharya, R.M.; Kulkarni, V.G. and Singh, M. 1980. *Canary Coloration of Indian wools.* Central Sheep and Wool Research Institute, Avikanagar Rajasthan.

Adam, C.E. 1961. Artificial insemination in rabbit. *J. Reprod. Fert.*, 2 : 521.

Adeleye, I.O.A. and Oguntona, E. 1975. Effects of age and sex on liveweight and body composition of the West African Dwarf sheep. *Nigerian J. Anim. Prod.*, 2 : 264-69.

Ademosun, A.A. 1970. Nutritive evaluation of Nigerian forages. 2. The effect of stage of maturity on the nutritive value of *Stylosanthes gracilis. Nigerian agric. J.*, 7 : 164- 73.

Adu, I.F. and Ngere, L.O. 1979. The indigenous sheep of Nigeria. *World Rev. Anim. Prod.*, 15 (3): 51-62.

Ahmed, S.I. 1955. Effct of glycine on storage of ram semen. *J. Agric. Sci.*, 46 : 164-67.

Ali, M.H.; Ranjhan, S.K. and Pathak, N.N. 1980. Note on the effect of supplement feeding on the performance of Muzaffarnagari lambs weaned at 60 and 80 days of age. *Indian* J. *Anim. Sci.*, 50 : 95-97.

Allcroft, W.M. 1963. Copper deficiency and molybdenum excess. In. *Animal Helath, Production and Pasture.* (Eds) A.H. Wordon, K.C. Sellers and D.E. Tribe. Longman, London. pp 662-80.

Anderson, J. 1945. The semen of animals and its use for artificial insemination. *Tech. Commun, Bur. Anim. Breed. Genet.(Edinb.), No.6:viii: 151.*

Anderson, J.J.B. 1970. *Parturient Hypocalcaemia.* Academic Press, New York, N.Y., London. P 276.

Andreevskii, V. Ja. 1940. Pricizy porocnosti spermy u baronov (Causes of semen abnormality in rams). *Trud. Lab. iskusst. Osemen. Zivoth.* (Mosk.), 1 : 36-45 (*Anim. Breeding Abstr.*, 13 : 36).

Andreoni, R. 1938. Notes on sheep breeding in Cyrenaica., Anim. Breeding Abstr., 7 : 19-20.

Andrewartha, K.A. and Caple, I.W. 1980. Res. *Vet. Sci.*, 28 : 101.

Andrewartha, K.A., Davies, W.D. and McDonald, J.W. 1980. Observations on serum Thyroxin concentration in lambs and ewes to assess iodine nutrition. *Aust. Vet. J.*; 56 : 18-21.

Anon. 1953. Sheep Farming in Rajasthan. *Indian Fmg.*, 3 (9) : 10.

Anon. 1972a. Progress Report A.I.C.R.P. on canary coloration of Indian wools. Central Sheep and Wool Research Institute, Avikanagar, Rajasthan.

Anon, 1972b. Blowflies and parasites: Costs to the sheep Industry 1969-70. *Bureau Agric. Economics, Wool Economics Research Report. No.* 22. *Australia Government* Publishing, Service, Canberra, p 99.

Anon, 1976. "Avikaline-a new carpet wool strain", evolved at Central Sheep and Wool Research Institute, Avikanagar, Rajasthan.

Anon, 1984. Annual Report. A.I.C.R.P. on goats. Central Sheep and Wool Research Institute, Avikanagar. (Rajasthan).

Antoine, D.; Ezakial Nepolean, R.; Kathiresan, D.; Pattabiraman, S.R., Padmanbhan, P.N.; Raman, K.S. and Kathperumal, V. 1992. Hormonal induction of estrum in anoestrus ewes. *Proc. 10th National convention Indian Soc. Study Anim. Reprod.* Madras Vety College. April 8-10, 1992.

Apgar, J. and Travis, H.F. 1979. Effect of a low zinc diet on the ewes during pregnancy and lactation. *J. Anim. Sci.*, 48 : 1234-38.

Armour, J. 1972. Facioliasis, Epidemiology, treatment and control. In : *Helminth Diseases of Cattle, Sheep and Horses in Europe.* (Eds.) G.M. Urguhart and J. Armour. Proc. Workshop, University of Glasgow. University Press, Glasgow. pp 110-105.

Armstrong, D.T. and Evans, S.G. 1983. Factors influencing success of embryo in sheep and goats. *Theriogenology,* 19 : 21-30.

Arora, C.L.; Acharya, R.M.; Badashiya, B.S. and Dass, N.C. 1975a. Characterisation of Malpura breed in Rajasthan and future prospects for its improvement. *Indian J. Anim. Sci.*, 45 : 843-48.

Arora, C.L., Acharya, R.M., Bhadashiya, B.S. and Dass, N.C. 1975b. Characterisation of Chokla breed in Rajasthan and future prospects of its improvement. *Indian J. Anim. Sci.*, 45 : 345-50.

Arora, C.L.; Acharya, R.M.; Ramadhar Singh and Buch, S.O. 1975c. Characterisation of Nali breed in Rajasthan and its prospects for improvement. *Indian J. Anim. Sci.*, 45 : 849-55.

Arora, C.L.; Acharya, R.M.; Sangoli, B.V. and Nagaraja, M.S. 1977. Characterisation of Sonadi breed in Rajasthan and future prospects for improvement. *Indian J. Anim. Sci.*, 47 : 20- 25.

Arora, C.L.; Singh, R.N.; Bohra, S.D.J.; Nivsarkar, A.E.; Gurbachan Lal; John, M.E. and Kumar, P.N. 1976. All India Coordinated Research Project on Sheep Breeding. 2. Sheep for Mutton Annual Report (1976). Central Sheep and Wool Research Institute, Avikanagar, Rajasthan.

Arora, R.G. and Oyar, P.K.R. 1973. Studies on the pathology of fascioliasis in sheep and goats. *Indian J. Anim. Sci.*, 43 : 720-23.

Ashmawy, G.M. and Al-Azawi, W.A. 1982a. A comparative study of fleece characteristics in Iraqi sheep. I. Greasy fleece weight, shrinkage percentage and fibre type ratio. *Egyptian J. Anim. Prod.*, 22 : 53-61.

Ashmawy, G.M. and Al-Azawi, W.A. 1982b. A comparative study of fleece characteristics in Iraqi sheep. II. Fibre diameter, medullated fibre percentage and fibre length. *Egyptian J. Anim. Prod.*, 22 : 63-71.

Ashton, W.M. and Yousef, I.M. 1961. A study of the composition of Clun Forest ewe's milk: mineral constituents. *J, agri. Sci.* (Camb.), 61 : 77-80.

Ashton, A., Owen, J.B. and Ingleton Jean, W. 1964. A study of the composition of Clun forest ewe's milk. *J. agri. Sci.* (Camb), 63 : 85-90.

Asker, A.A. and El-Khalisi, I.J. 1966. Some observations on commerical flocks of sheep in Iraq. *Ann. Agric. Sci. University, Aim Shams.* 10, No. 2. (1965) : 17-28, (Anim. Breeding Abstr. 35 : 439).

Aslanjan, M.M. and Camuha, M.D. 1950. Izbirateljnostj oplodotvore enija u ovec pri osemene nii ih smesjju semeni ohlazdennovo i sverago (Selective fertilization in sheep by insemination with mixed stored and fresh semen). *Sovetsk. Zootch,* 5 (12) : 65-70 (Anim. Breeding Abstr., 18 : 724).

Aswad, A., Abdou, M.S.S., Al-Bayaty, F. and El-Sawaf, S.A. 1976. The validity of the 'Ultra-sonic Method' for pregnancy diagnosis in ewes and goats. *Zentralblatt fur Veterinarmedizin* A, 23 : 467-74.

Aswad, A. Rebesko, B., Fathalla R. and Fahri, R. 1974. Sinhronizacija porodaja (Parturition synchronisation). *Veterinarski Clasnik,* 28 : 609-14.

Atmadilaga, D. 1958. Study on the milk yield of Indonesian sheep with special reference to the Priangan breed. *Hemera Zoa.*, 65 : 3-14.

Axtmayer, J.H.; Hernandez, G.R. and Cook, D.H. 1938. The nutritive value of some forage crops of Puerto Rico. *J. Agric. Univ. P. Rico* 22 : 455-81.

Ayalon, N. and Shemesh, M. 1979. Use of milk progesterone measurement for determining pregnancy and ovarian activity in cattle and sheep. *Rafuah Veterinarith,* 36 : 79.

Bachrach, H.L. 1968. Foot and mouth disease. *Ann. Rev. Microbiel,* 22 : 201.

Bali, H.S. 1972a. On the occurence of *Calicophoron calicophorun* (Fischoeder, 1901) Nasmark, 1937 in sheep of Jammu and Kashmir state. *Indian Vet. J.,* 49 : 432-33.

Bali, H.S. 1972b. On the morphology of a new trichostrongylid Nematodirusr foteda in species of sheep *(Ovis aries)* from Kashmir. *Punjab Agricultural University, J. Res.,* 9 (1) : 195-98.

Bali, H.S. 1973. Incidences of Helminth Parasites of sheep in Bihar. *J. Anim. Health Production.,* 1 : 35-39.

Bali, H.S. 1976. A survey of Helminth parasites of sheep *(Ovis aries)* in Jammu and Kashmir. J. Anim. Health Prod., 4 : 25-32.

Bali, H.S. and Fotedar, D.N. 1972. Study on a new tremetode belonging to the genus *Paramphistomum fischoeder,* 1901. from *Ovis aries* Jammu and Kashmir state. *Indian J. Anim. Sci.,* 42 : 231.

Bali, M.K. and Singh, R.P. 1976. *Indian J. Anim. Res.,* 10 (2) : 111-12.

Bali, M.K. and Singh, R.P. 1977. Efficacy of Fenbendazole suspention (Hoechst) in goat nematodasis. *Haryana Agricultural University J. Res.,* 7(3) : 155-57.

Bali, H.S.; Kawatra, M.S. and Fotedar, D.N. 1976. Studies on the pathology of some helminth borne diseases in sheep of Kashmir. *J. Anim. Health Prod.,* 4 : 9-14.

Bansal, M.P. and Malik, B.S. 1966. Isolation of bacterial and viral agents associated with ovine lesions. *Indian Vet. J.,* 43 : 954-60.

Barnicoat, C.R.; Logan, A.G. and Grant, A.I. 1949. Milk secretion studies with New Zealand Romney ewes. II Milk yields of ewes and factors influencing them, *J. agri. Sci.* (Camb.), 39 : 47- 55.

Barrau, J. 1953. Research in Queensland on tropical plants and animal industries. *5th Pacif-1 2 BPfc/1* Lm. *Tech. Paper, 43. Use ordinary Barsaul C.S. and Talapatra,* S.K. 1970. *The nutritive value of oat fodder as green, hay and silage. Indian Vet. J.,* 47 : 508-12.

Bath, G.F. 1979. *J. S. Afr. Vet. Assoc.,* 50(1) : 3.

Bawa. A.K.; Gupta, I.C. and Sharma, B.D. 1988. Natural resources and their management in range lands of western Rajasthan. *Curr. Agric.* 12 : 33-56.

Bawa, H.S. 1943. *Sorghum helepense* pers poisoning in cattle and sheep. *Indian. J. Vet. Sci.,* 13 : 326-27.

Bell, T.D. 1945. Breeding studies in Rambouillet ewes and rams. *Pr. Bull. N. Mex. Agric. Exp. Sta.* No. 1003 : 2, (Anim. Breeding Abstr., 14 : 154).

Bellaver, C, Arruda, F.de A.V. and Simplicio, A.A. 1980. Avaliacao da producao leiteira das racas deslanadas Morada Nova e Santa Ines. (Evalution of milk yield in Morade Nova and Santa Ines ewes). In

Pesquisa em ovinos no Brasil 1975-1979, P. 110. Empresa Brasileira de Pasquisa Agropecuaria, Bage, Brazil.

Berger, Y. and Ginisty, L. 1980. Bilan de 4 annees d'etude de la race ovine Djallonke Cote d'lvoire (Results of a 4-year study of the Djallonke breed of sheep in the Ivory Coast), *Revue d'Elevage et de Medicine Veterinaire des Pays Tropicaux,* 33 : 71-8.

Bermudez, L.A.; Caballos, E.Y. and Chaverta, H. 1968. Las *Leguminosas espentaneas* en el valle del Sinu. *Agricultura trop.,* 24 : 589-603.

Bhaik, N.L. and Kohli, I.S. 1980. Reproductive performance of *Chokla and Magra ewes in Mbtropical area of North Western* arid zone of Rajasthan. 2. Parturition and involution of uterus. *Indian Vet.* J., *57* : 327-33.

Bhandari, D.S.; and Gupta, M.L. 1972. Studies on the digestibility and nutritive value of Ardu *(Ailanthus excelsa* Roxb). *Indian Vet.* J. 49 : 512-6.

Bhandari, D.S.; Govil, H.N. and Hussain, A. 1979. Chemical composition and nutritive value of Khejri *(Prosopis cinearia)* tree leaves. *Ann. Arid Zone,* 18 (3): 170-3.

Bhattacharya, A.N. and Harb, M. 1973. Sheep production on natural pastures by roaming Bedouins in Lebanon. *J. Range Mgmt.,* 26 : 266-69.

Bhasin, N.R. and Desai, R.N. 1965. Studies on factors affecting characters concerning quality of wool fiber in Chokla flock of sheep. *Indian Vet. J.,* 42: 782.

Bhatia, D.R. and Munshi, S.K. 1972. The chemical composition and nutritive value of F-2-5 strain of Bajra × orientale hay. *Indian Vet. J.,* 49 : 274-77.

Bhatia, B.B. and Pande, Y.U.B. 1950. *Proc. Nat. Acad. Sci. India,* Sect. B., 30 : 3.

Bhatia, D.R.; Mohan, M. and Patnayak, B.C. 1971. Effect of supplementary feeding on growth and carcass yield of Malpura lambs. Annual Report. Central Sheep and Wool Research Institute, Avikanagar, (Rajasthan).

Bhattacharyulu, Y. and Pandey, B.P. 1969. On a collection of adult amphistomes in sheep a scientific evaluation. *Indian J. Anim. Sci.,* 39 : 321.

Bhimaya, C.P.; Kaul, R.N. and Ganguli, B.N. 1963. Species suitable for afforestation of different arid habitats of Rajasthan. *Ann. Arid Zone, 2* : 162-68.

Bhuvanakumar, C.K. and Subramanian, P. 1981. Effect of body weight on greasy fleece weight in Merino, Nilgiri and their crosses. *Cheiron,* 10 : 40-2.

Blackshaw, A.W. 1958. The effects of glycerol on the paravital staining of spermatozoa. *Aust. Vet. J.*, 34 : 71-76.

Beaz, T.G.; Forbes, J.M.; Kalkenny, J.B.; Mudd, C.H.; Robinson, J.J.; Russel, A J.F. and Treacher, T.T. 1973. *Feeding the ewes. Technical Report. Sheep Improvement service, meat and livestock commission No. 2. Blatchley, U.K., Meat and Livestock commission, pp. 52.*

Bogart, R. and Mayer, D.T. 1946. Environmental temperature and thyroid gland involvement in lowered fertility of ram. Res. *Bull. Mo. Agric. Exp. Sta.;* No. 402; 43. (Anim. Breeding Abstr., 14 : 235).

Bohra, S.D.J.; Nagarcenkar, R. and Sharma, K.N.S. 1979. Factors affecting preweaning body weights in Malpura sheep. *Indian Vet. J.*, 56 : 125-28.

Boland, M.P.; Crosby, T.F. and Gordon, I. 1978. Morphological abnormality of cattle embryos following superovulation using PMSG. *Theriogenelogy*, 10 :175.

Boray, J.C. 1969. Experimental fascioliasis in Australia. *Adv. Parasitol., 7 :* 95-210.

Bosc, M.J. 1972. The induction and synchronisation of lambing with the aid of dexamethasone. *J. Reprod. Fert.*, 29 : 347-57.

Boyazoglu, J.G. 1963. Aspects quantitatifs dela production lactiere des brebis. I. Mise all point bibliographique. *Annal Zootech.*, 12 : 237-96.

Bradford, G.F. and Kennedy, D.W. 1980. Genetic aspects of embryo transfer. *Theriogendogy*, 13 : 13-28.

Braithwaite, G.D. 1976. Calcium and phosphorus metabolism in ruminants with special reference to parturient paresis. *J. Dairy Res.*, 43 : 501-502.

Brandis, D. 1906. *Indian trees* (Reprinted, 1971). Bishan Singh, Mahendra Pal Singh; Dehradun, 767 p.

Britt, J.H. 1979. Prospects for controlling reproduction processes in cattle, sheep and swine from recent findings in reproduction. *J. Dairy Sci.*, 62 : 651-65.

Brown, D.E.; Cameron, C.W. and Ngere, L.O. 1972. Synchronisation of oestrus and early post-partum rebreeding of Nungua Blackhead sheep in Ghana. *Ghana J. agri. Sci.*, 5 : 183-7.

Bryner, J.H. 1976. Vibriofetus induced abortion. *Theriogenology*, 5 : 128.

Buch, B.B. and Jayraman, S., 1953. Culling a means for improvement of sheep. *Indian Fmg* 3(10) : 24-26.

Builow, S.V. and Dzhaparidze, T.G. 1975. Breeding Clun Forest sheep. Anim. Breeding Abstr., 44 : 4791.

Bush, E.C. 1965. A comparison of the milking capacity of Corriedale and Corriedale X Border Leicester ewes. *E. Afr. agric. For. J.*, 31 : 31-4.

Butterworth, M.H.; Houghton, T.R.; Macartney, J.C; Prior, A.J.; Middlemiss, C.P. and Edmond, D.E. 1968. Some observations on the lactation of Blackhead ewes and the growth of lambs: the composition and yield of milk. *J. agric. Sci.* (Camb.), 70 : 203-7.

Buvanendran, V.; Adu, I.F. and Oyejola, B. A. 1981. Breed and environmen-tal effects on lambs production in Nigeria. *J. agric. Sci.* (Camb.), 96 : 6-15.

Calderton cartes, J.F.; Robinson, J.J.; McHattie, J. and Fraser, C. 1977. The sensitivity of ewe milk to chages in dietary crude protein concentration. *Anim. Prod.*, 24 : 135.

Cameron, N.D. 1992. Correlated physiological responses to selection for cacass lean content in sheep. *Livestock Prod. Sci.* 30 : 53-58.

Carbonero Bravo, D. 1955. La insemination artificial en la Karakulizacion de algunas razas ovinas espanola. *Rev. Patron. Biol. Anim.*, (Madr.) 1: 199-246 (Anim. Breeding Abstr., 24 : 220).

Castellanos Ruelas, A. and Valancie, Z.M. 1982. Quantitative and qualitative study of milk yield in Pelibuey ewes. *Trop. Anim. Prod., 7* : 232-40.

Castillo Rojas, J.; Hernandez Ledezma, J.J., Berruecos, J.M. and Lopez Angeles, J.J. 1977. Comportamiento reproductive del borrego Tabasco mantenido en clima tropical. III. Puberted y duracion del estro (Reproductive performance of Tabasco lambs maintained in a tropical climate. III. Puberty and duration of oestrus), *Tecnica Pecuariaen Mexico,* 32 : 32-5.

Chakrabarti, A.; Basak, D.K.; Shorn, P.N. and Banergie, A.K. 1975. Studies on osteo dystophia fibrosa in Black Bengal goats. *Indian J. Anim. Health.*, 14(2): 95-97. Chakarvarty, S.C. and Haq, I. 1946. Hairy sheep of India. *Indian Fmg., 7 :* 15-16.

Chang, H.C. 1945. The sperm production of adult rams in relation to frequency of semen collection. *J. agric. Sci.* (Camb), 35 : 245-46.

Chang, M.C. and Marden, W.G.R. 1954. The aerial transport of fertilized mammalian ova. *J. Hered.*, 45 : 75.

Charyulu, B.; Deshmukh, A.P. and Patil G.D. 1988. Fleece production and wool quality of Deccani and their halfbreds with Dorset and Merino. *Indian Vet. J.*, 65 : 242-45.

Chellappa, D.J. and Gopal Kishan, CA. 1977. *Indian J. Anim. Res.*, 11 (2): 74-75.

Chatterjee, R. and Arora, R.K. 1974. Scope and practical importance of eliminating highly medullated fibers in Indian fleeces. *Text. Res. J.*, 44 : 325-26.

Chaturvedi, M.D. 1948. *The role of leaf fodder in the United Provinces. U.P. For. Dep. Bull.* No. 19. 9 p.

Chaturvedi, M.D. 1956a. Arua the tree of distinction. *Indian Fmg.*, 6(3) : 33-34.

Chaturvedi, M.D. 1956b. The friendly shisham. *Indian Fmg.*, 6 (2): 19-20.

Chaudhry, A.L. 1988. *Interospection of carpet wool availability and production strategy in India.* Paper presented at National Seminar on Carpet Wool Production and utilisation held at Central Sheep and Wool Research Institute, Avikanagar, Oct. 24-26, 1988.

Chippalkatti, H.R.; Sule, A.D.; Juneja, K.K. and Aggarwal, M.C. 1965. Some observation on canary coloration in Indian raw wools, *J. Text. Inst.*, 56 : T 649 - T 664.

Chorey, P.A. and Naithani, S.P. 1965. The effect of antibiotic feeding on growth of lambs and kids. *Indian Vet. J.*, 42: 433-38.

Chucko, C.L. 1954. The performance of indigenous and introduced grasses in Malabar. *Indian Vet. J.*, 31 : 187-93.

Church, R.B. and Shee, B. 1975. Some aspects of bovine embryo transfer. In: *Egg Transfer in cattle in the E.E.G. programme coordination of research on beef products.* Cambridge, *Anim. Breeding Abstr.* 54 : 168.

Clark, L; Carlisle, C.H. and Beasley, P.H. 1975. Observations on the pathology of bent leg of lambs in south-western Queensland. *Aust. Vet. J.*, 51 : 4-10.

Combellas, J. and Gonzalez, J.E. 1973. Rendimientoy valor nutritivo de forrajes tropicales. 5. Pasto para *(Brachiaria mutica* Stapf) Yield and nutritiv: value of tropical forages. 5. Para grass *(Brachiaria mutica* Stapf), *Agronomia trop.* (Venezuela), 23 : 277-86.

Combellas, J., Gonzalez, J.E and Trujillo, A. 1972. Rendimiento y valor nutritivo de forrajes tropicales. 1. Bermuda cv, coastal *(Cynodon dactylon* (L) Pers.) Yield and nutritive value of tropical forages. 1. Coastal Bermuda grass *(Cynodon dactylon* (L) pers.). *Agronomia trop.* (Venezuela), 22 : 231-8.

Combellas, J. de. 1978. Comparaction de los pesos al nacer, pesos al destete y ganancias en peso de corderos West African y corderos mestizos de Dorset Horn. (A comparison of body weight at birth and weaning and gains of West African and crossbred Dorset Horn lambs). In *Informe Anual.* Instituto de Produccion Animal, pp. 71-72. Universidad Central de Venezuela Maracay, Venezuela.

Combellas, J.De 1980. Determinacion de la production de leche en ovejas West African × Dorset Horn en su primera lactancia (Determination of milk yield of West African × Dorset Horn ewes in their first lactation). In *Informe Anual* 1979. Instituto de produccion Animal, pp. 69-70. Universidad Central de Venezuela, Maracay, Venezela.

Combellas, J.de, Rondon, Z. and Gullen, J. 1980. Sincronizacion del estro en ovejas. (Synchronisation of oestrus in ewes). In *Informe Anual 1979.* Instituto de produccion Animal, p. 71. Universidad Central de Venezuela, Maracay, Venezuela.

Comstock, R.E.; Green, W.W.; Winters, L.M. and Nordskog, A.N. 1945. Studies of semen and semen production. *Tech. Bull. Minn. Agric. Exp. Sta.* No. 162 : 55 *Animal Breeding Abstr.,* 13 : 37).

Coombe, J.B.; Wardrop, I.D. and Tribe, D.E. 1960. Astuoly of milk production of the grazing ewe with emphasis on experimental technique employed. J. Agri. Sci. (Camb.) 54 : 353-59.

Cupps, P.T.; MuGowan, B.; Rohlmann, D.F.; Reddon, A.R. and Weir, W.C. 1960. Seasonal changes in the semen of rams. *J. Anim. Sci.,* 19:208-13.

Crowder, L.V. 1960. Gramincasy leguminosas forrageras on Colombia. *DIA Beletin Technice,* 8. Bogota.

Daflapurkar, D.K.; Kaushik, S.N. and Kalpatal, B.C 1979. Comparative studies on wool production and wool quality traits in Patanwadi and Deccani crosses. *Indian Vet. J.,* 56 : 764- 67.

Dattilo, M. and Congiu, F. 1979. Utilizzazione energetica degli aimenti per la produczione del latte in ovini di razza Sarda (Utilisation of food energy for milk production in ewes of the Sardinian breed). *Rivista di Zootecnia e Veterinaria,* 3 : 183-90.

Daniel, J.C. Jr., Ed. 1971. *Methods in Mammalian Embryology.* San Francisco, W.H. Freeman P 532.

Das Kornel 1992. A note on mastitis in Corriedale ewes. *Indian Vet. J.* 69 : 357-58.

Davis, D.L. and Day, B.N. 1978. Cleavage and blastocyst formation by pig eggs *in vitro. J. Anim. Sci.,* 46 : 1043.

De Groot, L.J. and Stanbury, J.B. 1975. *The Thyroid and its Diseases.* Wiley, New York, NY, p 823.

Demiruren, A.S.; Beheshti, R.D.; Salimi, H.; Saleh, B.A. and Djafer, A. 1971. Comparison of the reproductive and productive capacities of sheep of the Kellakul, Kizil, Bakhtiari and Bakuchi breeds of Iran. *Tech. Rep. No. 1.* Anim. Husb. Res. Inst. Tehran.

Dennis, S.M. 1961. Vibriofetus infection of sheep. *Vet. Rev. Annotations, 7 :* 69-82.

Deshpande, B.R. 1970. Morphological changes in the semen picture and testicular tissue with artificially induced sub-fertility and sterility by scrotal insulation in rams. Ph. D. Thesis M.P.K.V. Rahuri (India).

Deshpande, K.S. and Sakhre, P.G. 1987. Comparative performance in wool quality traits of Deccani and its crosses. *Indian J. Anim. Sci,* 57 : 894-96.

Dhanda, O.P. 1970. Studies on oestrus pattern, hormonal synchronisation of oestrus and subsequent fertility in Nali sheep. M.Sc. Thesis Haryana Agricultural University; Hisar.

Dhanda, O.P. and Arora, K.L. 1977. A note on synchronisation of oestrus in Nali sheep. *Indian J. Anim. Sci.*, 47: 369-70.

Dhar, D.N. and Singh, K.S. 1970. Some effects of cortesone administration on the natural and acquired resistance in lambs to the nematode, *Oesophagostomum columbianum* (Curtice, 1890) (Stossich, 1899). *Indian J. Anim. Sci.*, 40 : 640-649.

Dhar, D.N.; Tiwari, H.C. and Rao, Y.V.B. 1972. Prevlence of lung worm infections in Kashmir. *Indian J. Anim. Sci.*, 42 : 112-14.

Dharmendra Nath, 1971. Observations on the treatment of amphistomiasis in sheep. *Indian Vet. J.*, 48 : 653-655.

Dhillon, R.S. 1980. Karakul lamb pelt: Evaluation and classification. *Wool & Woollens India.*, 17 (3): 32-35.

Dhillon, J.S.; Acharya, R.M. and Singh, R.N. 1979. Factors affecting libido and fertility in rams. *Indian J. Anim. Sci.*, 49 : 196-99.

Dierks, R.E. 1979. Rabies, pathogenesis and diagnosis. *J. Lab. Clin. Med.*, 94 :1-4.

Dotta, V.; Abate, 0.; Guglielmino, R.; Zimmerman, S. and Ponzetto C. 1978. *Schweizer Archiv'fur Tierheilkunde,* 120 (6): 291.

Dotta, V.; Abate, O.; Goglielmino, R.; Zimmerman, S. and Ponzetto, G 1979. *Clinica Veterinaria,* 102 (2): 166.

Dougall, H.W. and Bodgan, A.V. 1958. Browse plants of Kenya with special reference of those occurring in South Baringo. *E. Afr. agric. J.*, 23 : 236-45.

Dubey, S.C. and Kaushish, S.K. 1981. Enterotoxaemia - A dreaded sheep disease. *Farmers J.*, 1 : 33-34.

Dunn, A.M. 1948. *Veterinary Helminthology.* William Heinemann. Medical Books Ltd. London.

Dunn, I.G. 1980. Relationship of nutrition to successful embryo transplantations. Theriogenology, 13 : 27-39.

Dun, R.B. 1955. The cervix of the ewe its importance in artificial insemination of sheep. *Aust. Vet. J.*, 31 : 101-103.

Dun, R.B. 1956. Temporary infertility of rams associated with flooding. *Aust. Vet. J.*, 32 : 1-3.

Dun, R.B.; Ahmed, W. and Morrant, A.J. 1960. Annual reproductive rhythm in Merino sheep related to the choice of a mating time at Trangie, Central Western New South Wales. *Aust. J. agric. Res.*, 11 : 805-26.

D'Souza, B.A. 1949. Observations on the outbreak of the so called obscure sheep disease at Livestock Research Station, Hosur in 1946-47. *Indian Vet. J.*, 25: 321-30.

Dwarkanath, P.K. and Ghoshal, A.K. 1981. Trace minerals and productivity of Cattle ins esmi arid tracts of Rajasthan. Indiran J. Anim. Health. 20 (1): 127-29.

Dziuk, P.*J. 1971. Obtaining eggs and embryos from sheep and pigs. In:Methods in Mammalian Embryology.* (Ed.). J. C. Daniel, Jr. San Francisco. W.H. Freeman, pp 76-85.

Dziuk, P.J.; Graham, E.F.; Donker, J.D.; Marion, G.B. and Petersen, W.E. 1954. Some observations in collection of semen from bulls, goats, boars and rams by electrical stimulation. *Vet. Med.*, 48 : 455-58.

Dzhurov, A. 1979. *Obshchai Sravnitelna Patologia, 7* : 110.

Easley, G.T. 1951. A report of artificial insemination : Factors affecting - reproduction and semen studies in sheep of the highland pastures of Southern Peru. *J. Amer. Vet. Med. Ass.*, 119 : 278-282.

Edelsten, R.M. 1980. *Trop. Anim. Health Prod.*, 12 (2) : 69.

Egerton, J.R. and Roberts, D.S. 1969. The aetiology and pathogenesis of ovine foot rot. *J. Comp. Pathol. Ther., 79* : 207-15.

El-Homosi, F.F. and El-Hafiz, G.A.A. 1982. Reproductive performance of Ossimi and Saidi sheep under two prepubertal planes of nutrition. *Assiut Vet. Med. J.*, 10 (19) : 59-66.

Eliya, J. and Juma, K.H. 1970. Birth weight, weaning weight and milk production in Awassi sheep. Trap. *Agri.* Trin., 47 : 321-24.

El-oksh, H.A.; Kadry, A.E.H.; Ghanem, Y.S. and Galal, E.S.E. 1979. Estimates of some physical wool characteristics of Barki sheep under semi arid desert condition, *Beitrage Zur Tropischen Landwirtschaft und Veterinarmedizin,* 17 : 391- 95.

Elsden, R.P.; Lewis, S.; Cumming, I.A. and Lawson, R.A.S. 1974. Superovulation in the cow following treatment with PMSG and Prostaglan-din F_2. alpha *J. Reprod. Fert.*, 36 : 455-456.

El-Sherbiny, A. A. and Shesha, A. 1972. Effect of crossing Ossimi sheep with Merino on milk production and mothering quality *Agric. Res. Rev.*, 50(3) : 37-43.

El-Wishy, A.B. 1974. Some aspects of reproduction in fat tailed sheep in subtropics. IV. Puberty and sexual maturity. *Zeitschrift fur Tierzuchtung und Zuchtungsbiologie,* 91 : 311-16.

Emmens, C.W. 1959. Fertility in the male (Supplementary). Progresses *in the Physiology of Farm Animal.* (Ed.) J. Hammond. London : Butterworths Scientific Publications, pp. 1047- 1116.

Emmens, C.W. and Robinson, T.V. 1962. Artifical Insemination. In sheep. *The semen of Animals and Artificial Insemination* (Ed.) J.P. Maule Commonwealth Agricultural Bureaux. Farnham Royal, Bucks, England Engela, D.J. 1948. Fertility in rams. *Fmg. S. Afr.*, 23 : 218-82. (Anim. Breeding Abstr., 26 : 1025).

Engela, D.J. 1948. Use *of fodden trees and shruubs* Qd. Deptt. preimary Industries Div. Plt. Ind. Adirsory leaflat 1024

Everist, S.L. 1969. Use *of Fodder Trees and Shrubs.* Qd. Dept. Primary Industries, Div. Plt. Ind. Advisory leaflet 1024.

Everist, S.L.; Harvey, J.M. and Bell, A.T. 1958. Feeding sheep on mulge. Qd. Dept. Agr. and Stock, Reprint 7/4/58.

Eyal, E.; Lawi, A.; Folman, Y. and Morag, M. 1978. Lamb and milk production of a flock of dairy ewes under an accelerated breeding regime. *J. agric. Sci.* (Camb.), 91: 69-79.

Fall, A., Diop, M.; Sandford, J.; Wissooq, Y.J.; Durkin, J. and Trail, J.CM. 1982. Evaluation of the productivities of Djallonke sheep and N'-Dama cattle at the Centre de Recherches Zootecniques, Kolda, Senegal. Research *Report No.* 3. Intern. Livestock Centre for Africa, Addis Ababa, Ethiopia.

F.A. O, 1978 Production year Book. Vol. 31 FAO, Rome, P 291.

Farid, A.; Makarenchian, M. and Sefidbakht, N. 1976. Crossbreeding of Iranian fat-tailed sheep. 1. Preweaning growth performance of Karakul, Mehraban, Naeini and their reciprocal crosses. Iranian *J. agric. Res.*, 4 : 69-79.

Faure, AS.; Boshoff, D.A. and Burger, F.J.L. 1980. Comparison between the intravaginal sponge and subcutaneous implant for synchronisation of oestrus in Karakul ewes., 1976-1977. In *Agricultural Research 1979.* pp. 83-85. Division of Agricultural Information, Oretirua, South Africa.

Field, AC. 1969. Urinary calculi in animals. *Aust. Vet. J.*, 54 : 267- 70.

Figuiredo, E.A.P.; Simplicio, A.A. and Pant, K.P. 1982. Evaluation of sheep breeds for early growth in tropical North-east Brazil. *Trop. Anim. health Prod,* 14 : 219-23.

Fitch, L.W.N. 1954. Osteodystrophic diseases in sheep in New Zealand II Bowie and Bentleg. *NZ. Vet. J.*, 2 : 118-27.

Fletcher, I.C.; Lishman, A.W.; Thring, B. and Holmes, J.A. 1980. Plasma progesterone levels in lactating ewes after hormone- induced ovulation during the non-breeding season. *S. Afr. J. Anim. Sci,* 10 : 151-57.

Florich, G. 1931. Karakul and its breeding. Anim. Breeding Abstr., 1 : 58.

Foilmom, S.N.; Lunca, I.; Bratescu and Otel, V. 1956. Cercetari asupra diluarii Si Conservarii Spermei de berbec. si taur. (The dilution and storage of ram and bull semen). *Anal. Inst. Cera. Zootec.,* 14 : 231-41 (Anim. Breeding Abstr., 25 : 1943).

Fraser, A.H.M. 1954. Influence of nutrition on wool growth. *Nutr. Abstr. Rev.,* 4 : 9-13

Frazer, A. 1948. Cheviot sheep. *J. Minist. Agri.,* 55 : 277-83.

Fraser, I.E.B. and Mulcock, A.P. 1956. Staining of wool by bacterial pigment. *Nature, 177*: 628-29.

French M.H. 1935. The nutritive value of legume hays. *Ann. Rept. Dept. Vet. Sci. Anim. Husb., Tanganyika,* 86-89.

French M.H. 1937. Compartive nutritive value of legume hays. *Ann. Rept. Dept. Vet. Sci. Anim. Husb. Tanganyika,* 104-106.

French M.H. 1957. Nutritional value of tropical grasses and fodders. *Herb. Abstr.,* 21 : 1-9.

Funes, F. 1975. Digestibility and nutritive value of pangola grass *(Digitaria decumbens* Stent) in relation to nitrogen fertilisation and harvesting time. *Cuban J. Agric. Sci.,* 9 : 369-78.

Gaillard, Y. 1979. Caracteristiques de la reproduction de la brebis Oudah (Reproductive characteristics of the Uda ewe), *Revue d'Elevage et de Medecine Veterinaire des Pays Tropicaux.* 32 : 285-90.

Gandolfi, F.; Brevini, T.A.L., Richardson, L.; Brown, C.R. and Moss, R.M. 1989. Characterisation of proteins secreted by sheep ovi duct epithelial cells and their function in embryonic development. *Development,* 106 : 303-12.

Ganesakale, D. 1975. Effect of season of mating on the economic traits in sheep. *Cheiron,* 4 : 40-44.

Ganguli, B.N.; Kaul, R.N. and Nambiar, K.T.N. 1964. Preliminary Studies on a few top feed species. *Ann. Arid Zone,* 3 : 31- 37.

Gardner, D.E. 1973. Pathology of *Clostridium welchii* type D enterotoxaemia. *J. Comp. Pathol,* 83 : 499.

Gardner, R.W. and Hogue, D.E. 1964. Effects of energy intake and number of lambs suckled on milk yield, milk composition and energetic efficiency of lactating ewes. *J. Anim. Sci.,* 23 : 935-42.

Gaur, D.; Chopra, S.C; Balaine, D.S. and Chhikara, B.S. 1977. Preweaning and postweaning body weights of Nali and Corriedale × Nali lambs. *Indian J. Anim. Sci.,* 47 : 534-38.

Georgi, J.R. 1969. *Parasitology for Veterinarians,* W.B. Saunders and Co. Philadelphia.

Ghanekar, V.M. 1983. Deccani sheep. A study in retrospect part III. Studies on Daccani wool. *Wool & Woollens India,* 20 (3) : 21-40.

Ghoneim, K.E.; Kazzal, N.T.; Abdallah, R.K.; Thannoon, N. and Alatroshy, A.H. 1973. *Some wool characteristics of Karadi sheep in North of Iraq.* University of Mosul College of Agriculture and Forestry, Hamman Al-Alil,Iraq.

GIBCO. 1976-77. *Catalog,* Grand Island Biological Company, Grand Island, NY.

Gihad, E.A. 1976. Intake, digestibility and nitrogen utilisation of tropical natural grass hay by goats and sheep. *J. Anim. Sci.,* 43 : 879-83.

Gill, R.S. and Negi, S.S. 1971. Requirement of maintenance digestible crude protein for adult sheep. *Indian J. Anim. Sci.,* 41 : 980.

Godwin, K.O. 1975. The role and metabolism of selenium in animals. In : *Traceelements in Soil-Plant-Animal-Systems.* (Eds.) D.J.D. Nicholas and A.R. Egan. Academic Press. New York, NY., London, pp 259-70.

Goings, R.L.; Jacobson, N.L.; Beitz, D.C.; Wiledike, E.T. and wiggers, K.D. 1974. Prevention of parturient paresis by prepartum Calcium deficient diet. . *Dairy Sci.,* 57 : 1184-88.

Gonzole, C; Vijil, E. and Reguera, O. 1985. Study of milk yield, health condition of the mammary gland and udder shape throgh out lactation in machine milked Churra Sheep. *36th Ann. Meat. EAAP* Greece.

Gootwine, E.; Alef, B. and Gadeesh, S. 1980. Udder confirmation and heritability in Assaf (Alwassi × East Friesian) cross of dairy sheep in Israel. *Ann. Gen. Sci. Anim.,* 12 (1) : 9-13.

Gordon, H.Mc. D. 1958. *Proc. Aust. Soc. Anim. Prod.,* 2 : 59.

Gordon, H.Mc. D. 1967. The diagnosis of helminthiasis in sheep. *Vet. Med. Rev.,* 2/3 : 140-168.

Grumes, D.L. 1973. Grass tetany of cattle and sheep. In : *Antiquality components of Forages.* Crop Sci. Soc. Ann. Wisconsin, pp 113-140.

Guha, S.; Kohli, M.L. and Bhattacharya, P. 1951. Artificial insemination of sheep and goat at Izatnagar. *Indian). Vet. Sci. A.H.,* 21 : 171.

Gunn, R.M.C.; Sanders, R.N. and Granger, W. 1942. Studies in fertility in sheep. 2. Seminal changes affecting fertility in rams. *Bull. Coun. Sci. Industr. Res. Aust.,* No. 148-149. *(Anim. Breeding Abstr. 11 : 236).*

Gupta, B.R.; Reddy, K.K. and Munerathnam, D. 1987. Puberty and sexual maturity in Nellore sheep and its synthetics with Dorset. *Indian J. Anim. Sci.,* 57 : 1013-16.

Gupta, B.S.; Satpathy, N.; chabra, S.S. and Ranjhan, S.K. 1970. Chemical composition and nutritive value of a summer legume, Chakunda silage *(Cassia cirra)* with and without molasses for sheep. *Indian J. Dairy Sci,* 23 : 220.

Gupta, P.C.; Randhir Singh; Sangwan, D.C. and Pradhan, K. 1975. Chemical composition and *in vitro* nutrient digestibility of some of the tree leaves. *Indian Forester,* 101: 674-80.

Gupta, J.L.; Chopra, S.C. and Dhamale, S.P. 1974. Effect of mating seasons on growth of Deccani sheep and their crosses with Merino and Rambouillet. *Haryana Agricultural University J. Res.* 4 : 158-61.

Gurgis, R.A. 1973. The study of variability in some wool traits in a coarse wool breed of sheep. *J. agric. Sci.* (Camb.), 80 : 233-38.

Guyer, P.O. and Dyer, A.J. 1954. Study of factors affecting sheep production. *Missouri Agri Expt. Sta. Res. Bull.* No. 448 : 79 pp.

Gwatkin, R.B.L. 1972. Chemically defined media for mammalian eggs and early embryos, *In Vitro,* 8 : 59.

Gyang, F.D. and Sarror, D.I. 1979. *Bull. Anim. Health Prod. Afr.,* 27 (3): 195.

haberman, E. and Wellhorner, H.H. 1972. Studies on the active Tracers In *Microbial Immunology.* International Atomic Energy Agency, Vienna, pp 67-76.

Hafez, E.S.E.;Baderldin, A.L. and Darwisch, Y.H. 1955 Seasonal variations in semen characteristics of sheep in the subtropics. *J. agric. Sci.* (Camb), 45 : 283-92.

Hafs, H.D.; Purchas, R.W. and Pearson, A.M. 1971. A review: Relationship of some hormones to growth and carcass quality of ruminants. *J. Anim. Sci,* 33 : 64-71.

Haresign, W. 1973. Ovulation control in the sheep. In: *Control of Ovulation* (Eds) D.B. Crighton, N.B. Haynes, G.R. Foxcroft and G.E. Lamming. Butterworths London, pp 435-51.

Harlan, J.R. 1976. The plants and animals, that nouresh man. Sci. Amer. 235 : 89-97.

Hay, M.F. and Moor, R.M. 1978. Changes in the Graffian follicle population during the follicular phase of the oestrus cycle. In *Control of Ovulation* (Eds) D.B. Crighton, N.B. Haynes, C.R. Foxcroft and G.E. Lamming. Butterworths, London, pp 177-196.

Heape, W. 1890. Preliminary note on the transpalantation and growth of mammalian ova within a uterine foster mother. *Proc. Roy. Soc.* (Lond.), 48 : 457.

Heard, T.M. and Walker, S.K. 1992. premature ovulation and embryo collection in the ewe. *Proc. Ann. Conf. Intern. Embryo* Transfer held in Denver, Colorado. Jan 12-14, 1992.

Hendricksen, R.E.; Poppi, D.P. and Minson, DJ. 1981. The voluntary intake, digestibility and retention time by cattle and sheep of stem and leaf fractions of a tropical legume *(Lablab purpureus), Aust. J. agri. Res.,* 32 : 389-98.

Henderson, A.F. 1965. Cited by Narayan, S, 1969.

Hetzel, D.J.S.; Yates, N.G.; Obst, J.M.; Chaniago, T. and Bakrie, B. 1982. A preliminary report on the growth of Javanese Thin-tail sheep and their crosses with Suffolk, Polled Dorset and Wiltshire Horn rams. *Proc. Aust. Soc. Anim. Prod.*, 14 : 629.

Hiji, C.S.G. 1935. Preliminary notes on a disease of sheep and goats locally known as "Phet or pitto". *Indian Vet. J.*, 12 : 18-21.

Hill, J.R. (Jr.); Hurst, V. and Godleg, W.C. 1958. A comparison of reconstituted skim milk and egg yolk sodium citrate as extenders for ram semen. *Amer. J. Vet. Res.*, 19 : 132-134.

Hinkolman, L.A.; Bradford, C.E.; Pathak, E.T.; Anderson, G. and Cupps, P.T. 1979. An embryo transferstudy of birth weight in Finish Landrace and Targhee sheep. *J. Anim. Sci.*, 48 : 298-306.

Hiregoudar, L.S. 1970. Ananthelmintic trial with "Nilverm" against gastro intestinal nematodes of sheep. *Indian Vet. J.*, 47 : 15-18.

Hirsch, S. 1933. Sheep and goats in Palestine. Anim. Breeding Abstr., 2 : 400.

Hogue, D.E.; Reid, J.T.; Heffon, C.L.; Guterman, W.H. and Lisk, D.J. 1980. *Cornell Vet.*, 70 : 67.

Homfray, C.K. 1937. *Nursery and plantation notes for Bengal.* (4th Edition). Govt. of Bengal. Forest Department. Supdt. Govt. Printing, Alipore, Bengal, 210 p.

Honmode, J. 1969. Hormonal synchronization of oestrus in Malpura ewes. *Indian J. Anim. Sci.*, 39 : 95.

Honmode, J. 1970. Gestation period in Rambouillet, Chokla and Malpura ewes. *Indian J. Anim. Sci.*, 40 : 229.

Honmode, J. 1971. Effect of megasterol, megasterol acetate and progesterone on oestrus and fertility in cyclic Malpura and Chokla ewes. *Indian J. Anim. Sci.*, 41 : 972.

Honmode, J. 1972. Annual Report, Central Sheep Wool Research Institute, Avikanagar, Rajasthan.

Honmode, J. and Pachlag, S.V. 1973. Time of ovulation and chloride content of vaginal mucous during estrus in sheep. *Indian J. Anim. Sci.*, 43 : 136.

Honmode, J. and Tiwari, S.B. 1968. Artificial breeding of Rambouillet, Malpura, Chokla, R × C and R × M ewes during March/April. Annual Report, Central Sheep and Wool Research Institute, Avikanagar, Rajasthan, p 4.

Honmode, J. and Tiwari, S.B. 1973. Artificial breeding of exotic, indigenous and their crossbred ewes during March/April. Annual Report, Central Sheep and Woo Research Institute, Avikanagar, Rajasthan.

Honmode, J. and Tiwari, S.B. 1974. Effect of frequency of semen collection on quality and fertility of Malpura and Chokla rams. *Indian Vet. J.*, 51 : 100-104.

Honmode, J. Patil, B.D. and Pachlag, S.V. 1971a. Hormonal induction of superovulatory oestrus in early postpartum indigenous and crossbred ewes. *Indian J. Anim. Sci.*, 41 : 977.

Honmode, J., Bohra, S.D.J. and Patil, B.D. 1971 b. Oestrus synchronisation in anoestrus motegenous and exstic ewes-A note. *Indian J. Anim. Sci.*, 41 : 1066.

Honmode, J. Patil, B.D. and Tiwari, S.B. 1971c. Reproductive perforniance of Chokla rams as influenced by differential feeding. *Indian Vet. J.*, 48 : 1044-47.

Honmode, J.; Chatterjee, R. and Patil, B.D. 1971d. Investigation into the effect of season, hormone and plane of nutrition on reproduction and characteristics of wool of six indigenous sheep breeds. Annual Report, Central Sheep and Wool Research Institute, Avikanagar Rajasthan, p : 48.

Hulet, C.V.; Voigtlander, H.P. (Jr.); Pope, A.L. and Caside, L.E. 1956. The nature of early season infertility in sheep. *J. Anim. Sci.*, 15 : 607-16.

Hultz, F.S. 1943. The Wyoming systems for scoring Corriedale sheep. *Bull. Wyo. Agric. Exp. Sta.* No. 258 : 44. (*Anim. Breeding Abstr., 12:81).*

Hunt, W.L.; Addleman, D., and Bogart, R. 1971. Induction of multiovulation in the ewe following synchronisation of estrus. *J. Anim. Sci.*, 32 : 491-95.

Hunter, G.L. 1957. The maternal influence on size in sheep. *J. agric. Sci.* (Camb). 48: 36.

Hunter, G.L. 1968. Increasing the frequency of pregnancy in sheep. I. Some factors affecting rebreeding during the postpartum period. *Anim. Breeding Abstr.*, 36 : 347-78.

Hunter, G.L.; Adams, C.E. and Rowson, L.E.A. 1955. Interbred ovum transfer in sheep. *J. agric. Sci.* (Camb.), 46 : 143-49.

Hunter, R.H.F. 1980. *Physiology and Technology of Reproduction in Female Domestic Animal.* Academic Press, London, UK.

Hussain, B.S. 1977. Salt tolerance of species as determined by germination of seeds at different salinity levels. *Pakist. J. For.*, 22 (2) : 95-7.

International consulation on Ipil-Ipil Research. 1976. Summary Report. CEC, U.P.L.B. College, Loguna, Phillpines.

Istvan, S. 1956. Effektivnostj ispolizovanija moloka v kacestve razbavitelja sememni barana. (The effectiveness of using milk as a diluent for ram semen), *Ovcevodstvo,* 1956 (4) : 33-37. (*Anim. Breeding Abstr.*, 25, 231).

Ivanov, M.F. and Yudin, B.M. 1932. *'Karakul Fur pelts'* Publication of Moscow higher Zooleanical Institute.

Jackson, T.H. 1974. Effects of Nutrition on the Carcass. British Council Course 324. March-April, 1974 : Edinburgh.

Jain, P.C. and Kamalapur, S.K. 1971a. On the occurence of *Calicophoron canliorchis* (Steles and Goldberger, 1910. Nasmark, 1937 paramphistomidae : Trematoda) in sheep in India. *Indian Vet. J.*, 47 : 307-308.

Jain, P.C. and Kamalapur, S.K. 1971b. A note on the occurence of *Shriabinema ovis* (Skrjabin, 1915) wereschtchagin, 1926 (Syphaciinae : oxyuridae) in sheep in Madhya Pradesh, *Orissa Vet. J.*, 6 : 22-23. Jainudeen, M.R. and Hafez, E.S.E. 1974. Egg transfer, In : *Reproduction in Farm Animals.* (Ed.) E.S.E. Hafez. Lea and Febiger. Philadelphia.

Jawahar Lal; Sambasivarao, K.; Suresh Chandra; Naithani, R.C.; Chattopadhyay, S.K. and Sabir, M. 1976. Studies on the comparative efficacy of *Cedrus deodara* oil, Benzyl benzoate and tetraethyle thiuram monosulphide against *sarcoptic mange* in sheep. *Indian Vet. J.*, 53 : 543-45.

Jayal, M.N. 1966. A study on the nutritive value of bamboo (Dendrocalamus *strictus)* leaves. *Indian Vet. J.*, 40 : 625-28.

Jayal, M.N. and Kehar, N.D. 1962. A study of the nutritive value of Mulberry *(Morus indica)* tree leaves. *Indian J. Dairy Sci.*, 15 : 21-27.

Jayaramen, S. and Buch, B.B. 1953. Care and management of rams. *Indian Fmg.*, 3 (8) : 26-28.

Jenness, R. 1974. The composition of milk. In *Lactation* vol. III, (Eds) B.L., Larson, V.R., Smith, pp. 3-107. Academic Press, New York, USA.

Johnson, W.P. and Perdval, R.C. 1955. Tetracycline therapy of Experimentally Induced Anthrax in sheep. *J. Amer Vet. Med. Assoc*, 127 : 142.

Johri, C.B. and Talapatra, S.K. 1971. Growth studies with Jammunapari goats. II Jamunapari kids under browsing and stall feeding conditions. *Indian Vet. J.*, 48 : 495-503.

Johri, C.B.; Kulshreshtha, S.K. and Saxena, J.S. 1971. Chemical composition and nutritive value of green soyabean and soyabean straw. *Indian Vet. J.*, 48 : 938-40.

Joshi, D.C. 1966. Studies on high altitutde Himalayan pastures (Bugiyals) of Uttar Pradesh. *Indian Vet. J.*, 43 : 1019-27.

Johari, D.C. 1973. Studies on the semen quality of Polwarth, Rambouillet, Bikaneri and Rampur Bushair rams. (2) Semen quality in summer, rains and autumn seasons. *Indian J. Anim. Health*, 12 : 85-88.

Joshi, D.C. and Ludri, R.S. 1966. The chemical composition and nutritive value of Timla *(Ficus roxburghii)* and Kharik *(Celtis tetrenda)* tree fodders. *Indian Vet. J.*, 43 : 833-37.

Joshi, D.C. and Ludri, R.S. 1967. The chemical composition and nutritive value of Buffaloe *(Pennistum lacerdum* griseb) and *Panicum Kabulabula* grasses. *Indian Vet. J.*, 44 : 800-804.

Joshi, D.C. and Ludri, R.S. 1970. Studies on sheep nutrition. II. Plane of nutrition of grazing sheep during different months. *Indian Vet. J.*, 47 : 1089-94.

Joshi, J.D. and Singh, G. 1968a. Studies on ram semen diluents. I. Effect of egg yolk citrate series diluents on livability and morphology of ram spermatozoa. *Indian J. Vet. Sci*, 38 : 574.

Joshi, J.D. and Singh, G. 1968b. Studies on ram smen diluents III. Effect of skim-milk series diluents on the livability and morphology of rma spermatozoa. *Indian J. Vet. Sci.*, 38 : 583.

Joshi, J.D. and Singh, G. 1968c. Studies on ram semen diluents II. Effect of tomato juice series diluents on the livability and morphology of rma spermatozoa. *Indian J. Vet. Sci.*, 38 : 591.

Kabbali, A. 1977. Etude de la production laitiere et de la croissance des agneaux de brebis Timahdite et Beni-Hsen : influence due niveau energetique (Study of milk production and lamb growth in Middle Atlas and Beni Ahsen sheep : effect of energy level), *Hommes, Terre et Eaux.*, 6 (25) : 31-43.

Kadambi, K. and Dabral, S.N. 1955. Studies on the suitability of the different methods of artificially regenerating forest trees (Sowing, entire transplanting and stump planting). *Indian Forester*, 81 (2) : 129-51.

Kale, S.M. 1966. Transportation of sheep by rail and road. *Indian Vet. J.*, 43 : 354-357.

Kalita, C.C.; Gautam, O.P. and Banerjee, D.P. 1978. Fenbendazole against haemonchosis in sheep. *Indian Vet. J.*, 55 : 660- 62.

Kandasamy, N. and Pant, K.P. 1980. Gestational oestrus during pregnancy in sheep. *Indian Vet. J.* 57 : 434-35.

Kapoor, U.R.; Aggarwala, O.N.; Panchauri, P.C.; Nath, K. and Narayan, S. 1972. The relationship between diet, the copper and sulphur content of wool and fibre characteristics. *J. agric. Sci.* (Camb.), 79; 109-14.

Kardymowiez, O.; Brusson, F. du M. du; Wintenberger-torres, S. and Zind Zaplatal, H. 1976. The long distance transportation of fertilised sheep ova stored *in vitro* at a temperature above 0 degree Centigrade VIII. *Intern. Cong. Anim. Repod. A.I. Karakaw J.* 1976. Vol. I. Conm. Abst.

Karihalo, A.K. and Singh, O.N. 1967. Oestrus synchronization and its utility in sheep improvement. *Indian Vet. J.*, 44 : 1069.

Kaschanian, N. 1973. Vergileichende Untersuchungen uber die Milchleistung von Awassi, Schall-und Schall- X Awassi-Schaften im Iran (A comparison of milk, production in Awassi, Shal and Shal × Awassi-sheep in Iran), *Zuchtungskunde.* 45 : 270-75.

Katiyar, R.D. 1961a. Annual Report. ICAR scheme for the appointment of Asstt. Disease Investigation officers (Sheep and goats) for the year 1960-61.

Katiyar, R.D. 1961b. Final Report. ICAR scheme for the appointment of Assistant Disease Investigation Officer (Sheep and goats).

Katiyar, R.D. 1964. Annual Report. ICAR scheme for the appointment of Assistant Disease Investigation Officer (Sheep and goats).

Katiyar, R.D. 1962-1968. Annual Reports 1962-1968. ICAR scheme for the appointment of Asstt. Disease Investigation Officers (Sheep and Goats).

Katiyar, R.C. and Ranjhan, S.K. 1969a. Chemical composition of Makra *(Dactyloctenium aegypticum* linn) and indigenous monsoon grass. Indian J. Dairy Sci. 23 : 33-36.

Katiyar, R.C. and Ranjhan, S.K. 1969b. Chemical composition and nutritive value of kikuyngrass *(Pennisetum dandestinum). Indian J. Dairy Sci,* 22 : 42-45.

Katiyar, R.C; Ranjhan, S.K. and Shukla, S.K. 1970. Yield and nutritive value of *Clitoria ternatea* - A wild perennial legume for sheep. *Indian J. Dairy Sci,* 23 : 79-81.

Katiyar, R.D.; Shrivastava, V.K.; Khera, R.C. and Mahopatra, H.K. 1981. Description of Bonpala sheep. *Livestock Adv.,* 6 (2) : 57-59.

Kaul, R.N. and Ganguli, B.N. 1962. Khejri - the tree that does not mind drought. *Indian Fmg.,* 12 (5) : 7-8.

Kaura, R.L. 1941. Some common breeds of Indian sheep. *Indian Fmg.,* 2: 175-9.

Kaura, R.L. 1942. Some common breeds of sheep. *Indian Fmg.,* 3 : 122-25.

Kaushish, S.K. 1971. Studies on gestation length, blood constituents and process of parturition in sheep. M.Sc. Thesis, Haryana Agricultural University, Hisar.

Kaushish, S.K. 1980. Existing sheep managemental practices in Malpura tract of Rajasthan. *Wool & Woollens India,* 17(1) : 43-47.

Kaushish, S.K. 1989. Studies on reproduction in sheep : Relationship between gestation length, birth weight and weight at service. Paper presented at *V Annual Conf. SAPI.* Munnar (Keral) Dec, 27-30, 1989.

Kaushish, S.K. 1991. Improving sheep production through improved management practices. *Carpet-e-World, Intern.,* 13 : 47-50.

Kaushish, S.K. 1992. Carpet wool production and utilization in India. *Carpet-e-World, Mem., 14* : 70-74.

Kaushish, S.K. 1992. Bhedon avum Bakrion Ke Sankramak rog aur unki Roktham. (Contageous diseases of Sheep and Goats and their Prevention). *Fanners and Parliament.* 27 (2) : 37.

Kaushish S.K. 1993. Quality aspects of Indian wools as compaird to New Zeal and wools. Carpet er world Indian 15 : 39-41.

Kaushish, S.K. and Arora, K.L. 1973. Effect of weight of dam at seirce on gestattion length, parturition and wieth of lamb born to Nali sheep. *Indian J. Anim. Prod.,* 3 (1) : 41-50.

Kaushish, S.K. and Arora, K.L. 1974a. Studies on reproduction in sheep. 3. Factors affecting gestation length. *Indian J. Anim. Prod.,* 5 (1-4) : 85-90.

Kaushish, S.K. and Arora, K.L. 1974b. Studies on reproduction in sheep. 4. process of parturition. *Indian J. Anim. Sci.,* 44 : 667-71.

Kaushish, S.K. and Arora, K.L. 1975. Studies on relationship among weight of placenta, number of cotyledons and birth weight. *Indian J. Anim. Sci.,* 45 : 248-51.

Kaushish, S.K. and Arora, K.L. 1977. Studies on reproduction in sheep. 5. Blood and plasma cosntitutents before and after parturition in Nali sheep. *Haryana Vet.,* 16 (2) : 74-77.

Kaushish, S.K. and Jain, P.C. 1987. Economics of rearing mixed flocks of Malpura sheep in ORP villages. *Carpete-World Intern., 9* : 161.

Kauslish S.K. and sharma R.K. 1982 shearing and dopping in sheep, Haryana Farning (April) : 9.

Kaushish, S.K. and Sahni, K.L. 1975. A note on the time of parturition in Russian Merino ewes. *Indian Vet. J.,* 52 : 837-39.

Kaushish, S.K. and Sahni, K.L. 1976. Effect of feeding animal protein (Egg + milk) and trace elements and provision of cooler climate on libido, semen quality and certain physiological reactions of Russian Merino rams during summer season. *Indian J. Anim. Sci.,* 46 : 135-39.

Kaushish S.K. and Sahni, K.L. 1977a. Reproductive performance of merino sheep during autumn and spring seasons under semi-arid conditions. *Indian J. Anim. Sci.,* 47 : 18-22.

Kaushish, S.K. and Sahni, K.L. 1977b. Seasonal variation in reproductive behaviour and semen quality of Russian Merino rams under semi arid conditions. *Indian J. Anim. Sci.,* 47 : 189-92.

Kaushish, S.K. and Arora, K.L. 1983. Effect of weights and measurements on placenta weight and number of cotyledons in Nali and Lohi sheep. *Intern. J. Trop. Agri.* 1 (2) : 167-69.

Kaushish, S.K.; Arora, K.L. and Dhanda, O.P. 1973. Studies on reproduction in sheep 2. Time of parturition. *Indian J. Anim. Prod.,* 4(1) : 56-59.

Kaushish, S.K.; Bhatia, D.C. and Arora, K.L. 1976. Studies on adaptability sheep to subtropical climate and seasonal haematological attributes of Nali sheep. *Indian Vet. J.,* 53 : 760-65.

Kaushish, S.K.; Dubey, S.C. and Uppal, P.K. 1980. Pashuon devara manushyon men failne wale mukhya rog tatha unkebachav ke upaye. (Zoonotic diseases of animals and their prevention). *Pashu Vigyan,* (Oct-Dec.) 6 (4) : 14-17.

Kaushish, S.K.; Sahani, M.S.; Dubey, S.C. an Sharma, S.C. 1982. Performance of crossbred rams and their progenies under field conditions. *Wool & Woollens India.* (July-Sept.) 19 (3) : 39-44.

Kaushish, S.K.; Rawat, P.S. and Kalra, D.B. 1986. Studies on reproduction in sheep 6. Effect of gestation length and dams weight at service and lambing on preweaning body weight in exotic fine wool breeds. *Indian J. Anim. Reprod.,* 7 (1) : 92-94.

Kaushish, S.K.; Rawat, P.S. and Sharma, S.C. 1990. Performance of native sheep (Malpura) and its crosses with Avikaline under semi arid conditions. *World Rev. Anim. Prod.,* 25 (1) : 43-46.

Kaw, J.L. and Khera, S.S. 1970. Sheep pneumonia in Jammu and Kashmir state. 1. Incidence in four sheep farms. *Indian J. Anim. Sci.,* 40 : 511-17.

Keast, J.C. and Meeley, F.H.W. 1949. Some observations on artificial insemination of sheep. *Aust. Vet. J.,* 25 : 281- 87.

Kehar, N.D. and Rao, M.G. 1944. Observations on toxicity of *Rhododendron arboreum* to livestock. *Indian J. Vet. Sci.,* 14 : 177-78.

Kharat, ST.; Prasad, V.L.; Sobale, B.N.; Sane, M.S.; Joshi, A.L. and Rangnekar, D.V. 1980. Note on comparative evaluation of *Leucaena leucocephala, Desmanthus virgetus* & *Medicago sativa* for cattle. *Indian J. Anim. Sci,* 50 : 638-39.

Khot, S.S.; Mahal, G.S. and Lall, H.K. 1962. *Researches in Animal Husbandry* sheep and wool - A review, of work done duing 1929-54. ICAR, Publication.

Killeen, J.D. and Moore, N.W. 1971. The morphological apperande and development of sheep ova fertilized by surgical insemination. *J. Reprod. Fert.,* 24 : 63-70.

King, NJ.; Mungomery, R.W. and Hughes, C.G. 1965. *Manual of cane growing.* Sydney. Angus and Robertson.

Kirk, J.H.; Huffman, E.M. and Anderson, B.C. 1980. Mastitis and Udder abnormalities as related to neonatal lamb mortality in shed lambed range ewes. *J. Anim. Sci.,* 50 : 610-16.

Kishore, K.; Chopra, S.C.; Acharya, R.M. and Balaine, D.S. 1978. performance fo Rambouillet, Malpura and their crosses for wool traits. *Indian J. Anim. Sci*, 48 : 737-40.

Kishore, K; Gour, D.; Rawat, P.S. and Arora, C.L. 1980. Note on gestation length in cross-bred sheep. *Indian f. Anim. Sci.*, 50 : 565-67.

Kishore, K; Gour, D.; Rawat, P.S. and Malik, R.C. 1982. A study of some reproductive traits of Avikalin and Avivastra ewes and factors affect-ing them. *Cheiron*, 11 : 20-24.

Kishore, K.; Sethi, I.C. and Basuthakur, A.K. 1983. Study on milk production in Avikalin and Avivastra ewes. *Cheiron*, 12 : 257-60.

Koger, M. 1951. Storage, dilution and use of ram semen in artificial breeding of sheep. *Bull. Max. Agric, exp. Sta.* No. 366 : 12 pp. (Anim. Breeding Abstr., 21 : 798).

Kohli, R.K.; Jolly, M.S. and Khan, A.M. 1969. Foliar constitutents of the food plants of *tusar*silkworm *Antheraea mylitta* Drury. *Indian Forester*. 95: 614-17.

Krishnan, T.S. 1939. Nutrition of sheep and wool production. *Indian. J. Vet. Sci., 9* : 59-60.

Krishnan, A., Blagoveschensky, E.N. and Rakhecha, P. 1968. Water-balance of *Prosopis spicigera* community. *Indian J. Met. Geophys.* 19 (2) : 181-92.

Kronfield, K.S. 1972. Ketosis in pregnant sheep and lactating cows. A review. *Aust. Vet. J.*, 48 : 680-87.

Kulshrestha, S.K.; Upadhaya, R.B. and Saxena, J.S. 1971. Utility of Pangola grass *Digitaria decribens* for sheep. *Indian Vet. J.*, 40 : 1152-55.

Kunznetsov, M.P. 1934. The theoretical basis of the mothods of introducing sperm in artificial insemination of sheep. *Prob. Zhivotn*, IV : 122-125. (Anim. Breeding Abstr., 3 : 36, 37).

Labban, F. 1973. *A study on Wool Characteristics of the Barbary Sheep. Ministry* of Agriculture, Tripoli, Libyan Arab Republic.

Labban, F.M. and Ghali, A. 1969. A study on increasing lambing rate of sheep through management. *J. Anim. Prod. United Arab Republic, 9* : 285-93.

Labussiere, J.; Dotchewsbi, D. and Cembaild, J.P. 1981. Caracteristique morphologique de is mamelledes brebis Lacaune. *Ann. Zootech.*, 30 (1) : 115-36.

Lagrix, A. 1949. Sheep breeding in France, *Anim. Breeding Abstr.*, 18 : 286.

Lal, M. and Kalra, D.V. 1960. Lantana-poisoning in doemsticated animals. *Indian Vet. J.*, 37 : 263-69.

Lal, H.K. 1947. Some common breeds of Indian Sheep. *Anim. Breeding Abstr.*, 18 : 285.

Lamond, R. 1964. Synchronization of ovarian cycles in sheep and cattle. *Anim. Breeding Abstr.*, 269-285.

Lander, P.E. and Dharmani, L.C. 1925. A new fodder (siloed *shisham* leaves) for dairy cows. *Agric. Res. Inst. Pusa. Bull.* 158 : 11.

Lardy, H.A. and Phillips, P.H. 1939. preservation of spermatozoa. *Proc. Amer. Soc. Anim. Prod.*, 219-21.

Lasley, J.F., Easiloy, G.T. and McKenzie, F.F. 1942. A staining method for the differentiation of live and dead spermatozoa. I. Applicability to the staining of ram spermatozoa. *Anat. Rec.*, 82 : 167-74.

Lawlor, M.J.; Louca, A. and Mavrogenis, A. 1974. The effect of three suckling regimes on the lactation performance of Cyprus Fat-tailed, Chios and Awassi sheep and the growth rate of lambs. *Anim. Prod.*, 18 : 293-99.

Lawton, R.M. 1968. The value of browse in dry tropics. £. *Afr. Agric. For. J.*, 33 : 227.

Lea, L.C. 1954. *Proc. Pan Indian Ocean Sci. Cong.* Pt. II. Perth, W. Australia.

Lee, H.J. 1956. The toxicity of *Phalaris tuberosa* to sheep and cattle and preventive role of cobalt. In *Proc. 71 R Int. Grassl. Cong. Wellington* (Ed.) G.J. Neale, pp 387-95.

Lee, H.J. 1963. Cobalt deficiency. In *Animal Health production and Pasture* (Eds) A.N. Worden, K. Sellers, and D.E. Tribe, Longman London pp 662-80.

Lee, H.J. 1976. *Animal to Human Nutrition - Achievements and Prospects.* Commonwealth Scientific and Industrial Research Organigation. Melbourne pp 126.

Leone, E. 1953. Quaderi Sex. *Perugins Soc. ltal. Biol. Sper.*, 10 : 1.

Lewis, G.; Terlecki, S. and Parker, B.N.J. 1974. Observations on the pathogenesis of delayed sway back. *Vet. Rec*, 95 (14) : 313-16.

Lima, C.R.; Araujo, M.R. and Souto, S.M. 1972a. Valores nutritivos da silagem de sorgo forrageiro e capins elefante, coloniao, pangola e guatemala (Nutritive value of silages of forage sorghum and elephant, colonial, pangola and guatemala grasses), *Pesquisa Agropecuaria Brasileira, Zootecnia, 7* : 53-57.

Lima, C.R.; Souto, S.M.; Garcia, J.M.R. and Arujo, M.R. 1972b. Valores nutritivos do feno de siratro *(Phaseolus atropurpureus)* em diferentes estadios de crescimento (Nutritive value of siratro *(Phaseolus atropurpureus)* at different stages of growth), *Pesquisa Agropecuaria Brasileira, Zootecnia, 7* : 63-66.

Lincoln, R.E.; Walker, J.S.; Klein, F. and Haines, B.W. 1964. Anthrax. *Adv. Vet. Sci.*, 9 : 327.

Link, P. 1942. The development of Australian Merino breed. *Anim. Breeding Abstr.*, 10 : 72.

Little John, A.I. 1968. Psoroptic mange in the goat. *Vet. Rec,* 82 : 148-55.

Lohan, O.P.; Lall, D.; Pal, R.N. and Negi, S.S. 1979. Studies on tannins in fodder trees. *Fifteenth Dairy Industry conf.* Hyderabad, A.P.; Feb. 1979.

Lohani,D.N. 1979. Preliminary trials with high yielding variety of *Leucaena leucocephala. Indian Forester,* 105 : 199-202.

Lone, M.M.; Koul, T.N. and Sharma, D.D. 1972. Field trial studies on Nitroxynil (Trodex M & B) and Hexachlorophene (Distodin Pfizer) against fasciolisis in naturally infected sheep. *Indian Vet. J.*, 49 : 764-67.

Lonkar, P.S.; Uppal, P.K.; Belwal, L.M. and Mathur, P.B. 1983. Blue tongue in sheep in India. *Trop. Anim. Health. Prod.*, 15 (2): 86.

Louca, A. 1972. The effect of suckling regime on growth rate and lactation performance of the Cyprus Fat tailed and Chios sheep. *Anim. Prod.*, 15 : 53-59.

Lowson, R. A.S. and Rowson, L.E.A. 1972. The influence of breed of ewe and offspring on litter size after egg transfer in sheep. *J. Reprod. Fert., 28 :* 433-39.

Luck, P.E. 1965. *Dolichos lab-lab,* a valuable grazing crop. *Qd. agric. J.*, 91 : 308-309.

Macpherson, J.W. 1957. Goats milk as a semen diluent. *Can. J. Comp. Med. Vet. Sci,* 21 : 161-62.

Mahadevan, V. 1961. *Feed requirements of sheep.* ICAR Scheme, IVRI, Izatnagar.

Mahajan, J.M.; Acharya, R.M. and Dhillon, J.S. 1980a. Crossbreeding for the production of fine wool: effect on wool production. *Indian J. Anim. Sci.*, 50 : 553-57.

Mahajan, J.M.; Acharya, R.M. and Dhillon, J.S. 1980b. Crossbreeding for the production of fine wool : effect on wool quality traits. *Indian J. Anim. Sci,* 50 : 624-28.

Maheshwari, S.R. 1977. A comparative study in energy and protein requirement of the native and crossbred lambs. Ph. D. Thesis, Panjab University, Chandigarh. Maheshwari, S.R. and Mathur, C.S. 1970. Studies on nutritive value of sweet sudan 59-3. *Ann. Arid Zone,* 9 : 209-211.

Maheshwari, M.L. and Talapatra, S.K. 1967. Sampling of the pasture grasses with oesophageal fistula technique. *Indian J. Anim. Health.*, 6 : 75-77.

Mahoney, M. 1949. An experience with Elastration. *J.* Amer. *Vet. med. Ass.*, 115 : 352.

Majumdar, B.N. and Momin, S.A. 1960. Studies on tree leaves as cattle fodder. Part V. The nutritive value of young gular leaves *(Ficus glomerata). Indian J. Dairy Sci,* 37 : 16-19.

Majumdar, B.N.; Momin, S.A. and Kehar, N.D. 1967a. Studies on tree leaves as cattle fodder. I. Chemical composition as affected by the stage of growth. *Indian J. Vet. Sci.,* 37 : 217-23.

Majumdar, B.N.; Momin, S.A. and Kehar, N.D. 1967b. Studies on tree leaves as cattle fodder. II. Chemical composition as affected by the locality. *Indian J. Vet. Sci.,* 37 : 224-31.

Makkar, M.S.; Joshi, H.C. and Gupta Indra. 1974. *Indian J. Anim. Res.,* 11 (2): 77-78.

Malik, N.S. and Nath, K. 1970. Chemical composition and nutritive value of green pala *(Ziziphus nummularia)* leaves. *Indian J. Anim. Sci.,* 40 : 41-45.

Mallikeswaran, K.; Mani, V. and Raman, K.S. 1980. Milk production in ewes and its relationship with the growth of suckling lambs. *Indian Vet. J.,* 57 : 235-8.

Mamedjarv, S.E. 1952. Osemenenie over smesannym semenem vuh barranow, (Inseminating sheep with mixed semen from two rams) *Socialist, Zivotn.,* 14 (9): 71-73 *(Anim. Breeding Abstr.,* 21 : 275).

Mandrik, F.L; Daukev, I.V. and Yakub, G.G. 1974. *Veterinariya, Moscow.* 12 : 27.

Mani Mohan, Bohra, S.D.J. and Acharya, R.M. 1979. New apparel wool (Avivastra) and superior carpet wool (Avikalin) strain evolved at Central Sheep and Wool Research Institute Avikanagar (Rajasthan).

Mann, T. 1964. *The biochemistry of semen and of the male reproductive tract.* Indian Edn. Methuen & Co. Let; London.

Marin, J.S.; Dubey, S.C. and Kaushish, S.K. 1980. Effect of different managrerial conditions on the performance of local sheep - A field study. *Indian J. Anim.* Res., 14 (2): 101-104.

Maqsood, M. 1951. Seasonal variations in the testis histology of the ram. *Vet. Rec.,* 63: 597-98.

Marai, I.F.M. 1975. A study of some characteristics of two Egyptian breeds of sheep. *Rivista di Agricoltura Subtropicl e Tropicale.* 69 : 157-66.

Marston, H.R. 1952. Copper, cobalt and molybdenum in the nutrition of plants and animals. *Physiol. Rev.,* 32 : 66-121.

Marston, M.R. 1955. Nutritional factors involved in wool production by Merino sheep. *Aust. J. Sci. Res.* Series. 1 (3): 376-87.

Martemucci, G.; Ballitti, E.; Toteda, F. and Manchisi, A. 1983. Study of first signs of oestrus in ewe lambs according to genotype. *Annalidella Facolta di Agraria, University di Bari,* 31 : 185-200. *(Anim. Breeding Abstr.,* 5 : 4979).

Matharu, B.S. 1962. Transporation of animals by mechanical transport (Road transportation). *Indian Vet. J.,* 39 : 611- 16.

Matharu, B.S. 1963a. Transportation of animal on foot (By walking). *Indian Vet. J.,* 40:104-108.

Matharu, B.S. 1963b. Transportation of animals by sea. *Indian Vet. J.,* 40 : 623.

Mathauda, G.S. 1956. Indications about artificial regeneration of some tree species based on the work done at New Forest during the period 1928-55. *Proc. Ninth Silvic., Conf.* Part I. 198- 208. F.R.I, and College, Dehradun.

Mathur, A.K. 1985. Semen quality of exotic rams during summer. Annual Report, Central Sheep and Wool Research Institute Avikanagar. (Raj.) P 49.

Mathur, C.S. 1960. Common fodder grasses native to the desert soil of Rajasthan and their feeding value. *Indian J. Vet. Sci,* 30 : 69-103.

Mathur, C.S. 1973. Effect of protein nutrition on production and quality of wool. Ph. D. Thesis, University of Udaipur.

Matter, *H.E. 1974. Mating and conception during gestation in sheep,* Yearbook, Karakul Breeders Society of South Africa. 16 : 45-55.

Mathur, P.B. 1970. Variations in liverfluke egg counts within a day and the incidence of infection in sheep and intermediate snail host. *Indian J. Sci, Ind.,* Section B 4 (1): 11-15.

Mathur, P.B. and Dutt, S.C. 1970. Treatment of *Fasciola gigentica* infection in sheep and cattle. *Indian J. Sci,* Indian Section B. 4(1): 1-6.

Mathur, S.D. 1977. *Rajasthan men bhedon ki Mukhya Nasle.* Directorate of Sheep and Wool, Rajasthan.

Mathur; A.K.; Srivastava, R.S. and Kalra, D.B. 1987. Synchronization of oestrus and fertility in ewes treated with prostaglandin F_2. alpha *Indian J. Anim. Sci,* 57 : 709-10.

Mathur, A.K.; Srivastava, R.S.; Anil Joshi and Kalra, D.B. 1989. Pellet freezing of ram semen. Indian *J. Anim. Sci,* 59 : 1529.

Mavrogenis, A.P. 1982. Environmental and genetic factors influencing milk production and lamb output of Chios sheep. *Livestock Prod. Sci,* 8 : 519-27.

Mavrogenis, A.P.; Papachristoborou, Chr.; Lysan drides, P.H. and Roushias, A. 1985. Environmental and agenetic factors affecting udder characteristics and milk production in Chios sheep. *36th Ann. Meet. EAAP* Geece.

Mayo, G.M.E. and Mulhean, G.I. 1969. Inheritance of congenital goitre due to a thyroid defect in Merino sheep. *Aust. J. Agri. Res.,* 20 : 533-47.

Mc Dowell, R.E. 1977. Ruminant products, more than meat and milk. Winrock Report. Winrock Intern. Livestock Res. and Tr. Centre, A.R., p 28.

Mc Intosh, G.H. 1978. Urolithiasis in animals., *Aust. Vet. J.*, 54 : 267-70.

McKenzie, F.F. and Terrill, C.E. 1937. Oestrus, ovulation and related phenomena in the ewe. *Missouri, Agri. Expt. Sta. Res. Bull.* 264: 1-88.

Mc Leod, M.N. and Minson, D.J. 1974. Differences in carbohydrate fractions between *Lolium perenne* and two tropical grasses of similar dry matter digestibility. *J. agric. Sci.* (Camb.), 82 : 449-54.

Mc Mohan, P.R. and Speakman, J.B. 1941. The action of light on wool and related fibres, NZ. *J. Sci. Tech.*, 22 : 235-60.B

Mellanby, K. 1972. *Scabies.* E.W. Cassey Ltd., Hampton, p. 81.

Meyer, M.E. 1974. Advances in research on brucellosis. 1957-72. *Adv. Vet. Sci. Comp. Med.*, 18 : 231-50.

Mia, W.H.; Sahai, B.; Majumdar, B.N. and Kehar, N.D. 1960. Studies on tree leaves as cattle fodder. Part III. The nutritive value of barged leaves *(Ficus bengalensis) Indian J. Dairy Sci.*, 13 (1): 1-8.

Mies Filho, A. and De Almeida Ramos, A. 1955a. Efficiencia de differents techicas de inseminacao artificial en ovinos. *Bol. Insem. Artif.* (Riode J), 7 : 29-36 *(Anim. Breeding Abstr.*, 25 : 1946).

Mies Filho, A. and De Almeida Ramos, A. 1955b. Relacao entre alguns caracteres do semen de ovinos. *Veterinaria* (Rio de J. 9 (3/4) 37-41 *(Anim. Breeding Abstr.*, 25 : 1948).

Milford, R. 1967. Nutritive values and chemical composition of seven tropical legumes and lucerne grown in subtropical south eastern Queensland. *Aust. J. exp. Agric. Anim. Husb.*, 7 : 540-45.

Milford, R. and Minson, D.J. 1968. The effect of age and method of hay making on the digestibility and voluntary intake of the forage legumes *Dolichos lab-lab* and *Vigna sinensis. Aust. J. exp. Agric. Anim. Husb.*, 8 : 409-12.

Milhand, G. and Cazieuz, A. 1980. *Vet. Human Toxicol.*, 22 (2): 81. Milhand, G.; Zundenl, E. and Crombet, M. 1980. *Recucil de Medicine Veterinarire,* 156 (1): 37.

Miller, S.J. 1975. Personal communication from his data on A.I. in sheep in Australia.

Mikus, M. 1968. Studium vztahov medzi mnozstvom mlieka a, rezmermi vemena oviec pocas laktacnej periody. *Ved. Pr. Vysk. Ust V Trencine,* IV : 137-52.

Minson, D.J. 1972. The digestibility and voluntary intake by sheep of six tropical grasses. *Aust. J. expt. Agric. Anim. Husb.*, 12 : 27.

Minnett, F.C. 1948. Pathogenesis of black quarter, *J.* Comp. *Pathol,* 58 : 245-66.

Mishra, M.S. and Sengupta, B.P. 1965. Semen quality of ram and goat in autumn. *Indian Vet. J.*, 42 : 742.

Mishra, R.K. and Singh, V.K. 1979. Breeding sheep for wool and mutton in temperate region. *Wool & Woollens India.* 16 (1) : 28-31.

Mishra, H.N.; Mullick, S.G.; Rao, C.C.P. and Khera, S.S. 1970. Analysis of results of bacteriological examination, morbid materials from cases of sheep pneumonia during the period 1959- 68. *Indian Vet. J.*, 47 : 714-19.

Mittal, J.P. 1980. Seasonal changes in semen characteristics of Merino and Magra sheep of Indian desert. *J. agric. Sci,* (Camb.), 95 : 721.

Mittal, J.P. 1984. Sperm production in the ram. Relationship to tropical environment, *Anim. Breeding Abstr.*, 53 : 5024.

Mittal, J.P. 1985. A study on sexual season of Marwari sheep of Rajasthan desert. *Indian Vet. J.*, 62 : 247-48.

Mittal, J.P. and Ghosh, P.K. 1981. Male reproductive characteristics of sheep in Indian Arid Zone. *J. agric. Sci,* (Camb.), 97 : 313-18.

Mittal, J.P. and Pandey, M.D. 1975. Effect of age and season on wool yield in Bikaneri (Magra) ewes. *Indian J. Anim. Sci.*, 45 : 161-62.

Mittal, J.P. and Sengupta, B.P. 1981. Superovulation and multiple lambing in desert adapted Magra ewes with serum and chorionic gonadotraopins. *Ann. Arid Zone,* 20 : 247-52.

Moe, P.W.; Tyrell, H.F. and Flatt, W.P. 1971. *J. Dairy Sci.*, 54 : 1.

Mohan, R.N. 1948. Verminous pneumonia in animals in Bengal. *Indian J. Vet. Sci.*, 18 : 33-36.

Mohan, M.; Sheikh, Q.D. and Bhatia, D.R. 1971. Digestible energy content of the available pastures of Rajasthan. Annual Report, Central Sheep and Wool Research Institute, Avikanagar, Rajasthan pp 37-38.

Momin, S.A. and Ray, S.C. 1943. Tree leaves as cattle fodder. I. The seasonal variation in the composition of some edible tree leaves, *indian J. Vet. Sci.*, 13 : 183-90.

Moore, B.H. 1949. Characteristics of substrates and media essential for metabolism and motility of ram, boar and stallion spermatozoa. *Microfilm Abi.;* IX : 12-13. (*Anim. Breeding Abstr.* 19 : 30).

Moore, N.W. 1970. Preliminary studies on *in vitro* culture of fertilized sheep ova. *Aust. J. Biol.*, 23 : 721.

Moore, R.W. 1966. Genetic factors affecting the milk intake of lambs. *Aust. J. Agric. Res.*, 17 : 191-99.

Moore, B.H.; Mayer, D.T. and McKenzie, F.F. 1940. Factors influencing motility and metabolism in ram semen. *Proc. Amer. Sci. Anim. Prod.*, 210-15.

Moore, D.N. and Eppleston, J. 1979. Embryo transfer in goats. *Aust. J. Agri. Res.*, 30 : 973-81.

Moore, N.W. and Shelton, J.N. 1964a. Response of the ewe to horse anterior pituitary extract. *J. Reprod. Fert., 7 : 79-87.*

Moore, N.W. and Shelton, J.N. 1964b. Egg transfer in the sheep. Effect of degree of synchronization between doner and recipient, age of egg and site of transfer on the survival of transferred eggs. *J. Reprod. Fert.*, 7 : 145-52.

Moore, R.M. and Trounson, A.O. 1977. Hormonal and follicular factors affecting maturation of sheep oocytes *in vitro* and ther subsequent developmental capacity. *J. Reprod. Fert.*, 49 : 101-109.

More, T.; Rawat, P.S. and Sahni, K.L. 1981. Some observations on high and low potassium Chokla sheep given water intermittently under semi arid conditions. *J. agric. Sci.* (Camb.), 96 : 463-69.

Morley, F.H.W. 1948. The effect of flystrike on the scrotum on subsequent fertility in rams. *Aust. Vet. J.*, 24 : 94-95.

Morrant, A.J. and Dun, R.B. 1960. Artificial Insemination of sheep. II. Techniques and equipment used at Trangi Agricultural Experimental Station. *Aust. Vet. J.*, 36 : 1-7.

Morris, R.S.; Anderson, N. and McTaggart, I.K. 1977. An economic analysis of two schemes for the control of helminthiasis in breeding ewes. *Vet. Parasitol,* 3 : 349-63.

Murray, F.A. 1978. Embryo transfer in large domestic mammals. In : *Methods in mammalian Reproduction.* (Ed). J.C. Daniel. Jr New York. Academic Press, pp 285-305.

Murtagh, G.J. and Dougherty, A.B. 1968. Relative yields of *Lab-Lab* and Velvet bean. *Trap. Grasslds, 2* : 57-63.

Muthanna, K.D.; Arora, G.D. and Gyan Chand. 1976. Comparative performance of indigenous trees in arid zone under different soil working techniques. *Ann. Arid Zone,* 15 (1&2) : 67-76.

Nagarcenkar, R. 1961. Ph. D. Thesis, University of Poona, Poona. Nagar-Nagar cenkar, R. 1963. Canary coloration of wool. *Zootec. e. Vet.*, 18 : 314-20. *(Anim. Breeding Abstr,* 32 : 1114).

Nagarcenkar, R. 1964. Further studies on the canary coloration of wool. *Zootec. e. Vet.*, 19 : 15-20.

Nagarcenkar, R. and Bohra, S.D.J. 1983. Production of carpet wool with special reference to India. *Wool & Woollens India,* 20 (1): 28-35.

Negi, S.S.; Pal, R.N. and Ehrich, C. 1979. Tree fodders in Himachal Pradesh (India). An introduction to six most common fodder trees in the Himachal Pradesh State of India and their feeding value for cattle. German Agency for Technical Cooperation (GTZ), Eshborn, Federal Republic of Germany. 68 p.

Nanda, P.N. 1942. The Punjab. *Indian Fmg.*, 3 : 500-502.

Nanda, P.N. 1947. Improvement of quality and quantity of wool in India. *Indian Fmg.*, 8 : 4-7.

Narayan, S. 1969. *Proc. Conf. Interns, app. Anim. Prod.* ICAR, New Delhi.

Narayana, H. and Setty, E.V.S. 1977. Studies on the incorporation of mullberry leaves *(Morus indica)* in layers mash on health, production and egg quality. *Indian J. Anim. Sci.*, 47 : 212-15.

Narayan, S. and Sharma, R.S. 1970. Studies on fleece components, skin follicle characteristics and fleece staining in Nali sheep. *Indian Vet. J.*, 47 : 858-66.

Narayanaswamy, M. and Balaine, D.S. 1976. A note on the oestrus cycle in Bannur sheep. *Indian Vet. J.*, 53 : 235-36.

Narayanaswamy, M; Balaine, D.S. and Chopra, S.C. 1975. A note on gestation length in Mandya sheep. *Indian J. Anim. Sci.*, 45 : 915-17.

Nath, K. and Malik, N.S. 1968. Nutritive value of *Indigofora cardifolia* and *Indigofera enneaphylla.* Annual Report, Central Sheep and Wool Research Institute, Avikanagar, (Raj.) pp 7-8.

Naude, J.E. and Grant, J.L. 1979. An investigation into an eight month reproduction cycle for sheep. In Division of Livestock and Pastures Annual Report for the year ended 30 September 1978. pp. 223-24. Department of Research and Specialist Services, Salisbury, Rhodesia.

Newson, I.E. and Thorpe, F. 1938. Lamb diseases in Colorado feed lots. *Colorade Agri. Expt. Sta. Bull* No. 448.

Newcomb, R. and Rowson, L.E.A. 1975. Conception rate after uterine transfer of cow eggs in relation to synchronization of estrus and age of eggs. *J. Reprod. Fert.*, 43 : 539.

Niekerk, C.H. Van 1979. Limitations to female reproductive efficiency. In *Sheep Breeding.* 2nd edn, pp. 303-13.. (Eds) G.L. Tomes, D.E. Robertson, RJ. Lightfoot, Butterworths, London, UK.

Norton, G.P. and Nicholls, CH. 1960. The effect of heating wool containing alkali. J. *Text. Inst.*, 51 : ,183-92.

Ortavant, R. 1958. Le cycle spermatogenotique Chez le belier. Thesis, University of Paris.

Otero, J. de R. 1958. Informacose sebre algumes plantas forrageires. Directora de Estatistiecs produccao, Ru die Janeiro.

Osikowski, M. 1965. Prezebieg rui a ewiec i optymalnyl termin pokrycis. (The stage of oestrus in sheep and the optimum time for service). *Przegl. hottowl,* 24 (5). 33-35 *(Anim. Breeding Abstr.,* 24 : 1729).

Osman, A.H. and El-Shafie, S.A. 1967. Carcass characteristics of Sudan Desert sheep. *Sudani. Vet. Sci.,* 8 : 115-19.

Osman, A.H.; El-Shafie, S.A. and Khattab, A.G.H. 1970. Carcass composition of fattened rams and wethers of Sudan Desert sheep *J. agric. Sci.* (Camb), 75 : 257-63.

Ostgard, O. 1957. Gestation period, birth weight and sex ratio in sheep. *Anim. Breeding Abitr.,* 26 : 272.

Over, H.J. and Dykstra, J. 1975. Infection rhythms in Fascioliasis. In: *Facts and Reflections.* 11. Workshop on Fasciolisis Central Diergemeeskundig Instituint. Lelyzted The Netherland Central Diergeneeskundig Institute, Lelystaci. pp 9-17.

Owen, J.B. 1972. *Beef and sheep systems on arable farms.* Br. Grassld. Soc. Occasional Symposium No. 7, 67-72.

Owen, J.B. and Ingleton, J.W. 1963. A study of food intake and production in grazing ewes. 2. The interrelationships between food intake and productive output. *J. agric. Sci.* (Camb), 61 : 329-40.

Pachauri, V.C. 1970. Effect of stilbesterol feeding on the utilization of nutrients is adult sheep. *Indian J. Anim. Health,* 9 : 223-27.

Pachauri, V.C. Negi, S.S. 1979. A note on the utilization by yearling Gaddi ewes of mixed *Zea mays* and *Sorghum vulgare* greens supplemented with three levels of linseed cake. *Indian J. Anim. Sci.,* 49 : 22.

Pachauri, V.C; Aggarwala, O.N. and Mahadevan, V. 1970. Nutritive value of doob *(Cynodo dactylon)* hay for adult sheep. *Indian Vet. J.,* 47 : 782-85.

Pacho, H.N. 1973. Diagnostico de gestacion por laparotomia en ovejas (Pregnancy diagnosis by laparotomy in ewes), *Veterinaria* (Mexico), 4 : 161-65.

Pal, R.N. Dogra, K.K.; Singh, L.N. and Negi, S.S. 1979. Chemical composition of some fodder trees. *Himachal Pradesh Forage Res.,* 5 : 109-15.

Pandey, M.D.; Seth, O.N.; Pant, H.C. and Roy, A. 1972. Augmentation of lambing percentage in Bikaneri (Magra) ewes through serum gonadotrophin treatment. *Indian J. Anim. Sci.,* 42 : 28.

Pant, K.P.; Kulshresta, T.S. and Bhatia, R.K. 1980. Studies on the wool quality of Nali and Chokla sheep and their crosses with Rambouillet and Soviet Merino. *Indian Vet. J.,* 57 : 915-18.

Parbery, D.P. 1967. Pasture and fodder crop plant introduction at Kimberley Research Station, W.A. 1963-64. Part I. Perennial legumes. CSIRO *Aust. Div. Land Res., Tech. Mem.* 67/6.

Patel, B.H. and Patel, P.S. 1957. Fodder value of tree and vegetable leaves in Kaira district. *Indian J. agric. Sci., 27* : 307-15.

Patel, B.M.; Patel, P.S. and Shukla, P.C. 1962. Nutritive value of neem leaves, mangolds and comfrey. *Indian J. Dairy Sci.,* 15(4) : 139-45.

Patel, B.M.; Shah, B.C and Mistry, V.V. 1958. A study of fodder of Hissar district in the Punjab. *Indian J. agri Sci.,* 28 : 597-606.

Patil, J.S. and Bhosrekar, M. 1978. Collection, storage and transplantation of embryo. *Livestock Adv.,* 3 (9) : 5-13.

Patnayak, B.C.; Mohan, M. and Bhatia, D.R. 1979. A note on the nutritional value of cowpea, moth (Dew gram) and Cluster bean fodders fed as hay to sheep. *Indian J. Anim. Sci.,* 49 : 546.

Patnayak, B.C.; Mohan M. and Rai, A.K. 1971. The source and level of suplementary feeding on the performance of lambs. Annual Report, Central Sheep and Wool Research Institute, Avikanagar, Rajasthan pp 44-47.

Patterson, D.P.S.; Foulkes, J.A.; Sweasey, D.; Glancy, E.M.; and Terlecki, S. 1974. J. Neuro-Chem., 23 : 1245.

Paynter, D.I. 1979. Glutathion peroxidase and selenium in sheep. 1. Effect of intra ruminal selenium pellets on tissue gluta thion peroxidase activity *Aust. J. Agric. Res.,* 30 : 695-702.

Peart, J.M.; Edwards, R. A. and Donaldson, E. 1972. The yield and composi-tion of the milk of Finnish Landrace x Blackface ewes. *J. agric. Sci.* (Camb), 79 : 303-313.

Phillips, T.O. 1947. Cheviots sheep in Scotland. *Anim. Breeding Abstr.,* 16 : 327.

Phillips, P.H. and Lardy, H.A. 1940. A yolk buffer pabulam for preservation of bull semen. *J. Dairy Sci.,* 23 : 390-404.

Pissaloux, R. 1943. The fat lambed Barbery sheep. *Anim. Breeding Abstr.,* 16 : 328.

Plant, J.W. 1980. Pregnancy diagnosis in the ewe. *World Anim. Rev.,* 36 : 44-47.

Pollott, G.E. and Ahmed, F.A. 1978. The effect of weaning age on the growth of Watish (Sudan Desert type) lambs under feed lot conditions. *Research Bulletin No. 2. Um Banein Livestock Res. St. Singa,* Blue Nile Province, Sudan.

Popoff, M. 1978. *Bull. Mansuel de la Societe Vetarinair Pratique de France,* 62 (4) : 203.

Pottor, B.J. 1969. The effects of sodium chloride in the drinking water of stock. *Water Res. Found. Aust. Rep.* No. 2981 : 1-10.

Powley, G.; Care, A.D. and Johnson, C.L. 1977. *Res. Vet. Sci.,* 23 : 43.

Poynter, D. and Selway, S. 1968. Diseases caused by Lungworms. *Vet. Bull.*, 36 : 539-55.

Prasad, D.A.; Reddy, K.J. and Subba Reddy, C.S. 1974. Chemical composition and nutritive value of anjun *(Hardwickia binata* Roxb) leaves. *Indian J. Anim. Sci.*, 44 : 150-52.

Pritchard, C.J.R.; Pennell, A.E. and Williams, G.L. 1975. A note on the wool characteristics of sheep at the Hothuf Agricultural Research Centre. *Publication No. 61.* Joint Agricultural Research and Development Project University College of North Wales, UK. and Ministry of Agriculture and Water, Saudi Arabia. Purushotam, CH. 1978. Com-parative study of the production and reproduction traits of Nellore and Mandya sheep. Thesis Abst. Haryana Agricultural University, 4 : 99.

Qureshi, M.J. 1968. Influence of docking fat tailed lambs on fat deposition. *Agric. Pakist.*, 19 : 97-100.

Qureshi, M.J. and Shaw, A.O. 1968. Influence of docking fat tailed lambs on breeding performance. *Agri. Pakist*, 19 : 93-96.

Rai, A.K.; Kaushish, S.K. and Honmode, J. 1976. A note on physical and biochemical studies on semen of Chokla, Malpura and crossbred rams. *Indian J. Anim. Sci.*, 46 : 658-59.

Ram Ratan : Kundu, S.S. and Bhatia, D.R. 1982. Note on nutritive value of *Clitoria ternatea* hay for sheep. *Indian Anim. Sci.*, 52 : 265-67.

Ram, S.; Singh, B.; Balaine, D.S. and Chopra, S.C. 1978. A note on fleece characteristics of Nali purebred and its crosses with Coriedale. *Indian Vet. J.*, 55 : 407-10.

Ramadan, M.Y.A.;Lucas, I.A.M. and Farnworth, J. 1977. Aspects of Najdi ewe and lamb performance in Saudi Arabia: summer and winter lamb production from Najdi ewes. *Publication No. 84.* Joint Agricultural Research and Development Project, University College of North Wales. UK, and Ministry of Agriculture and Water, Saudi Arabia.

Ramamurty, A. 1963. Breeding performance of Nilgiri, Romney Marsh and Kuibishow rams. *Indian Vet. J.*, 40 : 621.

Ramamurty, A. 1964. Performance of crossbred progency of Nilgiri ewes with exotic rams. *Indian Vet. J.*, 41 : 201-5.

Ranjhan, S.K. 1980. *Animal Nutrition in the Tropics. Vikas* Publishing House Pvt. Ltd. Delhi.

Ranjhan, S.K.; Singh, C.P.; Nadger, S.R. and Talpatra, S.K. 1959. Studies on some high protein green feeds of Uttar Pradesh. 1. Chemical composition and nutritive value of the green feed karr (*Carthamus tictorius* Linn). *Indian Vet. J.*, 36 : 267-75.

Rao, A.S.; Rao, P.N.; Govindappa, S. and Ramachraiah, S.V. 1987. Certain biochemical profiles of the oviduct and vagina of ewes (*Ovis aries*) during oestrus cycle. *Indian J. Anim.* Reprod., 8 : 138-44.

Rao, L.R.; Purushotham, C. and Reddy, K.K. 1978. Studies on gestation period in Mandya and Nellore breeds of sheep. *Livestock* Adv., 3 (6): 9-12.

Rastogi, R.K.; Youssef, F.G.; Keens-Dumas, M.J. and Davis, D. 1979. Note on early growth rates of lambs of some tropical breeds. *Trop. Agric. Trin.*, 56 : 259-61.

Rathore, A.K. 1969. A note on the effect of scrotal wool cover on morphological changes in ram spermatozoa due to heat stress. *Anim. Prod.,* 11: 561-63.

Rathore, A.K. 1970. A note on the relation between scrotal wool cover and fertility in Merino rams. *Anim. Prod.,* 12 : 165- 67.

Rathore, D.S. 1981. Migration of sheep. Proc. *First National Seminar sheep goat Prod. Utiliz.,* held at Jaipur. April 3-4.

Ravindran, V.; Rajamahendran, R.; Nadrajah, K. and Goonewardene, L.A. 1983. Production characteristics of indigenous sheep under tradition-al management systems in northern Sri Lanka. *World Rev. Anim. Prod.,* 19 (3): 47-52.

Rawal, C.V.S.; Singh, R. and Luktuke, S.N. 1986a. Certain biochemical studies on oestrual vaginal mucus of Muzaffarnagri sheep. *Indian J. Anim. Reprod., 7*: 84-87.

Rawal, C.V.S.; Singh, R.; Garg, R.C. and Luktuke, S.N. 1986b. A study on post partum oestrus interval in Muzaffarnagri sheep. *Indian J. Anim. Reprod., 7* : 95-98.

Rawal, C.V.S.; Singh, R.K. and Luktuke, S.N. 1987. Thyroid status during different phases of oestrus cycle in Muzaffarnagri sheep. *Indian J. Anim. Sci.,* 57 : 31-32.

Reid, J.T.; Bensadown, A.; Bull, L.S.; Burton, J.M.; Gleeson, P.A.; Han, I.K.; Joo, Y.D.; Johnson, D.E.; Mc Manus, W.R.; Paladines, O.L.; Stroud, J.W.; Tyrrell, H.F.; Van Neikerk B.D.H.; Welliugton, G.H. and Wood, J.D. 1968. Changes in body composition and meat characteristics of cattle pigs and sheep during growth. Proc. Cornell Meat Conf. 1968. New York, Oct, 1968. pp 183-87.

Richardson, C. 1972. Pregnancy diagnosis in the ewe: a review, *Vet. Record,* 90 : 264-75.

Recordeau, S. and Flament, J.C. 1969. Cross breeding between Prealpes du sud and East Friesland sheep. 3. Milk yield. *Anim. Zootech.,* 18 : 151-68. *(Anim. Breeding Abstr,* 38 : 3703).

Robinson, J.J.; Fraser, C.L.; Gill, J.C. and McHatel, I. 1974. The effect of dietary crude protein concentrate and time of weaning on milk production and body weight changes in the ewes. *Anim. Prod.*, 19 : 331-39.

Robertson, HA. and Sarda, I.R. 1971. A very early pregnancy test for mammals : its application to the cow, ewe and sow. *J. Endocrinol.*, 49 : 407-19.

Roux, P.J. le, Wyk, L.C. van 1977. Studies on reproduction in Karakul sheep : effect of shortened length of gestation after flametha zone treatment on the quality of Karakul lamb pelts. 1971-1975. I*n Agri. Res.* 1976. pp. 86-87. Department of Agricultural Technical Servies, Pretoria, South Africa.

Rose Innes, R. and Mabey, C.L. 1964. Studies on browse plants in Ghana. 1. Chemical composition. *Emp. J. exper. Agric.*, 32 : 114-24.

Rowson, L.E.A. and Moor, R.M. 1966. Embryo transfer in the sheep : the significance of synchronizing oestrus in donor and recipient animal. *J. Reprod. Fert.*,1 : 207-12.

Rowson, L.E.A.; Moor, R.M. and Lowson, R.A.S. 1969. Fertility following egg transfer in the cow : effect of method, medium and synchronization of oestrus. *J. Reprod. Fert.*, 18 : 517.

Ruxton, I.B. 1975. A preliminary report on the proximate composition of forages grown at the animal production and agricultural research centre, Hofhuf, Saudi Arabia. *Publication No. 59.* Joint Agricltural Research and Development Project, University College of North Wales, UK and Ministry of Agriculture and Water, Saudi Arabia.

Roy, A.; Datta, I.C.; Sahni, K.L.; Singh, B. and Sengupta, B.P. 1962. Studies on certain aspects of sheep and goat husbandry. Artificial breeding, telescoping the breeding seasons and certain reproductive phenomena in sheep and goats. *Indian J. Vet. Sci. & Anim. Husb.*, 32 : 269-75.

Roy, A.; Gupta, H.C.; Shrivastava R.K. and Pandey, M.D. 1956. Storage of ram spermatozoa *in vitro.* Pt. I. Preservation in glycine-egg yolk medium. *Indian Vet. J.*, 33 : 18-20.

Ryley,J.W. 1966. Nitrogenous fertilisers and animal production. *Proc. Trop. Grasslds Soc, Aust.*, 7 : 10-12.

Sagi, R. and Morag, M. 1974. Udder confirmation, milk yield and milk fractionation in the dairy ewe. *Ann. Zootech.*, 23 (2) : 185-92.

Sagreiya, K.P. 1940. On the desirability of cultivating fodder yielding trees in the Central Provinces. *Indian Forester.* 66 : 14-16.

Sahani, M.S.; Dubey, S.C. and Sharma, S.C. 1981. Annual economics analysis of 100 Marwari sheep under field conditions. *Wool & Woollens India.* 17 (3) : 42-48.

Sahani, M.S. and Pant, K.P. 1978. Breed differences in the duration of pregnancy in sheep. *Indian Vet. J.*, 55 : 99-102.

Sahani, M.S.; Uppal, P.K. and Sahni, K.L. 1975. The milk production of native and corssbred under semiarid conditions. *Indian Vet. J.*, 52 : 832-36.

Sahebi, A. 1974. *J. Vet. Faculty, University*, Tehran, 30 : 49.

Sahni, K.L. 1967. Studies on quality, preservation and fertility of semen of sheep and goat. Ph. D. Thesis Agra University Agra.

Sahni, K.L. and Roy, A. 1967a. A study on the sexual activity of Bikaneri sheep *(Ovis aries)* and conception rate through artificial insemination. *Indian J. Vet. Sci.*, 37 : 327.

Sahni, K.L. and Roy, A. 1967b. A note on summer sterility in Rommney Marsh rams under tropical conditions. *Indian J. Vet. Sci.*, 37 : 335-38.

Sahni, K.L. and Roy, A. 1969. Influence of season on semen quality of rams and effects of diluters and dilutions on *in vitro* preservation. *Indian J. Anim. Sci.*, 39 : 1-14.

Sahni, K.L. and Roy, A. 1972a. A note on occurance of abnormal spermatozoa in different breeds of sheep and goat under tropical conditions. *Indian J. Anim. Sci.*, 42 : 501.

Sahni, K.L. and Roy, A. 1972b. The influence of colour additives on the keeping quality and identification of sheep and goat semen. *Indian J. Anim. Sci.*, 42 : 214.

Sahni, K.L. and Roy, A. 1972c. A note on seasonal variation in semen production of Corriedale and Corriedale × Bikaneri (Magra) rams under tropical conditions. Indian J. Anim. Sci., 42 : 99.

Sahni, K.L. and Roy, A. 1972d. Post partum conception in Bikaneri sheep. Indian J. Anim. Sci., 42 : 1038-39.

Sahni, K.L. and Sahani, M.S. 1973. Annual Report. Central Sheep and Wool Research Institute, Avikanagar. Rajasthan, p. 6.

Sahni, K.L. and Tiwari, S.B. 1972. Reproductive efficiency of young ewes of various sheep breeds. Annual Report, Central Sheep and Wool Research Institute, Avikanagar, Rajasthan.

Sahni, K.L. and Tivvari, S.B. 1973. Artificial insemination of sheep with diluted and stored semen in milk diluent. *Indian Vet. J.*, 50 : 536-39.

Sahni, K.L. and Tiwari, S.B. 1974a. The artificial insemination of sheep under semi-arid condition. *Wool & Woollens India,* XI (ii): 1.

Sahni, K.L. and Tiwari, S.B. 1974b. Effect of early rebreeding on certain aspects of sheep production. *Indian. J. Anim. Sci.*, 44 : 767.

Sahni, K.L. and Tiwari, S.B. 1975a. The effect of two efaculation frequencies on senmen characterestion of Rembuillet raws in diferrednt seasowns. *Indian Ve.t J.* 52. 421-25.

Sahni, K.L.; Tiwari, S.B. 1975b. Studies on milk production and preweaning growth of lambs in different sheep breeds under semiarid contions. *Indian J. Anim. Sci,* 45 : 252-55.

Sahni, K.L.; Tiwari, S.B. and Sahani, M.S. 1975. The artificial inseminatin of sheep with short distance transported semen of Rambouillet rams under semi-arid conditions. *Indian Vet. J.,* 52 : 1-5.

Sahni, K.L.; Tiwari, S.B. and Sahani, M.S. 1976. Effect of season on the occurence of oestrus and fertility in different breeds of sheep under semi-arid conditions. *Indian Vet. J.,* 53 : 515-22.

Said, A.N.; Slagsvold, P.; Bergh, H. and Laksesvela, B. 1977. *Nordisk Veterinaer median.* 29 (4/5) : 172.

Salamon, S. and Robinson, TJ. 1962. Studies on the artificial insemination of Merino sheep. I and II. *Aust. J. Agri. Res.* 13.

Salevro,A. and Malossini, F. 1968. Milk production in Seprauis Sana, Wurttem berg, re-de-France and Berrichon sheep with particular reference to the first month of lactation. *Annali. 1st. sper-Zootec.* 1 : 59-61 *(Anim. Breeding Abstr.,* 38 : 1424).

Salisbury, G.W. 1946. The glycolysis, livability and fertility of bovine spermatozoa as influenced by their concentration. *The problem of fertility.* Eds. E.T. Engle. P. Princeton, NJ.; Princeton. University Press, London; Geoffery Cumberlege. Oxford University Press, pp. 134-152 discussion; 152-53. *(Anim. Breeding Abstr.,* 15, 252).

Salisbury, G.W.; Fuller, H.K. and Willett, E.J. 1942. Preservation of bovine spermatozoa in yolk citrate diluents and field results from its use. J. Dairy. Sci., 26 : 69-78.

Salisbury, G.W.; Van Demark, N.L. and Lodge, J.R. 1978. *Physiology of Reproduction and Artificial Insemination of Cattle.* W.H. Freeman & Co. San Francisco.

Sane, C.R.; Gokhale, D.R. and Marathe, MR. 1952. Unpublished records of A.I. centre, Poone.

Saraswath, K.C.; Seth, O.N. and Roy, A. 1968. Effect of season of fertility in an experimental flock of Bikaneri (Magra) and live weight gain of lambs. *Indian J. Vet. Sci.,* 38 : 778.

Sarles, M.P. 1944. Effects of experimental nodulerworm *(Oesophagostomum columbianum)* infection in sheep. U.S. *Deptt. Agric. Tech. Bull.* 375.

Sarson, M. 1972. L'elevage du mouton de la race Barbarine au Centre d'Qusseltia, Tunisie Centrale (Raising Barbary sheep at the Qusseltia Centre in Central Tunisia). Decuments Techniques No. 55. Institute National de Recherche Agronomique de Tunisie, Tunis, Tunisia.

Sastry, M.S. and Mahadevan, V. 1963. Investigation on the toxicity of *Lantana camara* Linn. *Indian Vet. J.,* 40 : 47-52.

Sattar, A.; Raizada, B.C. and Pandey, M.D. 1978. A note on some wool fibre characteristics of Merino and crossbred lambs. *Vet. Res. Bull.* 1. : 148-49.

Saxena, V.B. and Tripathi, S.S. 1987. Seasonal affect on sperm morphology of Nali Rams. *Indian J. Anim. Sci.,* 57 : 294.

Saxena, V.B. and Singh, G. 1967. Preservation of ram spermatozoa. I. Effects on motility, livability and morphology of ram spermatozoa in four semen diluents. *J.Anim. Morph. Physiol.,* 14 : 114-20 (*Anim. Breed-ing Abstr.,* 37 : 2662).

Saxena, J.S.; Kulshreshtha, S.K. and Johri, C.B. 1971. Studies on exotic legume fodder siratro (*Phaseolus atropurpurens*) cultivation and nutritive value for sheep. *Indian Vet. J.,* 48 : 849-53.

Saxena, J.S.; Kulshreshtha, S.K. and Upadhyaya, R.B. 1972. Studies on the assessment of palatability and nurtitive value of green panic (*Panicum maximum* var trichoglume). *Indian Vet. J.,* 49 : 177-79.

Saylor, W.W. and Leach, R.M. Jr. 1980. *J. Nutri.,* 110 : 448.

Schuler, J.A. 1937. Note on the sheep breeding situation in Syria. *Anim. Breeding Abstr.,* 5 : 281.

Schiloberg, E.M. 1979. *In vivo* investigations in sheep on the influence of blood magnesium level on magnesium absorption from rumen. *Inagural dissertation, Tierarythihe Hochschule, Hanaover.*

Sedky, A.R. 1948. The Egyptian sheep : an attempt to improve quality and quantity. *Emp. J. exp. Agric,* 16 : 79-89.

Sehgal, J.P. and Kaushish, S.K. 1989. Bhedon Ko Kab aur Kaisa Ahar den. *Kheti* (April): 17-18.

Sen, K.C. and Ray, S.N. 1971. *Nutritive values of Indian cattle feeds and the feeding of animals.* Indian Council for Agricultural Research, New Delhi, 133 p.

Sengupta, B.P.; Mittal, J.P.; Mittal, P.C.; Pant, H.C. and Pandey, M.D. 1976. Studies on factors influencing lambing performance in PMSG treated Magra ewes. *Indian J. Anim. Sci.,* 46 : 23-28.

Sengupta, B.P.; Mittal, P.C.; Mittal, J.P. and Pandey, M.D. 1978. Fertility in young and parous Magra ewes superovulated with different cobinations of serum and chorionic gonadotrophins. *Indian J. Anim. Sci.,* 48 : 889-90.

Serra, J.A. and De Matos, R.M.A. 1951. Nature of the canary coloration of wool. *J. Text. Inst.,* 42 : T 329-T 331.

Seth, O.N.; Pandey, M.D. and Roy, A. 1972. A note on live weight, growth and mortality in lambs as singles, twins or triplets. *Indian J. Anim. Sci.,* 42 : 695-98.

Seth, S.K.; Raizada, M.B. and Waheed Khan, MA. 1962. *Trees for vana mahotsava.* F.R.I. & College, Dehradun, 130 p.

Shahi, H.B. 1941. Sheep breeding research in India. *Indian Fmg.* 2 : 61-65.

Sharma, B.M. 1966. Ecological studies on desert plants. *Prosopis spicigera* Linn. *Trop. Ecol.,* 7 (1) : 54- 66.

Sharma, Deonani, V.P. 1967. Observations on the pathogenicity of filaria infection and report on treatment trials with Dicklycidecyana cethydra zide. *Indian J. Vet. Sci.,* 37 : 39-46.

Sharma, D.D.; Gill, R.S.; Chander, S. and Negi, S.S. 1966a. Chemical composition of some fodder tree leaves in the Kangra district. *Punjab Agricultural University, J. Res.* 3 : 438- 42.

Sharma, D.D.; Negi, S.S. and Sidhu, G.S. 1966b. Chemical composition and nutritive vlue of Biul (*Grewia Punjab oppositifolia* Rox B) tree leaves. *Punjab Agricultural University J. Res.,* 3 : 178-85.

Sharma, D.D.; Chawala, M.A. and Negi, S.S. 1968. Chemical composition and nutritive value of bamboo *(Bamboosa arundinacea)* and Kachnar *(Bauhinia variegata)* tree leaves. *Punjab Agricultural University J. Res.,* 5 : 253-58.

Sharma, K.M.; Upadhyaya, R.B. and Saxena, J.S. 1972. *Setaria sphacelata-* a promising fodder for sheep. 1. Studies on palatability and nutritie value, *Indian Vet. J.,* 49 : 1137-40.

Sharma, R.P.; Tiwari, S.B. and Roy, A. 1969. Effect of frequency of semen collection from ram on seminal attributes and fertility. *Indian f. Anim. Sci,* 39 : 15.

Sharma, V.K.; Gupta, R.C.; Khurana, N.K. and Khar, S.K. 1992. Effect of body traits on super ovulatory response and embryo recovery in cross bred ewes. *Proc. 10th National conv. Indian Soc. Study Anim. Reprod.* Madras Vety. College April 8-10, 1992.

Sharma, V.V. and Murdia, P.C. 1974. Utilization of berseem hay by ruminants. J. *agric. Sci,* (Camb.), 83 : 289-93.

Shelton, J.N. and Moore, N.W. 1967. Evaluation of alternative methods of administration of progestagens in the entire cyclic. In : *The control of the ovarian cycle in the sheep.* Syndey University press, 59-75.

Shergin, N.P. 1935. *Probl. Zhivotn.,* 12 : 100 *(Anim. Breeding Abstr.,* 4 : 270).

Shiga, A.; Miyachi, T.; Sato, M.; Shinozaki, K.I.; Sacki, T. and Hayashi, M. 1979. *Faculty Agri. Iwate, University.* 14 (3) : 173.

Shukla,D.D. and Bhattacharya, P. 1952. Seasonal variation in reaction time and semen quality. *Indian, J. Vet. Sci.,* 22 : 109.

Shukla, D.C. and Pal, A.K. 1972. Ratio of dietary protein to energy on body composition of sheep. *Indian J. Anim. Sci.,* 42 : 223-26.

Shukla, K.S. and Ranjhan, S.K. 1969. Chemical composition and nutritive value of Gokhru. *Indian Vet. J.*, 46 : 715-18.

Shukla, K.S.; Ranjhan, S.K. and Katiyar, R.C. 1970a. *Rhynchosia minima* as a feed for sheep. *Indian* J. *Dairy Sci.*, 23 : 82-84.

Shukla, K.S.; Ranjhan, S.K. and Katiyar, R.C. 1970b. Nutritional studieson Kasni *(Chicorium intybus)* A most common weed of Rabi leguminous fooders. 1. Chemical composition and nutritive value of Kasni for *sheep. Indian J. Anim Sci.* 40 : 356.

Sinclair, A.N. 1957. Effect of variation of time of mating, frequency and semen dose rate on conception in Merino sheep. Aust. Vet. J., 33 : 88-91.

Sinclair, T. 1949. The latent phase of sheep scab : its nature and relation to the eradication of the disease.;. *Comp. Pathol.* 59 : 305-18.

Singh, Bharat, 1963. The first irrigated plantation of Rajasthan. *Indian Forester.* 89 : 690-700.

Singh, N.P. 1975. Chemical composition and nutritive value of Anjan *(Cenchrus ciliaris)* grass of sheep. *Indian Vet. J.*, 52 : 435-8.

Singh, R.V. 1985. *Fodder trees of India.* Oxford & IBH Publishing Co., New Delhi, Bombay, Calcutta.

Singh, C.S. and Joshi, D.C. 1957. Goja *(Amaranthus spinosa* Linn) A drought resistant evergreen useful feed for sheep. *Indian Vet. J.*, 34 : 191-96.

Singh, H.K. and Mudgal, V.D. 1967. Chemical composition and nutritive value of *Leucaena glauca* (white popinac). *Indian J. Dairy Sci.*, 20 : 191-95.

Singh, M. and Mahadevan, V. 1968. Studies on protein metabolism in adult sheep I. Endogenous urinary nitrogen determined graphically and dry direct method based on nitrogen free ration. *Indian Vet. J.*, 45 :659-65.

Singh, M. and Mahadevan, V. 1969. Studies on protein metabolism in adult sheep 11 Determination of metabolic faecal nitrogen and its expression on the basis of dry matter voided in faeces. *Indian Vet. J.*, 41 : 151-54.

Singh, M. and Mahadevan, V. 1970. A note on the protein requirement for maintenance. *Anim. Prod.*, 12 : 135-89.

Singh, N.P. and Patnayak, B.C. 1977a. Nutritive value of *Ailanthus excelsa* Roxb (Ardu) leaves for sheep. *Indian Vet. J.*, 54 : 198-201.

Singh, N.P. and Patnayak, B.C. 1977b. Nutritive value of *Vigna sinensis* (Cowpea) silage with or without molasses. *Indian J. Anim. Sci.*, 47 : 377-82.

Singh, V.K. and Rathore, D.S. 1987. Breeding sheep for carpet wool production. A lead paper presented at National Seminar on "Small Ruminants Production" held at Central Sheep and Wool Research Institute, Avikanagar, Rajasthan. Jan. 5-7, 1987.

Singh, B. and Roy, A. 1963. Studies on certain aspects of sheep and goat husbandry. VI. Seasonal variation in the semen quality of Corriedale and Bikaneri rams. *Indian J. Vet. Sci. & Anim. Husb.,* 33 : 211-14.

Singh, G and Saxena, V.B. 1965. Effect of various dilutors on preservability and morphology of ram spermatozoa. *Agric. Res.,* 5:215.

Singh, N. and Taneja, G.C. 1979. Effect of prolonged salt intake on the reproduction of ewes and weight of lamb born to Marwari sheep. *Indian J. Anim. Res.,* 13 (2) : 71-74.

Singh, B.; Kalra, S.; Ram, S. and Chopra, S.C. 1981. Fleece characteristics of Nali lambs and its crosses with Corriedale and Russian Merino. *Livestock Adv.* 6 (6) : 29-32.

Singh, S.R.; Singh, G.S. and Singh, S.N. 1965. Studies on para grass *(Panicum barbnode* or *Brachiaria mutica* stapf). Chemical composition, digestibility and nutritive value of paragrass when fed as green and hay. *Indian J. Dairy Sci.,* 16 : 136-49.

Singh, B.; Singh, G.S. and Singh, S.N. 1963. Effect of hybridization on the chemical composition and nutritive value of the Napier grass *(Pennisetum purpureum). Indian J. Vet. Sci. Anim. Husb.,* 33 : 301-309.

Singh, M.; Patnayak, B.C.; Singh, N.P.; Arya, D.L. and Sidiqui, S.S. 1977. Observations on the *effect* of feeding level on canary coloration of Indian wools. *Indian J. Anim. Sci.,* 47 : 34- 38.

Singh, Tej.; Handa, K.L. and Sobti, S.N. 1959 Chemical composition of some legumes of Jammu and Kashmir. *Indian J. Vet. Sci. 29* (2&3) : 32-39.

Singh, Sham; Krishnamurthi, S. and Katyal, S.L. 1967. *Fruit culture in India.* Indian Council for Agricultural Research, New Delhi. 225-29.

Singh, N.P.; Singh, M. and Patnayak, B.C. 1979. Effect of energy and protein levels on nutrient digestibility and wool production. *Indian J. Anim. Sci,* 49 : 277.

Sinha, N.K. 1978. Fertility in indigenous sheep bred artificially with Suffolk and Dorset rams. Annual Report, Central Institute for Re arch on Goats, Makhdoom.

Sinha, A.K. and Jolly, M.S. 1971. Foliar constituents of the food plants of tasar silkworm *Antheraea mylitta* D. *Indian Forester.,* 97 (5) : 261-63.

Sinha, N.K. and Khan, B.U. 1989. Reproductive ability of ewes of different ages. Paper presented at V *Ann. Conf. SAP1,* Munnar (Kerala) Dec. 27-30, 1989.

Sinha, N.K.; Wani, G.M. and Sahni, K.L. 1979. Observations on reproductive behaviour and semen quality of rams reared under tropical conditions. *Indian J. Anim. Sci.,* 49 : 916.

Sinha, N.K.; Wani, G.M. and Sahni, K.L. 1980. Note on the incidence of gestational oestrus in Muzaffarnagri ewes. *Indian J. Anim. Sci.,* 50 : 1009.

Skerman, P.J. 1977. *Tropical forage legumes.* FAO Plant Production and Protection series No.2. Food and Agriculture Organisation, UN, ROME.

Skolovskaya, I.L.;Drozdova, L.P.; Goluisheva, M.G.;Korotkov, A.I.; Maximov, U.V. and Lebedeva, V.A. 1956. Improvement of diluents for semen of farm animals. Proc *Lenin Acad. Agric. Sci.,* XXI : 17-24 *(Vet. Bull XXVII : 937).*

Smith, R.M. and Gcwthorn, J.M. 1975. The biochemical basis of defidences of zinc, manganese, copper and cobalt in animals. In : *Trace Elements in Soil-Plant-Animal-Systems.* (Ed). A.R. Egan, Academic Press, New York, N.Y. London, pp 243-58.

Smorag, Z.; Wierzbowaski, S. and Wierzchos, E. 1978. Results of transplanting 3, 6 and 7 day old frozen sheep embryos. *Scientes Biologigues.,* 26 (4) : 273-75.

Soulsby, E.J.L. 1965. *Text book of Veterinary Clinical Parasitology* Vol. I. Helmiths Blackwell, Scientific Publication, Oxford P 1120.

Soulsby, E.J.L. 1968. *Helminths, Arthropods and Protozoa of Domesticated Animals.* Balliers Tindal and Cassell. London p 824.

Speeding C.R.W. and Brown T.H. 1957. *J. agric. Sci.* (Camb), 49 : 229.

Speeding C.R.W.; Brown, R.H. and Wilson, I.A.N. 1953. *Nature,* (Lond) 181 : 168.

Srivastava, R.K. and Chaturvedi, M.L. 1971. Determination of pasture composition -comparison of different methods of sampling pasture grass. *Indian Vet. J.,* 48 : 958-62.

Srivastava, V.K. and Roy, A. 1969. Influence of season on the fertility of ewes and growth of lambs. *Indian J. Anim. Sci,* 39 : 185.

Stojanovskaja, V.I. and Mesjacev, A.S.I 957. Puti povysenija produktivnosti karakuljskin ovec neznoi konstitucii (Methods of increasing the prodcutivity of karakul ewes with a delicate constitution). *Karakulevodstvo i zverovodstovo,* 10 (4) : 10-14. (Anim. *Breeding Abstr.,* 26 : 276).

Stoynovskaya, S. 1966. Cited from *ovstevods vo, psrt II* Edited by G.R. Litovehenko, Moskova, KOLOS, 1972. pp 410.

Suckling, G.W.; Catress, T.W.; Healy, W.B. and Mattinglet, J. 1974. NZ *J. Agri. Res.,* 17 : 311.

Surreaux, P.G.; Thompson, D. de M. and Correa, T.B. 1980. Verficacao do desenvolvimento ponderal am cordeiros Fl Ile de France × Corriedale ate os 90 dias (Growth of Fl Ile-de-France × Corriedale lambs to 90 days). In *Pesquisa em ovinos no Brasil* 1975-1979, p. 117. Empresa Brasileira de Pesquisa Agropecuaria, Bage, Brazil.

Sutton, R.; Nancarrow, CD.; Walker, A.L.C. and Rigby, N.W. 1984. Identification of anoestrus associated glycoprotein in oviductal fluid of sheep. *J. Reprod. Fert.* 72 : 415-22

Sutton, W.G. 1935. Some studies of the yolk in New Zealand wools II-A study of yolk in the fleece during one year. *J. Text. Inst.,* 24 : 341-350.

Tait, R.M. and Dubeski, P.L. 1979. *Can. J. Anim. Sci.* 59 : 609.

Takahashi, M. and Ripperton, J.C. 1949. Koe haole *(Leucaena glauca)* : Its establishment, culture and utilization as a forage crop. *Hawaii Agric. exp. Sta.* Bull. 100.

Talpatra, S.K. and Bansal, C.S. 1957. Studies on indigenous creepers : A new rainfed high protein phosphorous rich plant. *Indian J. Vet. Sci.,* 27 : 45-60.

Taneja, G.C. 1966. Fertility in sheep. I. Conception rate in relation to lamb survival. *Indian Vet. J.,* 43 : 727.

Taparia, A.L. 1972. Breeding behaviour in Sonadi sheep. *Indian J. Anim. Sci.,* 42 : 576-79.

Tervit, H.R. and Glood, PG. 1978. The culture of sheep embryos in either a bicarbonate buffered medium or a phosphate buffered medium enriched with serum. *Theriogenology,* 10 : 251.

Tervit, H.R.; Whittinghan, D.G. and Rowson, L.E.A. 1972. Successful culture *in vitro* of sheep and cattle ova. *J. Reprod. Fert.,* 30 : 493.

Teixeira, F.J.L.; Bessa, MN.; Souza, A.A. de; Freitas, J.P. de.; and Mendonca, H.L. 1980. Contributicao ao estudo da efciencia reproductiva de ovelhasdeslanades Morada Nova-Variedade Branca (Reproductive efficiency of White Morada Nova sheep), *In Pesquisa em ovina no Brasil* 1975-1979. p. 55, Empresa Brasileira de Pesquisa Agropecuaria, Bage, Brazil.

Terrill, C.E. 1974. Sheep. In Hafez, *Reproduction in Farm Animals,* (Ed.) Hafez, E.S.E. 3rd edn. pp. 265-74. Lea and Febiger, Philadelphia, USA.

Thwaites, C.J. 1981. Development of ultrasonic techniques for pregnancy diagnosis in the ewe. *Anim. Breeding Abstr,* 49 : 427-34.

Tiwari, S.B. 1978. Studies on semen production of native and cross bred rams as influenced by age, body size, breed, seasons and levels of feeding. Ph. D. Thesis, Agra University, Agra. Tiwari, S.B. and Honmode, J. 1968. Age of matrutiy of young rams of various breeds with special reference to semen production. Annual Report, Central Sheep and Wool Research Institute, Avikanagar, Rajasthan pp 4.

Tiwari, S.B. and Kalra, D.B. 1982. Sexual behaviour and semen quality of exotic rams (Rambouillet and Soviet Merino) as influenced by age under semi arid condition. *Indian Vet. J.*, 59 : 992-94.

Tiwari, S.B. and Sahni, K.L. 1974a. A note on the effect of underground housing on semen production in Rambouillet rams under semi-arid condition. *Indian J. Anim. Sci.*, 44 : 206-208.

Tiwari, S.B. and Sahni, K.L. 1974b. The extent of reproductive wastage in Rambouillet rams under semi-arid conditions. *Indian Vet. J.*, 51 : 497.

Tiwari, S.B. and Sahni, K.L. 1975. Semen quality of Rambouillet rams and their crosses (with native) during hotter part of the year, *Indian Vet. J.*, 52 : 614-19.

Tiwari, S.B. and Sahni, K.L. 1981. Puberty and reproductive performance in ram lambs of indigenous exctic and cross-bred sheep under semi-arid conditions. Indian *J.* Anim. Sci., 51 : 634-38.

Tiwari, S.B. and Sahni, K.L. 1982. Semen production in relation to age, body weight and testis dimension in crossbred rams (Rambouilet × Malpura) under semi-arid condition. *Indian Vet. J.*, 52 : 995-97.

Tiwari, S.B.; Ahmed, N. and Sahni, K.L. 1973. Meat characteristics of Malpura lambs weaned at two ages. *Indian Vet. J.*, 50 : 1014-19.

Tiwari, S.B.; Sharma, R.P. and Roy, A. 1968. Effect of dilution on preservation of ram and goat spermatozoal survivalbity. *Indian J. Vet. Sci.*, 38 : 567.

Tiwari, S.B.; Sharma, R.P. and Roy, A. 1969c. Process of parturition in sheep and goat and morphological characteristics of their placenta. *Indian Vet. J.*, 46 : 576.

Tiwari, S.B.; Honmode, J. and Sahni, K.L. 1969b. A study on the effect of post-partum interval on efficiency of rebreeding in sheep. Annual Report, Central Sheep and Wool Research Institute, Avikanagar, Rajasthan, p 30.

Tiwari, S.B.; Singh, V.K. and Mathur, P.B. 1969c. A study on the reporductive performance of different breeds of sheep during spring season. Annual Report, Central Sheep and Wool Research Institute, Avikanagar, Rajasthan, p 16.

Tiwari, S.B.; Honmode, *J.*; Singh, V.K.; Singh, L.B. and Sahni, K.L. 1969d. Studies on oestrus activity and artificial insemination in sheep during autumn season. Annual Report, Central Sheep and Wool Research Institute, Avikanagar, Rajasthan, p 23.

Tiwari, S.B.; Srivastava, R.S. and Kalra, D.B. 1978. Effect of stage of heat on conception rate. Annual Report. Central Sheep and Wool Research Institute, Avikanagar, Rajasthan.

Tiwari, S.N.; Singh, G.D.N.; Jha, G.L. and Sinha, B.K. 1978. *Indian. J. Anim. Health. 17* (2): 141.

Tomar, S.S. 1979. Time of parturition in sheep. *Indian J. Anim. Res.,* 13 : 68-70.

Trapp, M.J. and Slyter, A.L. 1983. Pregnancy diagnosis in the *ewe. J. Anim. Sci. 57 :* 1-5.

Tontis, A.; Konig, H. and Luginbuhl, H. 1980. *Schweizer Azchiv fur Tierheikunde,* 122(2): 107.

Towns, K.M.; Boston, R.C. and Leaver, D.D. 1978. The effect of intravenous administration of phosphorus on phosphorus and calcium metabolism. *Aust. J. Agri. Res.,* 29 : 587.

Tressol, J.L. and Lamand, M. 1979. *Annals de Recherches veterinarais,* 10 (1) :7I.

Trail, J.C.M. and Sacker, CD. 1966. Production records of lambs from East Africa Black head ewes and Dorset Horn rams. *E. Afr. Agric. For. J.,* 32 : 133-36.

Tripathi, S.S. 1977. Synchronization of oestrus in sheep with melengestrol acetate (MGA), *Pantnagar J. Res.,* 2 : 195-96.

Trounson, A.O. and Moore, N.W. 1974. Fertilization in the ewe following multiple ovulation and uterine insemination. *Aust. J. Bio. Sci, 27 :* 301-304.

Trounson, A.O. and Rowson, LE.A. 1976. Research on embryo transfer in domestic animals. *J. Roy. Soc. England,* 137 : 77-85.

Trounson, A.O. and Willadsen, S.M. 1977. Reproductive function in prepubertal lambs : ovulation, embryo development and ovarian steroidogenesis. *J. Reprod. Fert.,* 49 : 69-75.

Troup, R.S. 1921. *The silviculture of Indian Trees.* Vol. I, II and III. Clarendon Press, Oxford.

Tyagi, J.C. and Lavania, J.P. 1965. Breeding season of indigenous and crossbred ewes and role of temperature and bright sunshine hours. *Indian J. Anim. Sci.,* 39: 91.

Ugalde Orta, J. 1978. Analisis de algunos factores geneticos y ambientales que afectan el peso al nacer y crecimiento hasta los tres meses de borregas Romney Marsh (Analysis of some genetic and environmental factors affecting birth weight and growth to 3 months in Romney Marsh lambs), *Veterinaria* (Mexico), 9 : 120.

Udall, D.H. 1964. *The Practice of Veterinary Medicine.* Baitheir tindel and Cox, 7 and 8 Henrietha O sheet W.C.R. p 69.

Udall, R.H. and Chenchow, F. H. 1969. The aetiology and control of urolithiasis. *Adv. Vet. Sci.,* 13 : 29-57.

Underwood, E.J. 1977. *Trace Elements in Human and Animal Nutrition.* Academic Press. London.

Upadhyaya, R.B.; Kulshreshtha, S.K. and Saxena, S.K. 1971. The role of Krishna Neel *(Anagallis arvensis)* in the dietary of sheep. *Indian Vet. J.,* 48 : 1059-62.

Uppal, P.K. Dubey, S.C., Kaushush, S.K. Rawat, P.S. 1981. Efficacy of panacure (Fenbendazole) against naturally infected takewor infemc-tions in sheep and gosts under field Conditions. Indian

Valencia, J.; Barron, C; Ferandez-Baca, S. 1977. Pubertad en corderos Tabasco × Dorset (Puberty in Tabasco × Dorset lambs), *Veterinaria* (Mexico), 8 : 127-30.

Van Nieckerk, B.D.H. and Mulder, A.H. 1965. Duration of pregnancy and of post partum anestrus in autumn mated Dorper, Dohne Merino and Merino ewes. *Proc. S. Afr. Soc. Anim. Prod.,* 4 : 205 *(Anim. Breeding Abstr.,* 36 : 1593).

Van Rensburg, H.J. 1967. *Pasture legumes and grasses in Zambia.* Govt. Printer, Lusaka.

Van Rensburg, H.J. 1968. *Palatability of plants in Zambia.* Govt. Printer Lusaka.

Vavilov, N.I. 1951. The origin, variation, immunity and breeding of cultivated plants. *Chronica Botanica,* 13 : 1-364.

Venkataramany, P. 1968. Silviculture of genus *Albizia* species. The silviculture of Indian trees, No. 22. Manager of publications, Delhi, p 54.

Vankatchalam. 1960. Chemical composition and nutritive value of buffalo grass *(Brachiaria mutica). Indian Vet. J.,* 37 : 260-62.

Villegas, V. and Cruz, E.C. 1960. Ewes for dairy purposes, *Philippine J. Anim. Indust.,* 21 : 39-42.

Volcani, R. 1953. Seasonal variations in spermatogenesis of some farm animals under the climatic conditions of Israel. *Bull.* Res. *Coun. Israel,* 3 : 123-26 *(Anim. Breeding Abstr.,* 22 : 1279).

Wadhwani, B.B. 1953. *Prosopis spicigera* (Linn.) *Indian Forester,* 79 : 432-35.

Wallace, L.R. 1948. The growth of lamb before and after weaning in relation to level of nutrition. *J. Agri. Sci.,* 38 : 93-153.

Walker, S.K.; Obst, J. M.; Smith, D.H.; Hall, G.P. and Flavel, A. B. 1979. *Aust.* J. *Biol.* Sci., 32 : 221.

Walkar, S.K.; Seamark, R.F.; Quinn, P.; Warnes, G.M.; Ashman, RJ.; Smith, D.H. and Ancall, P. 1988. Culture of pronuclear embryo of sheep in a simple medium. *Proc. 11th Intern. Cong. Anim. Reprod. A.I.* Dublin, Ireland 4 : 483-85.

Wani, G.M. 1981. Ultrasonic pregnancy diagnosis in sheep and goats-a review. *World Rev. Anim. Prod.,* 17 (4): 43-8.

Wani, C.M. and Sahni, K.L. 1981. Ultrasonic pregnancy diagnosis in ewe under tropical field conditions. *Indian J. Anim. Sci.,* 51 : 194-97.

Wanner, M. and Bose, P.H. 1978. *Schweizer Achiv fur* Tierheilkunde, 120 : 369.

Waston, R.H. 1952. Seasonal variation in fertility in Marino sheep. *Pastoral Rev.,* 62 : 324 (*Anim. Breeding Abstr.,* 21 : 272).

Waston, R.H.; Sapsford, C.S. and McCance, I. 1956. The development of the testis, epididymis and penis in the young Merino ram. *Aust. J. Agric. Res., 7* : 574-90.

Webster, GM. and Haresign, W. 1981. A note on the use of dexamethasone to induce parturition in the ewe. *Anim. Prod.,* 32 : 341-44.

Webster, W.M. 1952. Infertility in rams, *Proc. N.Z. Sci. Anim., (Anim. Breeding Abstr.,* 20 : 1136).

Weisman, E. 1967. Epizootiology, diagnosis and control of the leptospirosis of cattle, sheep and pigs. *Bull, deioffice. Intern, des Epizootics.,* 68 : 15.

West, O. 1950. Indigenous tree crops for Southern Rhodesia. *Rhod. Agric. J.,* 47 : 204-217.

Westerheim, O. 1937. Lungworm in goats in Ryfylke, *Norway, Abs. Vet. Bull.* Z. 74.

Whanger, P.D.; Weswing, P.H.; Schmitz, J.A. and Oldfield, J.E. 1977a. *J. Nutr.,* 107 : 1298.

Whanger, P.D.; Weswing, P.H.; Schmitz, J.A. and Oldfield, J.E. 1977b. *J. Nutr.,* 107 : 1288.

Whilock, J.H. 1960. *Diagnosis of Veterinary Parasitism.* Lea Febiger, Philadelphia. P.A. p 236.

White, C.L. and Somers, M. 1977. *Aust. J. Biol. Sci.,* 30 (1-2) : 47.

Wiggins, E.L.; Terril, C.E. and Emik, L.O. 1953. Relationships between libido, semen characteristics and fertility in range rams. *J. Anim. Sci.,* 12 : 684-%.

Willardsen, S.M. and Polge, C. 1980. Embryo transplantation in the large domestic species. Implications and prospects in the light of recent experiments with egg and embryos. *J. Royal Agric. Sci.* (Camb.), 341 : 119-26.

Willardsen, S.M. and Polge, C. 1980. Embryo transplantation in the domestic species : applications and perspectives in the light of recent experiments with eggs and embryos, *J. Roy. Soc. England,* 141 : 115-26.

Willardsen, S.M.; Polge, C; Traunson, A.O. and Rowson, E.E.A. 1977. Transplantaiton of sheep and cattle embryo after storage at - 196°C in animal research station Cambridge, U.K. Pub. Amsterdam, Netherland, Elsevier, 198-201.

Williams, S.M.; Garrigus, U.S.; Norton, H.W. and Nalbandov, A.V. 1956. The occurrence of estrus in pregnant ewes. *J. Anim. Sci.,* 15 : 978-83.

Wilmut, I. and Sales, O.I. 1981. Effect of a synchronous environment on embryonic development. *J. Reprod. Fert.,* 61 : 174-84.

Wilson, J.G. 1975. Hypothyroidism in ruminants with special reference to foetal goitre. *Vet. Rec, 97* : 161-64.

Wilson, J.G. and Bredon, R.M. 1963. Nutritional value of some common

cattle browse and fodder plants of Karamoja, Northern province, Uganda. *E. Afr. Agric. Fer J.,* 31 : 204-205.

Wilson, R.T.; 1976. Studies on the livestock of southern Darfur, Sudan. III. production traits of sheep. *Trop. Anim. Health Prod.,* 8 : 103-14.

Wilson, R.T. and Durkin, J.W. 1983. Livestock production in central Mali : weight at first conception and ages at first and second parturitions in traditionally managed goats and sheep. *J. agric. Sci.* (Camb.), 100 : 625-28.

Wilson, R.T.; Peacock, C.P. and Sayers, A.R. 1981. A study of goat and sheep production on the Masai group ranch at Elangata Wusa, Kajiado district, Kenya. *Working Document* 24. International Livestock Centre for Africa, Addis Ababa, Ethiopia.

Wollny, C. and Niznikowski, R. 1985. Relationship between milk yield body confirmation and udder measurement. *35th Ann. Meet. EAAP* Greece.

Yenikoye, A.; Andre, D.; Ravault, J.P. and Mariana, J.C. 1981. Etude de quelques caracteristiques de reproduction chez la brebis Peulth, du Niger (Some reproductive characteristics of the Fulani ewe is Niger). *Reprod. Nutr. Develop.,* 21 : 937-51.

Yenikoye, A.; Pelletier, J.; Andre, D. and Mariana, J.C. 1982. Anomalies in ovarian function of Peulh ewes. *Theriogenology,* 17 : 355-64.

Yoshicka, Z.Y. Inudo and Torizuka, T. 1951. Experiments on the storages of semen and the insemination of the stored semen in sheep and goats. *Bull. Nat. Agric. Sci.,* (Cuba) G. No. 1, 53-60 (Anim. *Breeding Abstr.,* 20 : 245).

Young, W. 1950. Border Leicesters in South West Scotland. Scot. Agri., 30 : 18-21 *(Anim. Breeding Abstr.,* 18 : 412).

Younis, A.A. and Afifi, E.A. 1979. On the occurrence of oestrus in pregnant Barki and Merino ewes. *J, agric. Sci.* (Camb.) 92 : 505-506.

Younis, A.A. and El-Gaboory, I.A.H. 1978. On the diurnal variation in lambing and time for placenta expulsion in Awassi ewes. *J. agri. Sci.* (Camb.), 91 : 757-60.

Younis, A.A.; El-Gaboory, I.A.; El-Tawil, E.A. and El-Shobokshy, A.S. 1978. Age at puberty and possibility of early breeding in Awassi ewes. *J. agric. Sci.* (Camb.), 90 : 255-60.

Zanwar, S. 1981. Embryo transfer in sheep. *Indian J. Anim. Reprod.,* 1 : 102-104.